Get Certified

A Guide to Wireless Communication
Engineering Technologies

Get Certified

A Guide to Wireless Communication
Engineering Technologies

Edited by
Syed A. Ahson
Mohammad Ilyas

CRC Press
Taylor & Francis Group
Boca Raton   London   New York

CRC Press is an imprint of the
Taylor & Francis Group, an **Informa** business

AN AUERBACH BOOK

CRC Press
Taylor & Francis Group
6000 Broken Sound Parkway NW, Suite 300
Boca Raton, FL 33487-2742

First issued in paperback 2017

ISBN 13: 978-1-138-11816-4 (pbk)
ISBN 13: 978-1-4398-1226-6 (hbk)

Library of Congress Cataloging-in-Publication Data

Get certified : a guide to wireless communication engineering technologies / Syed A. Ahson and Mohammad Ilyas.
   p. cm.
"IEEE WCET exam."
Includes bibliographical references and index.
ISBN 978-1-4398-1226-6
   1. Wireless communication systems. 2. Wireless communication systems--Examinations--Study guides. 3. Electronic data processing personnel--Certification--Study guides. I. Ahson, Syed. II. Ilyas, Mohammad, 1953- III. Institute of Electrical and Electronics Engineers.

TK5103.2.G493 2009
621.384076--dc22   2009027070

Visit the Taylor & Francis Web site at
http://www.taylorandfrancis.com

and the CRC Press Web site at
http://www.crcpress.com

Contents

v

Contributors

Syed Ahson
Microsoft
Seattle, Washington, U.S.A.

Jesus Alcober
Polytechnic University of Catalonia
Barcelona, Spain

Huseyin Arslan
University of South Florida
Tampa, Florida, U.S.A.

Mohamad Assaad
Ecole Supérieure d'Electricité (Supélec)
Gif-sur-Yvette, France

Watit Benjapolakul
Department of Electrical Engineering
Chulalongkorn University
Bangkok, Thailand

Jalel Ben-Othman
PRiSM Laboratory
Versailles Saint-Quentin-en-Yvelines
   University
Versailles, France

Janez Bester
Faculty of Electrical Engineering
University of Ljubljana
Ljubljana, Slovenia

Qi Bi
Alcatel Lucent
New York, New York, U.S.A.

Thomas M. Chen
Department of Electrical Engineering
Southern Methodist University
Dallas, Texas, U.S.A.

José Tomás Entrambasaguas
Communications Engineering
Department, University of Malaga
Malaga, Spain

Ben Falchuk
Telcordia Technologies, Inc.
Piscataway, New Jersey, U.S.A.

Dave Famolari
Telcordia Technologies, Inc.
Piscataway, New Jersey, U.S.A.

Gábor Fodor
Ericsson Research
Stockholm, Sweden

Sahar Ghazal
PRiSM Laboratory
Versailles Saint-Quentin-en-Yvelines
   University
Versailles, France

Lal C. Godara
School of Electrical Engineering
University College, University
of New South Wales
Australian Defence Force Academy
Canberra, Australian Capital
Territory, Australia

Gerardo Gómez
Communications Engineering
Department
University of Malaga
Malaga, Spain

Steven D. Gray
Nokia Research Center
Espoo, Finland

Paul Hoffmann
Brooklyn Polytechnic University
New York, New York, U.S.A.

Marisol Hurtado
i2CAT Foundation
Barcelona, Spain

Mohamed Ibnkahla
Queen's University
Kingston, Ontario, Canada

Mohammad Ilyas
Florida Atlantic University
Boca Raton, Florida, U.S.A.

Tornaz Javornik
Ljubljana, Slovenia

Gorazd Kandus
Department of Communication
Systems
Jozef Stefan Institute
Ljubljana, Slovenia

Andrej Kos
Faculty of Electrical Engineering
University of Ljubljana
Ljubljana, Slovenia

Shoshana Loeb
Telcordia Technologies, Inc.
Piscataway, New Jersey, U.S.A.

F. Javier López-Martínez
Communications Engineering
Department
University of Malaga
Malaga, Spain

David Morales-Jiménez
Communications Engineering
Department
University of Malaga
Malaga, Spain

Nhut Nguyen
Network Systems Lab,
Samsung Telecommunications
America
Richardson, Texas, U.S.A.

Tero Ojanperä
Nokia Group
Espoo, Finland

Antoni Oller
Polytechnic University of Catalonia
Barcelona, Spain

Bernd-Peter Paris
George Mason University
Fairfax, Virginia, U.S.A.

Arogyaswami J. Paulraj
Stanford University
Palo Alto, California, U.S.A.

Sreco Plevel
Department of Communication
  Systems
Jozef Stefan Institute
Ljubljana, Slovenia

Teresa C. Piliouras
TCR
Weston, Connecticut, U.S.A.

András Rácz
Ericsson Research
Budapest, Hungary

Quazi Mehbubar Rahman
Queen's University
Kingston, Ontario, Canada

Norbert Reider
Department of Telecommunications
  and Media Informatics
Budapest University of Technology
  and Economics
Budapest, Hungary

Andre Rios
Polytechnic University of Catalonia
Barcelona, Spain

Sebastia Sallent
i2CAT Foundation
Barcelona, Spain

Juan J. Sánchez
Communications Engineering
  Department
University of Malaga
Malaga, Spain

Gordon L. Stüber
Georgia Institute of Technology
Atlanta, Georgia, U.S.A.

Mitja Stular
Faculty of Electrical Engineering
University of Ljubljana
Ljubljana, Slovenia

András Temesváry
Department of Computer Science and
  Information Theory
Budapest University of Technology
  and Economics
Budapest, Hungary

Kornel Terplan
Terplan Consulting
New York, New York, U.S.A.

Saso Tomažič
Faculty of Electrical Engineering
University of Ljubljana
Ljubljana, Slovenia

Mojca Volk
Faculty of Electrical Engineering
University of Ljubljana
Ljubljana, Slovenia

James B. West
Rockwell Collins—Advanced
  Technology Center

Yang Yang
Alcatel Lucent
New York, New York, U.S.A.

Andrianus Yofy
Department of Electrical
  Engineering
Chulalongkorn University
Bangkok, Thailand

Djamal Zeghlache
Institut National des
    Télécommunications
Evry, France

Qinqing Zhang
John Hopkins University
Washington, D.C., U.S.A.

Chapter 1

Introduction

Syed Ahson

The demand for broadband services is growing exponentially. Mobile users are demanding higher data rates and higher-quality mobile communication services. The Generation Mobile Communication System (third generation, 3G) is an outstanding success. The conflict of rapidly growing users and limited bandwidth resources requires that the spectrum efficiency of mobile communication systems be improved by adopting some advanced technologies. It has been proved, in both theory and practice, that some novel key technologies, such as MIMO (multiple-input, multiple-output) and OFDM (orthogonal frequency-division multiplexing), improve the performance of current mobile communication systems. Since the first deployment of the commercial wireless systems, wireless mobile networks have evolved from the first-generation analog networks to the second-generation (2G) digital networks. Along with the rapid growth of the wireless voice service, 3G wireless mobile networks are deployed to offer more efficient circuit-switched services that utilize many advanced techniques to more than double the spectral efficiency of the 2G systems.

Fixed-mobile convergence and voice–data networks have merged next-generation value-added applications, and integrated multimedia services, combining Web browsing, instant messaging, presence, voice-over Internet Protocol (VoIP), videoconferencing, application sharing, telephony, unified messaging, multimedia content delivery, and the like on top of different network technologies. The convergence of the communications networks is motivated by not only the need to support many forms of digital traffic but also amortization of implementation and operational costs of the underlying networks. Historically, the approach to build and deploy multimedia services has focused on single-point solutions. These solutions worked well to address the specific needs of the intended service or related

1

set of services; however, they possess shortcomings in extensibility to cater to the newer and emerging multimedia services. A more pragmatic approach is to develop a single consolidated platform that is capable of supporting a wide variety of multimedia services over several communication networks.

Many countries and organizations are researching next-generation mobile communication systems, such as the ITU (International Telecommunication Union); European Commission FP (Framework Program); WWRF (Wireless World Research Forum); Korean NGMC (Next-Generation Mobile Committee); Japanese MITF (Mobile IT Forum); and China Communication Standardization Association (CCSA). International standard organizations are working for standardization of the E3G (enhanced 3G) and 4G (fourth generation mobile communication system), such as the LTE (long-term evolution) plan of the Third Generation Partnership Project (3GPP) and the AIE (air interface of evolution)/UMB (ultra mobile broadband) plan of 3GPP2.

The Institute of Electrical and Electronics Engineers (IEEE) Communications Society (IEEE ComSoc) has designed the IEEE WCET (wireless communication engineering technologies) certification program to address the worldwide wireless industry's growing and ever-evolving need for qualified communication professionals who can demonstrate practical problem-solving skills in real-world situations. Individuals who achieve this certification are recognized as having the required knowledge, skill, and ability to meet wireless challenges in various industry, business, corporate, and organizational settings. This book provides technical information about all aspects with which communication professionals should be familiar and offers suggestions for further information and study. The areas covered range from basic concepts to research-grade material, including future directions. This book captures the current state of wireless cellular technology and serves as a source of comprehensive reference material on this subject. It has a total of 22 chapters authored by 50 experts from around the world.

The book provides the following specific salient features:

- A general overview of the evolution of wireless technologies, their impact on the profession, and common professional best practices
- An invaluable resource for keeping pace with evolving standards for experienced wireless professionals
- presentation of accurate, up-to-date information on a broad range of topics related to wireless technologies
- presentation of material authored by experts in the field
- presentation of information in an organized and well-structured manner

Multiple access in wireless radio systems is based on insulating signals used in different connections from each other. The support of parallel transmissions on the uplink and downlink is called *multiple access*, whereas the exchange of information in both directions of a connection is referred to as *duplexing*. Chapter 2, "Access

Methods," describes methods that facilitate the sharing of the broadcast communication medium.

Chapter 3, "Adaptation Techniques for Wireless Communication Systems," discusses adaptation algorithms for improving wireless mobile radio system performance and capacity. First, an overview of adaptive resource management and adaptive transmission technologies is given. Then, parameter measurements that make many of these adaptation techniques possible are discussed in detail.

Coding techniques in general are of two types: source coding and channel coding. The source-coding technique refers to the encoding procedure of the source information signal into digital form. On the other hand, channel coding is applied to ensure adequate transmission quality of the signals. Channel coding is a systematic approach for the replacement of the original information symbol sequence by a sequence of code symbols in such a way that permits its reconstruction. Chapter 4, "Coding Techniques," focuses on the channel-encoding technique.

Modulation is the process by which the message information is added to the radio carrier. Most first-generation cellular systems, such as the Advanced Mobile Phone Services (AMPS) use analog frequency modulation (FM) because analog technology was very mature when these systems were first introduced. Digital modulation schemes, however, are the obvious choice for future wireless systems, especially if data services such as wireless multimedia are to be supported. Digital modulation can also improve spectral efficiency because digital signals are more robust against channel impairments. Spectral efficiency is a key attribute of wireless systems that must operate in a crowded radio-frequency (RF) spectrum. Chapter 5, "Modulation Methods," provides a discussion of these and other modulation techniques that are employed in wireless communication systems.

Chapter 6, "Cellular Systems," presents fundamental concepts of cellular systems by explaining various terminologies used to understand the working of these systems. The chapter also provides details on some popular standards.

Chapter 7, "An Overview of cdma2000, WCDMA, and EDGE," provides an overview of the air interfaces of these key technologies. Particular attention is given to the channel structure, modulation, and offered data rates of each technology. A comparison is also made between cdma2000 and WCDMA (wideband code-division multiple access) to help the reader understand the similarities and differences of these two code-division multiple access (CDMA) approaches for the third generation.

Chapter 8, "Universal Mobile for Telecommunication Systems" describes Universal Mobile Telecommunications System (UMTS) services, their quality-of-service (QoS) requirements, and the general architecture of a UMTS network. The chapter focuses on the universal terrestrial radio access network (UTRAN) entities and protocols and presents the radio interface protocol architecture used to handle data and signaling transport among the user, UTRAN, and the core network.

High-speed downlink packet access (HSDPA) introduces first-adaptive modulation and coding, retransmission mechanisms over the radio link and fast packet scheduling, and multiple transmit-and-receive antennas. Chapter 9, "High-Speed Downlink Packet Access," describes the HSDPA system and some of these related advanced radio techniques.

Chapter 10, "IEEE 802.16-Based Wireless MAN," presents the IEEE 802.16 standard and the different versions developed since its appearance in 2001. IEEE 802.16 media-access control (MAC), physical layers, QoS, and mobility issues are discussed, concentrating on QoS scheduling methods and mechanisms proposed by some authors.

The term *radio resource management* is generally used in wireless systems in a broad sense to cover all functions that are related to the assignment and the sharing of radio resources (e.g., mobile terminals, radio bearers, user sessions) among the users of the wireless network. The type of the required resource control, the required resource sharing, and the assignment methods are primarily determined by the basics of the multiple-access technology, such as frequency-division multiple access (FDMA), time-division multiple access (TDMA), or CDMA and the feasible combinations thereof. Likewise, the smallest unit in which radio resources are assigned and distributed among the entities (e.g., power, time slots, frequency bands/carriers, or codes) also varies depending on the fundamentals of the multiple-access technology employed on the radio interface.

Chapter 11, "3GPP Long-Term Evolution," discusses the radio resource management (RRM) functions in LTE as well as presents a detailed description of the LTE radio interface physical layer focusing on the physical resources structure and the set of procedures defined within this layer.

Mobile IP (Request for Comments 3344), described in Chapter 12, "Mobile Internet Protocol," is an Internet standard that improves the IP to support mobility. It allows a mobile host to move from one network to another without breaking the ongoing communication by assigning two IP addresses to the mobile host. Mobile IP resides in the network layer in providing Internet mobility. Thus, this makes Mobile IP not depend on the media where the communication takes place. A mobile host can roam from one type of medium to another without breaking its connectivity.

The explosive growth of IP networks capable of providing sophisticated convergent services based on multimedia applications has increased the development of new services as an alternative to traditional communication systems. In this way, the services convergence is growing under in IP-based provisioning framework and is controlled mainly by lightweight application-level protocols like Session Initiation Protocols (SIPs). Chapter 13, "SIP: Advanced Media Integration," explains the SIP capabilities of building advanced media services and the possibilities of performing advanced services in an SIP environment.

Chapter 14, "VoIP in a Wireless Mobile Network," presents the challenges of supporting VoIP service in wireless mobile data networks and the various techniques that can be used to meet these challenges. With careful design and optimization, the 3G/4G mobile data networks are able to provide VoIP service with not only a similar service quality as in the existing 2G circuit voice network but also a significantly higher RF capacity for the application. By supporting voice service using VoIP techniques, the 3G/4G mobile data networks offer an integrated solution that has high spectral efficiency and high mobility for multimedia applications.

Global evolution and convergence of communication systems have substantially redefined the concept of design, operation, management, and usage of networks and services, called the next-generation networks (NGNs) concept. Throughout the years, in pursuit of this concept, new technologies have emerged, and real-world experiences have identified suitable methods and enablers to provide the anticipated network and service convergence in a uniform and mobile manner. As a result, various reference architectures and proposals have emerged that reflect contemporary research and development trends in different fields of telecommunication systems and apply the latest technologies available. One such standardized reference architecture is the IP Multimedia Subsystem (IMS) which is discussed in Chapter 15, "IP Multimedia Subsystem (IMS)."

Location-based services (LBSs), presented in Chapter 16, "Location-Based Services," allow service providers to target customers and offer them services specifically tailored to where they are and what they need at both a given moment in time and a given location in space. LBSs have the power to transform mobile services, making interactions more relevant, timely, and personal. Many factors contribute to successful LBSs, including positioning technologies, service policies, and content adaptation and personalization. Major communication providers, equipment providers, and application developers are actively supporting LBS standards development to encourage major rollouts of LBSs.

Chapter 17, "Authentication and Privacy in Wireless Systems," covers cryptographic techniques to enable privacy and authentication in the prevalent wireless networks, namely, IEEE 802.11 wireless local-area networks (LANs), 802.16 wireless metropolitan-area networks (MANs), 2G Global System for Mobile Communications (GSM) and CDMA cellular networks, and 3G networks. Privacy is typically protected by symmetric encryption. Authentication is a more difficult problem and involves more elaborate procedures in wireless systems.

Chapter 18, "Network Management Tools," discusses the need for automated tools to support complex network design, planning, and network management tasks. It also reviews the major network management protocols, with particular emphasis on the Simple Network Management Protocol (SNMP) and Remote Networking Monitoring (RMON) because these protocols are widely used and implemented in vendor products.

The time for proprietary solutions is over. Telecommunication service providers (TSPs) can no longer afford to maintain numerous and different support, documentation, and management systems. Standards bodies and industry associations can provide help in streamlining processes and organizational structures. Three applicable solutions—the enhanced telecommunications operations map (eTOM), the telecommunication management network (TMN), and the control objectives for information and related technology (CobiT)—are addressed in Chapter 19, "Blueprints and Guidelines for Telecommunication Service Providers."

Chapter 20, "Antenna Technology," is a brief overview of contemporary antenna types used in cellular, communication links, satellite communication, radar, and other microwave and millimeter wave systems.

Diversity is a commonly used technique in mobile radio systems to combat signal fading. The basic principles of diversity are described in Chapter 21, "Diversity." If several replicas of the same information-carrying signal are received over multiple channels with comparable strengths, which exhibit independent fading, then there is a good likelihood that one or more of these received signals will not be in a fade at any given instant in time, thus making it possible to deliver an adequate signal level to the receiver.

Chapter 22, "Multiple-Input Multiple-Output Systems in Wireless Communications," reviews systems utilizing multiple transmit and multiple receive antennas, commonly known as multiple-input multiple-output (MIMO) systems. This wireless networking technology greatly improves both the range and the capacity of a wireless communication system. MIMO systems pose new challenges for digital signal processing given that the processing algorithms are becoming more complex, with multiple antennas at both ends of the communication channel.

It is not required that chapters be read in any particular order. Communication professionals are encouraged to focus on those chapters in which they feel they are not up to date or lack a good command of the technology discussed. This book should not be viewed as a study guide for a wireless certification exam as it does not address all the topics that may be covered on such an exam. It is rather an outline of the technical areas with which a communication professionals employed in industry should be familiar and offers suggestions in the form of comprehensive references regarding where to turn for further information and study.

Chapter 2

Access Methods

Bernd-Peter Paris

Contents

2.1 Introduction

The radio channel is fundamentally a broadcast communication medium. Therefore, signals transmitted by one user can potentially be received by all other users within range of the transmitter. Although this high connectivity is useful in some applications,

7

like broadcast radio or television, it requires stringent access control in wireless communication systems to avoid, or at least to limit, interference between transmissions.

Throughout, the term *wireless communication systems* is taken to mean communication systems that facilitate two-way communication between a portable radio communication terminal and the fixed network infrastructure. Such systems range from mobile cellular systems to personal communication systems (PCSs) to cordless telephones.

The objective of wireless communication systems is to provide communication channels on demand between a portable radio station and a radio port or base station that connects the user to the fixed network infrastructure. Design criteria for such systems include capacity, cost of implementation, and quality of service. All of these measures are influenced by the method used for providing multiple-access capabilities. However, the opposite is also true: The access method should be chosen carefully in light of the relative importance of design criteria as well as the system characteristics.

Multiple access in wireless radio systems is based on insulating signals used in different connections from each other. The support of parallel transmissions on the uplink and downlink is called *multiple access*, whereas the exchange of information in both directions of a connection is referred to as *duplexing*. Hence, multiple access and duplexing are methods that facilitate the sharing of the broadcast communication medium. The necessary insulation is achieved by assigning to each transmission different components of the domains that contain the signals. The signal domains commonly used to provide multiple-access capabilities include the following:

> *Spatial domain:* All wireless communication systems exploit the fact that radio signals experience rapid attenuation during propagation. The propagation exponent ρ on typical radio channels lies between $\rho = 2$ and $\rho = 6$, with $\rho = 4$ a typical value. As signal strength decays inversely proportional to the ρth power of the distance, far away transmitters introduce interference that is negligible compared to the strength of the desired signal. The cellular design principle is based on the ability to reuse signals safely if a minimum reuse distance is maintained. Directional antennas can be used to enhance the insulation between signals. We do not focus further on the spatial domain in this treatment of access methods.
>
> *Frequency domain:* Signals that occupy nonoverlapping frequency bands can be easily separated using appropriate bandpass filters. Hence, signals can be transmitted simultaneously without interfering with each other. This method of providing multiple-access capabilities is called *frequency-division multiple access* (FDMA).
>
> *Time domain:* Signals can be transmitted in nonoverlapping time slots in a round-robin fashion. Thus, signals occupy the same frequency band but are easily separated based on their time of arrival. This multiple-access method is called *time-division multiple access* (TDMA).
>
> *Code domain:* In code-division multiple access (CDMA), different users employ signals that have very small cross correlation. Thus, correlators can be used

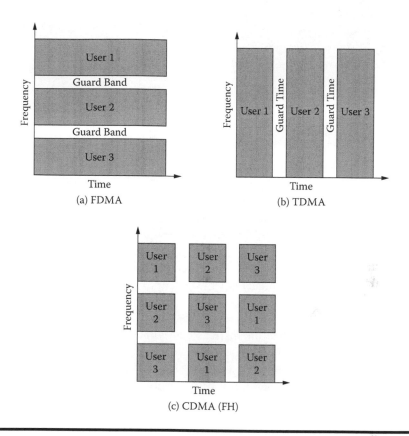

Figure 2.1 Multiple-access methods for wireless communication systems.

to extract individual signals from a mixture of signals even though they are transmitted simultaneously and in the same frequency band. The term code-division multiple access is used to denote this form of channel sharing. Two forms of CDMA are most widely employed and are described in detail: frequency hopping (FH) and direct sequence (DS).

System designers have to decide in favor of one, or a combination, of the last three domains to facilitate multiple access. The three access methods are illustrated in Figure 2.1. The principal idea in all three of these access methods is to employ signals that are orthogonal or nearly orthogonal. Then, correlators that project the received signal into the subspace of the desired signal can be employed to extract a signal without interference from other transmissions.

Preference for one access method over another depends largely on overall system characteristics. No single access method is universally preferable, and system considerations should be carefully weighed before the design decision is made. Before going into the detailed description of the different access methods, we discuss briefly the salient features of some wireless communication systems. This allows us later to assess the relative merits of the access methods in different scenarios.

2.2 Relevant Wireless Communication System Characteristics

Modern wireless radio systems range from relatively simple cordless telephones to mobile cellular systems and PCSs. It is useful to consider such diverse systems as cordless telephone and mobile cellular radio to illustrate some of the fundamental characteristics of wireless communication systems [1].

A summary of the relevant parameters and characteristics for cordless telephone and cellular radio is given in Table 2.1. As evident from the table, the fundamental differences between the two systems are speech quality and the area covered by a base station. The high speech quality requirement in the cordless application is the consequence of the availability of tethered access in the home and office and the resulting direct competition with wire-line telephone services. In the mobile cellular application, the user has no alternative to the wireless access and may be satisfied with lower, but still acceptable, quality of service.

In cordless telephone applications, the transmission range is short because the base station can simply be moved to a conveniently located wire-line access point

Table 2.1   Summary of Relevant Characteristics of Cordless Telephone and Cellular Mobile Radio

Characteristic or Parameter	Cordless Telephone	Cellular Radio
Speech quality	Toll quality	Varying with channel quality; possibly decreased by speech pause exploitation
Transmission range	<100 m	100 m–30 km
Transmit power	Milliwatts	Approx. 1 W
Base station antenna height	Approx. 1 m	Tens of meters
Delay spread	Approx. 1 µs	Approx. 10 µs
Complexity of base station	Low	High
Complexity	Low	High

(wall jack) to provide wireless network access where desired. In contrast, the mobile cellular base station must provide access for users throughout a large geographical area of up to approximately 30 km (20 miles) around the base station. This large coverage area is necessary to economically meet the promise of uninterrupted service to roaming users.

The different range requirements directly affect the transmit power and antenna height for the two systems. High-power transmitters used in mobile cellular user sets consume far more power than even complex signal-processing hardware. Hence, sophisticated signal processing, including speech compression, voice activity detection, error correction and detection, and adaptive equalization, can be employed without substantial impact on the battery life in portable handsets. Furthermore, such techniques are consistent with the goals of increased range and support of large numbers of users with a single, expensive base station. On the other hand, the high mobile cellular base station antennas introduce delay spreads that are one or two orders of magnitude larger than those commonly observed in cordless telephone applications.

Clearly, the two systems just considered are at extreme ends of the spectrum of wireless communications systems. Most notably, the PCSs fall somewhere between the two. However, the comparison highlights some of the system characteristics that should be considered when discussing access methods for wireless communication systems.

2.3 Frequency-Division Multiple Access

As mentioned in Section 2.1, in FDMA nonoverlapping frequency bands are allocated to different users on a continuous time basis. Hence, signals assigned to different users are clearly orthogonal, at least ideally. In practice, out-of-band spectral components cannot be completely suppressed, which leaves signals not quite orthogonal. This necessitates the introduction of guard bands between frequency bands to reduce adjacent channel interference (i.e., inference from signals transmitted in adjacent frequency bands; see also Figure 2.1a).

It is advantageous to combine FDMA with time-division duplexing (TDD) to avoid simultaneous reception and transmission that would require insulation between receive and transmit antennas. In this scenario, the base station and portable take turns using the same frequency band for transmission. Nevertheless, combining FDMA and frequency-division duplexing (FDD) is possible in principle, as is evident from the analog systems based on frequency modulation (FM) deployed throughout the world since the early 1980s.

2.3.1 Channel Considerations

In principle, there exists the well-known duality between TDMA and FDMA. In the wireless environment, however, propagation-related factors have a strong influence on the comparison between FDMA and TDMA. Specifically, the duration of a transmitted symbol is much longer in FDMA than in TDMA. As an immediate consequence, an equalizer is typically not required in an FDMA-based system because the delay spread is small compared to the symbol duration.

To illustrate this point, consider a hypothetical system that transmits information at a constant rate of 50 kbps. This rate would be sufficient to support 32-kbps adaptive differential pulse code modulation (ADPCM) speech encoding, some coding for error protection, and control overhead. If we assume further that some form of quadrature phase-shift keying (QPSK) modulation is employed, the resulting symbol duration is 40 μs. In relation to delay spreads of approximately 1 μs in the cordless application and 10 μs in cellular systems, this duration is large enough that only little intersymbol interference is introduced. In other words, the channel is frequency nonselective (i.e., all spectral components of the signal are affected equally by the channel). In the cordless application, an equalizer is certainly not required; cellular receivers may require equalizers capable of removing intersymbol interference between adjacent bits. Furthermore, it is well known that intersymbol interference between adjacent bits can be removed without loss in signal-to-noise ratio (SNR) by using maximum likelihood sequence estimation (e.g., [2, p. 622]).

Hence, rather simple receivers can be employed in FDMA systems at these data rates. However, there is a flip side to the argument. Recall that the Doppler spread, which characterizes the rate at which the channel impulse response changes, is given approximately by $B_d = v/cf_c$, where v denotes the speed of the mobile user, c is the propagation speed of the electromagnetic waves carrying the signal, and f_c is the carrier frequency. Thus, for systems operating in the vicinity of 1 GHz, B_d will be less than 1 Hz in the cordless application and typically about 100 Hz for a mobile traveling on a highway. In either case, the signal bandwidth is much larger than the Doppler spread B_d, and the channel can be characterized as slowly fading. Whereas this allows tracking of the carrier phase and the use of coherent receivers, it also means that fade durations are long in comparison to the symbol duration and can cause long sequences of bits to be subject to poor channel conditions. The problem is compounded by the fact that the channel is frequency nonselective because it implies that the entire signal is affected by a fade.

To overcome these problems, time diversity, frequency diversity, or spatial diversity could be employed. Time diversity can be accomplished by a combination of coding and interleaving if the fading rate is sufficiently large. For very slowly fading channels, such as the cordless application, the necessary interleaving depth would introduce too much delay to be practical. Frequency diversity can be introduced simply by slow frequency hopping, a technique that prescribes users to change the carrier frequency periodically. Frequency hopping is a form of spectrum spreading because the bandwidth occupied by the resulting signal is much larger than the symbol rate. In contrast to direct sequence spread spectrum (discussed in Section 2.5), however, the instantaneous bandwidth is not increased. The jumps between different frequency bands effectively emulate the movement of the portable and thus should be combined with the just-described time diversity methods. Spatial diversity is provided by the use of several receive or transmit antennas. At carrier frequencies exceeding 1 GHz, antennas are small, and two or more antennas can be accommodated even in the handset. Furthermore, if FDMA is combined with TDD, multiple antennas at the base station can provide diversity on both uplink and downlink. This is possible because the channels for the two links are virtually identical, and the base station, using channel information gained from observing the portable's signal, can transmit signals at each antenna such that they combine coherently at the portable's antenna. Thus, signal-processing complexity is moved to the base station, extending the portable's battery life.

2.3.2 *Influence of Antenna Height*

In the cellular mobile environment, base station antennas are raised considerably to increase the coverage area. Antennas mounted on towers and rooftops are a common sight, and antenna heights of 50 m above ground are no exceptions. Besides increasing the coverage area, this has the effect that frequently there exists a better propagation path between two base station antennas than between a mobile and the base station (see Figure 2.2).

Assuming that FDMA is used in conjunction with TDD as specified at the beginning of this section, then base stations and mobiles transmit on the same frequency. Now, unless there is tight synchronization between all base stations, signals from other base stations will interfere with the reception of signals from portables at the base station. To keep the interference at acceptable levels, it is necessary to increase the reuse distance (i.e., the distance between cells using the same frequencies). In other words, sufficient insulation in the spatial domain must be provided to facilitate the separation of signals. Note that these comments apply equally to cochannel and adjacent channel interference.

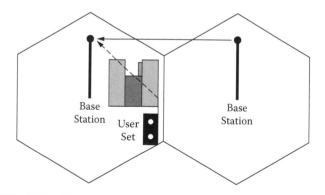

Figure 2.2 High base station antennas lead to stronger propagation paths between base stations than between a user set and its base stations.

This problem does not arise in cordless applications. Base station antennas are generally of the same height as user sets. Hence, interference created by base stations is subject to the same propagation conditions as signals from user sets. Furthermore, in cordless telephone applications there are frequently attenuating obstacles, such as walls, between base stations that reduce intracell interference further. Note that this reduction is vital for the proper functioning of cordless telephones since there is typically no network planning associated with installing a cordless telephone. As a safety feature, to overcome intercell interference, adaptive channel management strategies based on sensing interference levels can be employed.

Example 2.1: CT2

The CT2 standard was originally adopted in 1987 in Great Britain and improved with a common air interface (CAI) in 1989. The CAI facilitates interoperability between equipment from different vendors, whereas the original standard only guarantees noninterference. The CT2 standard is used in home and office cordless telephone equipment and has been used for telepoint applications [3].

CT2 operates in the frequency band 864–868 MHz and uses carriers spaced at 100 kHz. FDMA with TDD is employed. The combined gross bit rate is 72 kbps, transmitted in frames of 2-ms duration, of which the first half carries downlink and the second half carries uplink information. This setup supports a net bit rate of 32 kbps of user data (32-kbps ADPCM encoded speech) and 2-kbps control information in each direction. The CT2 modulation technique is binary frequency shift keying.

2.3.3 *Further Remarks*

From the discussion, it is obvious that FDMA is a good candidate for applications like cordless telephones. In particular, the simple signal processing makes it a good choice for inexpensive implementation in the benign cordless environment. The possibility of concentration of signal-processing functions in the base station strengthens this aspect.

In the cellular application, on the other hand, FDMA is inappropriate because of the lack of built-in diversity and the potential for severe intercell interference between base stations. A further complication arises from the difficulty of performing handovers if base stations are not tightly synchronized.

For PCSs, the decision is not as obvious. Depending on whether the envisioned PCS application resembles more a cordless private branch exchange (PBX) than a cellular system, FDMA may be an appropriate choice. We will see later that it is probably better to opt for a combined TDMA/FDMA or a CDMA-based system to avoid the pitfalls of pure FDMA systems and still achieve moderate equipment complexities.

Finally, there is the problem of channel assignment. Clearly, it is not reasonable to assign a unique frequency to each user as there are insufficient frequencies, and the spectral resource would be unused whenever the user is idle. Instead, methods that allocate channels on demand can make much more efficient use of the spectrum. Such methods are discussed in the description of TDMA systems.

2.4 Time-Division Multiple Access

In TDMA systems, users share the same frequency band by accessing the channel in nonoverlapping time intervals in a round-robin fashion [4]. Since the signals do not overlap, they are clearly orthogonal, and the signal of interest is easily extracted by switching the receiver on only during the transmission of the desired signal. Hence, the receiver filters are simply windows instead of the bandpass filters required in FDMA. As a consequence, the guard time between transmissions can be made as small as the synchronization of the network permits. Guard times of 30–50 µs between time slots are commonly used in TDMA-based systems. As a consequence, all users must be synchronized with the base station to within a fraction of the guard time. This is achievable by distributing a master clock signal on one of the base station's broadcast channels.

TDMA can be combined with TDD or FDD. The former duplexing scheme is used, for example, in the Digital European Cordless Telephone (DECT) standard and is well suited for systems in which base-to-base and mobile-to-base propagation paths are similar, that is, systems without extremely high base station antennas. Since both the portable and the base station transmit on the same frequency, some signal-processing functions for the downlink can be implemented in the base station, as discussed for FDMA/TDD systems.

In the cellular application, the high base station antennas make FDD the more appropriate choice. In these systems, separate frequency bands are provided for uplink and downlink communication. Note that it is still possible and advisable to stagger the uplink and downlink transmission intervals such that they do not overlap to avoid the situation that the portable must transmit and receive at the same time.

With FDD, the uplink and downlink channels are not identical; hence, signal-processing functions cannot be implemented in the base station. Antenna diversity and equalization have to be realized in the portable.

2.4.1 Propagation Considerations

In comparison to a FDMA system supporting the same user data rate, the transmitted data rate in a TDMA system is larger by a factor equal to the number of users sharing the frequency band. This factor is 8 in the pan-European Global System for Mobile Communications (GSM; originally Groupe Spécial Mobile) and 3 in the digital advanced mobile phone service (D-AMPS) system. Thus, the symbol duration is reduced by the same factor, and severe intersymbol interference results, at least in the cellular environment.

To illustrate, consider the example in which each user transmits 25 thousand symbols per second. Assuming eight users per frequency band leads to a symbol duration of 5 µs. Even in the cordless application with delay spreads of up to 1 µs, an equalizer may be useful to combat the resulting interference between adjacent symbols. In cellular systems, however, the delay spread of up to 20 µs introduces severe intersymbol interference spanning up to five symbol periods. As the delay spread often exceeds the symbol duration, the channel can be classified as *frequency selective*, emphasizing the observation that the channel affects different spectral components differently.

The intersymbol interference in cellular TDMA systems can be so severe that linear equalizers are insufficient to overcome its negative effects. Instead, more powerful, nonlinear decision feedback or maximum-likelihood sequence estimation equalizers must be employed [5]. Furthermore, all of these equalizers require some information about the channel impulse response that must be estimated from the received signal by means of an embedded training sequence. Clearly, the training sequence carries no user data and, thus, wastes valuable bandwidth.

In general, receivers for cellular TDMA systems will be fairly complex. On the positive side of the argument, however, the frequency selective nature of the channel provides some built-in diversity that makes transmission more robust to channel fading. The diversity stems from the fact that the multipath components of the received signal can be resolved at a resolution roughly equal to the symbol duration, and the different multipath components can be combined by the equalizer during the demodulation of the signal. To further improve robustness to channel fading, coding, and interleaving, slow frequency hopping and antenna diversity can be employed as discussed in connection with FDMA.

2.4.2 *Initial Channel Assignment*

In both FDMA and TDMA systems, channels should not be assigned to a mobile on a permanent basis. A fixed assignment strategy would be either extremely wasteful of precious bandwidth or highly susceptible to cochannel interference. Instead, channels must be assigned on demand. Clearly, this implies the existence of a separate uplink channel on which mobiles can notify the base station of their need for a traffic channel. This uplink channel is referred to as the *random-access channel* because of the type of strategy used to regulate access to it.

The successful procedure for establishing a call that originates from the mobile station is outlined in Figure 2.3. The mobile initiates the procedure by transmitting a request on the random-access channel. Since this channel is shared by all users in range of the base station, a random-access protocol, like the ALOHA protocol, has to be employed to resolve possible collisions. Once the base station has received the mobile's request, it responds with an immediate assignment message

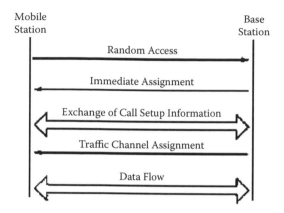

Figure 2.3 **Mobile-originating call establishment.**

that directs the mobile to tune to a dedicated control channel for the ensuing call setup. On completion of the call setup negotiation, a traffic channel (i.e., a frequency in FDMA systems or a time slot in TDMA systems) is assigned by the base station, and all future communication takes place on that channel. In the case of a mobile-terminating call request, the sequence of events is preceded by a paging message alerting the base station of the call request.

Example 2.2: GSM

Named after the organization that created the system standards (Groupe Spécial Mobile), this pan-European digital cellular system has been deployed in Europe since the early 1990s [6]. GSM uses combined TDMA and FDMA with FDD for access. Carriers are spaced at 200 kHz and support eight TDMA time slots each. For the uplink, the frequency band 890–915 MHz is allocated, whereas the downlink uses the band at 935–960 MHz. Each time slot is 577 μs, which corresponds to 156.26-bit periods, including a guard time of 8.25-bit periods. Eight consecutive time slots form a GSM frame of duration 4.62 ms.

The GSM modulation is Gaussian minimum shift keying with a time–bandwidth product of 0.3; that is, the modulator bandpass has a cutoff frequency of 0.3 times the bit rate. At the bit rate of 270.8 kbps, severe intersymbol interference arises in the cellular environment. To facilitate coherent detection, a 26-bit training sequence is embedded into every time slot. Time diversity is achieved by interleaving over 8 frames for speech signals and 20 frames for data communication. Sophisticated error-correction coding with varying levels of protection for different outputs of the speech coder is provided. Note that the round-trip delay introduced by the interleaver is on the order of 80 ms for speech signals. GSM provides slow frequency hopping as a further mechanism to improve the efficiency of the interleaver.

2.4.3 Further Remarks

In cellular systems, such as GSM or the North American D-AMPS, TDMA is combined with FDMA. Different frequencies are used in neighboring cells to provide orthogonal signaling without the need for tight synchronization of base stations. Furthermore, channel assignment can then be performed in each cell individually. Within a cell, one or more frequencies are shared by users in the time domain.

From an implementation standpoint, TDMA systems have the advantage that common radio and signal-processing equipment at the base station can be shared by users communicating on the same frequency. A somewhat more subtle advantage of TDMA systems arises from the possibility of monitoring surrounding base stations and frequencies for signal quality to support mobile-assisted handovers.

2.5 Code-Division Multiple Access

CDMA systems employ wideband signals with good cross-correlation properties [7]. That means the output of a filter matched to one user's signal is small when a different user's signal is input. A large body of work exists on spreading sequences that lead to signal sets with small cross correlations [8]. Because of their noise-like appearance, such sequences are often referred to as *pseudonoise* (PN) sequences, and because of their wideband nature, CDMA systems are often called *spread-spectrum systems.*

Spectrum spreading can be achieved mainly in two ways: through frequency hopping as explained or through direct sequence spreading. In direct sequence spread spectrum, a high-rate, antipodal pseudorandom spreading sequence modulates the transmitted signal such that the bandwidth of the resulting signal is roughly equal to the rate of the spreading sequence. The cross correlation of the signals is then largely determined by the cross-correlation properties of the spreading signals. Clearly, CDMA signals overlap in both time and frequency domains but are separable based on their spreading waveforms.

An immediate consequence of this observation is that CDMA systems do not require tight synchronization between users as do TDMA systems. By the same token, frequency planning and management are not required as frequencies are reused throughout the coverage area.

2.5.1 Propagation Considerations

Spread spectrum is well suited for wireless communication systems because of its built-in frequency diversity. As discussed, in cellular systems the delay spread measures several microseconds; hence, the coherence bandwidth of the channel is smaller than 1 MHz. Spreading rates can be chosen to exceed the coherence bandwidth such that the channel becomes frequency selective; that is, different spectral components are affected unequally by the channel, and only parts of the signal are affected by fades.

Expressing the same observation in time domain terms, multipath components are resolvable at a resolution equal to the chip period and can be combined coherently, for example, by means of a RAKE receiver [2]. An estimate of the channel impulse response is required for the coherent combination of multipath components. This estimate can be gained from a training sequence or by means of a so-called pilot signal.

Even for cordless telephone systems, operating in environments with submicrosecond delay spread and corresponding coherence bandwidths of a few megahertz, the spreading rate can be chosen large enough to facilitate multipath diversity. If the combination of multipath components described is deemed too complex, a simpler, but less powerful, form of diversity can be used that decorrelates only the strongest received multipath component and relies on the suppression of other path components by the matched filter.

2.5.2 Multiple-Access Interference

If it is possible to control the relative timing of the transmitted signals, such as on the downlink, the transmitted signals can be made perfectly orthogonal, and if the channel only adds white Gaussian noise, matched filter receivers are optimal for extracting a signal from the superposition of waveforms. If the channel is dispersive because of multipath, the signals arriving at the receiver will no longer be orthogonal and will introduce some multiple-access interference, that is, signal components from other signals that are not rejected by the matched filter.

On the uplink, extremely tight synchronization between users to within a fraction of a chip period, which is defined as the inverse of the spreading rate, is generally not possible, and measures to control the impact of multiple-access interference must be taken. Otherwise, the near–far problem (i.e., the problem of very strong undesired users' signals overwhelming the weaker signal of the desired user) can severely decrease performance. Two approaches are proposed to overcome the near–far problem: power control with soft handovers and multiuser detection.

Power control attempts to ensure that signals from all mobiles in a cell arrive at the base station with approximately equal power levels. To be effective, power control must be accurate to within about 1 dB and fast enough to compensate for channel fading. For a mobile moving at 55 mph and transmitting at 1 GHz, the Doppler bandwidth is approximately 100 Hz. Hence, the channel changes its characteristic drastically about 100 times per second, and on the order of 1000 bps must be sent from base station to mobile for power control purposes. As different mobiles may be subject to vastly different fading and shadowing conditions, a large dynamic range of about 80 dB must be covered by power control. Notice that power control on the downlink is really only necessary for mobiles that are about equidistant from two base stations, and even then neither the update rate nor the dynamic range of the uplink is required.

The interference problem that arises at the cell boundaries where mobiles are within range of two or more base stations can be turned into an advantage through the idea of soft handover. On the downlink, all base stations within range can transmit to the mobile, which in turn can combine the received signals to achieve some gain from the antenna diversity. On the uplink, a similar effect can be obtained by selecting the strongest received signal from all base stations that receive a user's signal. The base station that receives the strongest signal will also issue power control commands to minimize the transmit power of the mobile. Note, however, that soft handover requires fairly tight synchronization between base stations, and one of the advantages of CDMA over TDMA is lost.

Multiuser detection is still an emerging technique. It is probably best used in conjunction with power control. The fundamental idea behind this technique is to model multiple-access interference explicitly and devise receivers that reject or cancel the undesired signals. A variety of techniques has been proposed, ranging from optimum maximum likelihood sequence estimation via multistage schemes, reminiscent of decision feedback algorithms, to linear decorrelating receivers. An excellent survey of the theory and practice of multiuser detection is given by [9].

2.5.3 Further Remarks

CDMA systems work well in conjunction with FDD. This arrangement decouples the power control problem on the uplink and downlink, respectively. Signal quality-enhancing methods, such as time diversity through coding and interleaving, can be applied just as with the other access methods. In spread-spectrum systems, however, coding can be built into the spreading process, avoiding the loss of bandwidth associated with error protection. In addition, CDMA lends itself naturally to the exploitation of speech pauses that make up more than half the time of a connection. If no signals are transmitted during such pauses, then the instantaneous interference level is reduced, and the total number of users supportable by the system can be approximately doubled.

2.6 Comparison and Outlook

The question of which of the access methods is best does not have a single answer. Based on the preceding discussion, FDMA is suited only for applications such as cordless telephone with very small cells and submicrosecond delay spreads. In cellular systems and for most versions of PCSs, the choice reduces to TDMA versus CDMA.

In terms of complexity, TDMA receivers require adaptive, nonlinear equalizers when operating in environments with large delay spreads. CDMA systems in turn need RAKE receivers and sophisticated power control algorithms. In the future, some form of multiple-access interference rejection is likely to be implemented as well. Time synchronization is required in both systems, albeit for different reasons. The additional complexity for coding and interleaving is comparable for both access methods.

An often-cited advantage of CDMA systems is the fact that the performance will degrade gracefully as the load increases. In TDMA systems, in turn, requests will have to be blocked once all channels in a cell are in use. Hence, there is a hard limit on the number of channels per cell. There are proposals for extended TDMA systems, however, that incorporate reassignment of channels during speech pauses.

Not only would such extended TDMA systems match the advantage of the exploitation of speech pauses of CDMA systems, they would also lead to a soft limit on the system capacity. The extended TDMA proposals would implement the statistical multiplexing of the user data (e.g., by means of the packet reservation multiple-access protocol) [10]. The increase in capacity depends on the acceptable packet loss rate; in other words, small increases in the load lead to small increases in the packet loss probability.

Many comparisons in terms of capacity between TDMA and CDMA can be found in the recent literature. Such comparisons, however, are often invalidated by making assumptions that favor one access method over the other. An important exception is the article by Wyner [11]. Under a simplified model that nevertheless captures the essence of cellular systems, he computed the Shannon capacity. Highlights of his results include the following:

- TDMA is distinctly suboptimal in cellular systems.
- When the SNR is large, CDMA appears to achieve twice the capacity of TDMA.
- Multiuser detectors are essential to realize near-optimum performance in CDMA systems.
- Intercell interference in CDMA systems has a detrimental effect when the SNR is large, but it can be exploited via diversity combining to increase capacity when the SNR is small.

More research along this avenue is necessary to confirm the validity of the results. In particular, incorporation of realistic channel models into the analysis is required. However, this work represents a substantial step toward quantifying capacity increases achievable with CDMA.

Glossary

Capacity: Shannon originally defined capacity as the maximum data rate that permits error-free communication in a given environment. A looser interpretation is normally employed in wireless communication systems. Here, capacity denotes the traffic density supported by the system under consideration normalized with respect to bandwidth and coverage area.

Code-division multiple access (CDMA): Systems use signals with very small cross correlations to facilitate sharing of the broadcast radio channel. Correlators are used to extract the desired user's signal while simultaneously suppressing interfering, parallel transmissions.

Duplexing: The exchange of messages in both directions of a connection.

Frequency-division multiple access (FDMA): Simultaneous access to the radio channel is facilitated by assigning nonoverlapping frequency bands to different users.

Multiple access: The support of simultaneous transmissions over a shared communication channel.

Random-access channel: This uplink control channel is used by mobiles to request assignment of a traffic channel. A random-access protocol is employed to arbitrate access to this channel.

Time-division multiple access (TDMA): Systems assign nonoverlapping time slots to different users in a round-robin fashion.

References

1. Cox, D.C., Wireless network access for personal communications. *IEEE Commun. Mag.,* 96–115, 1992.
2. Proakis, J.G., *Digital Communications,* 2nd ed. New York: McGraw-Hill, 1989.
3. Goodman, D.J., Second generation wireless information networks. *IEEE Trans. Veh. Technol.,* 40(2), 366–374, 1991.
4. Falconer, D.D., Adachi, F., and Gudmundson, B., Time division multiple access methods for wireless personal communications. *IEEE Comm. Mag.,* 33(1), 50–57, 1995.
5. Proakis, J.G., Adaptive equalization for TDMA digital mobile radio. *IEEE Trans. Veh. Technol.,* 40(2), 333–341, 1991.
6. Hodges, M.R.L., The GSM radio interface. *Br. Telecom Tech. J.,* 8(1), 31–43, 1990.
7. Kohno, R., Meidan, R., and Milstein, L.B., Spread spectrum access methods for wireless communications. *IEEE Commun. Mag.,* 33(1), 58, 1995.
8. Sarwate, D.V., and Pursley, M.B., Crosscorrelation properties of pseudorandom and related sequences. *Proc. IEEE,* 68(5), 593–619, 1980.
9. Verdu, S., Multi-user detection. In *Advances in Statistical Signal Processing—Vol. 2: Signal Detection.* JAI Press, Greenwich, CT, 1992.
10. Goodman, D., Trends in cellular and cordless communications. *IEEE Commun. Mag.,* 31–40, 1991.
11. Wyner, A.D., Shannon-theoretic approach to a Gaussian cellular multiple-access channel. *IEEE Trans. Inform. Theory,* 40(6), 1713–1727, 1994.

Chapter 3

Adaptation Techniques for Wireless Communication Systems

Huseyin Arslan

Contents

3.1 Introduction

Wireless communication systems have evolved substantially over the last two decades. The explosive growth of the wireless communication market is expected to continue in the future as the demand for all types of wireless services is increasing. There is no doubt that the second generation of cellular wireless communication systems was a success.

These systems, however, were designed to provide good coverage for voice services so that a minimum required signal quality could be ensured over the coverage area. If the received signal quality is well above the minimum required level, the receivers do not exploit this. The speech quality does not improve much as the quality is mostly dominated by the speech coder. On the other hand, if the signal quality is below the minimum required level, a call drop will be observed. Therefore, such a design requires the use of strong forward error correction (FEC) schemes, low-order modulations, and many other redundancies at transmission and reception. In essence, the mobile receivers and transmitters are designed for the worst-case channel and received signal conditions. As a result, many users experience unnecessarily high signal quality from which they cannot benefit. While reliable communication is achieved, the system resources are not used efficiently.

New generations of wireless mobile radio systems aim to provide higher data rates and a wide variety of applications (such as video and data) to mobile users while serving as many users as possible. However, this goal must be achieved under spectrum and power constraints. Given the high price of spectrum and its scarcity, the systems must provide higher system capacity and performance through better use of the available resources. Therefore, adaptation techniques have been becoming popular for optimizing mobile radio system transmission and reception at the physical layer as well as at the higher layers of the protocol stack.

Traditional system designs focus on allocating fixed resources to the user. Adaptive design methodologies typically identify the user's requirements and then allocate just enough resources, thus enabling more efficient utilization of system resources and consequently increasing capacity. Adaptive channel allocation and adaptive cell assignment algorithms have been studied since the early days of cellular systems. As the demand in wireless access for speech and data has increased, link and system adaptation algorithms have become more important.

For a given average transmit power, adaptation allows the users to experience better signal qualities. Adaptation reduces the average interference observed from other users as they do not transmit extra power unnecessarily. As a result, the received signal quality will be improved over a large portion of the coverage area. These higher-quality signal levels can be exploited to provide increased data rates through rate adaptation. For a desired received signal quality, this might also translate into less transmit power, leading to improved power efficiency for longer battery life. On the other hand, for a desired minimum signal quality, this might lead to an increased coverage area or better frequency reuse. In addition, adaptive receiver designs allow the receiver to work with reduced signal quality values; that is, a desired bit error rate (BER) or frame error rate (FER) performance can be achieved with lower signal quality. Adaptive receivers can also enable reduced average computational complexities for the same quality of service, which again implies less power consumption. As can be seen, adaptation algorithms lead to improved performance, increased capacity, lower power consumption, increased radio coverage area, and eventually better overall wireless communication system design.

Many adaptation schemes require a form of measurement (or estimation) of various quantities (parameters) that might change over time. These estimates are then used to trigger or perform a multitude of functions, like the adaptation of the transmission and reception. For example, Doppler spread and delay spread estimations, signal-to-noise ratio (SNR) estimation, channel estimation, BER estimation, cyclic redundancy check (CRC) information, and received signal strength (RSS) measurement are some of the commonly used measurements for adaptive algorithms. As the interest in the adaptation schemes increases, so does the research on improved (fast and accurate) parameter estimation techniques.

In this chapter, an overview of commonly used adaptation techniques and their applications for wireless mobile radio systems is given. Some of the commonly used parameters and their estimation using baseband signal-processing techniques are explained in detail.

3.2 Overview of Adaptation Schemes

In wireless mobile communication systems, information is transmitted through a radio channel. Unlike other guided media, the radio channel is highly dynamic. The transmitted signal reaches the receiver by undergoing many effects, corrupting the signal, and often placing limitations on the performance of the system.

Figure 3.1 illustrates a wireless communication system that includes some of the effects of the radio channel. The RSS varies depending on the distance relative to the transmitter, shadowing caused by large obstructions, and fading due to reflection, diffraction, and scattering. Mobility of the transmitter, receiver, or scattering objects causes the channel to change over time. Moreover, the interference conditions in the system change rapidly. Most important, the radio channel is highly random, and the statistical characteristics of the channel are environment dependent. In addition to these changes, the traffic load, types of services, and mobile user characteristics and requirements might also vary in time. Adaptive techniques can be used to address all these changing conditions.

The adaptation strategy can be different depending on the application and services. Constant BER constraint for a given fixed transmission bandwidth and constant throughput constraint are two of the most popular criteria for adaptation. In constant BER, a desired average or instantaneous BER is defined to satisfy the

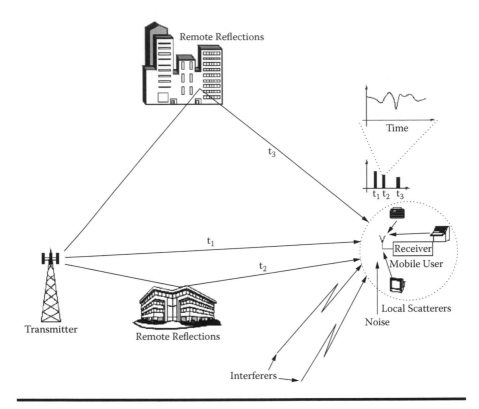

Figure 3.1 Illustration of some of the effects of radio channel. Local scatterers cause fading; remote reflectors cause multipath and time dispersion, leading to intersymbol interference (ISI); mobility of user or scatterers cause time varying channel; reuse of frequencies, adjacent carriers cause interference.

acceptable quality of service. Then, the system is adapted to the varying channel and interference conditions so that the BER is maintained below the target value. To ensure this for all types of channel and interference conditions, the system changes power, modulation order, coding rate, spreading factor, and so on. Note that this changes the throughput as the channel quality changes. On the other hand, for the constant-throughput case, the adaptations are done to make sure that the effective throughput is constant where the BER might change.

In general, it is possible to classify the adaptation algorithms as link and transmitter adaptation, adaptation of system resource allocation, and receiver adaptation. In the following sections, these adaptation techniques are briefly discussed.

3.2.1 Link and Transmitter Adaptation

A reliable link must ensure that the receiver is able to capture and reproduce the transmitted information bits. Therefore, the target link quality must be maintained all the time in spite of the changes in the channel and interference conditions. As mentioned, one way to achieve this is to design the system for the worst-case scenario so that the target link quality can always be achieved.

If the transmitter sends more power for a specific user, the user benefits from it by having better link quality, but the level of interference for the other users increases accordingly. On the other hand, if the user does not receive enough power, a reliable link cannot be established. To establish a reliable link while minimizing interference to other users, the transmitter should continuously control the transmitted power level. Power control is a simple form of adaptation that compensates for the variation of the received signal level due to path loss, shadowing, and sometimes fading.

Numerous studies on power control schemes have been performed for various radio communication systems (see [1] and the references listed therein). In code-division multiple-access (CDMA) systems, signals having widely different power levels at the receiver cause strong signals to swamp out weaker ones in a phenomenon known as the *near–far effect*. Power control mitigates the near–far problem by controlling the transmitted power.

It is possible to trade off power for bandwidth efficiency; that is, a desired BER (or FER) can be achieved by increasing the power level or by reducing the bandwidth efficiency. One way of establishing a reliable link is to add redundancy to the information bits through FEC techniques. With no other changes, this would normally reduce the information rate (or bandwidth efficiency) of the communication. In the same way, high-quality links can be obtained by transmitting the signals with spectrally less-efficient modulation schemes, like binary phase-shift keying (BPSK) and quaternary phase-shift keying (QPSK).

On the other hand, new-generation wireless systems aim for higher data rates made possible through spectrally efficient higher-order modulations. Therefore, a reliable link with higher information rates can be accomplished by continuously controlling the coding and modulation levels. Higher modulation orders with less-powerful coding rates are assigned to users that experience good link qualities, so that the excess signal quality can be used to obtain higher data rates. Recent designs have exploited this with adaptive modulation techniques that change the order of the modulation [1,2], as well as with adaptive coding schemes that change the coding rate [3,4].

For example, the Enhanced General Packet Radio Service (EGPRS) standard introduces both Gaussian minimum-shift keying (GMSK) and 8-PSK (phase shift key) modulations with different coding rates through link adaptation and hybrid automatic repeat request (ARQ) [5]. The channel quality is estimated at the receiver, and the information is passed to the transmitter through appropriately defined messages. The transmitter adapts the coding and modulation based on this channel quality feedback. Similarly, variable spreading and coding techniques are present in third-generation CDMA-based systems [3], cdma2000 and wideband CDMA (WCDMA, or UMTS [Universal Mobile Telecommunications System]). Higher data rates can be achieved by changing the spreading factor and coding rate, depending on the perceived communication link qualities.

Adaptive antennas and adaptive beam-forming techniques have also been studied extensively to increase the capacity and to improve the performance of wireless communication systems [6]. The adaptive antenna systems shape the radiation pattern in such a way that the information is transmitted (e.g., from a base station) directly to the mobile user in narrow beams. This reduces the probability of another user experiencing interference in the network, resulting in improved link quality, which can also be translated into increased network capacity. Although adaptive beam forming is an excellent way to utilize multiple-antenna systems to enhance the link quality, recently different flavors of the usage of multiantenna systems have gained significant interest. Space–time processing and multiple-input multiple-output (MIMO) antenna systems are some new developments that will allow further usage of multiple-antenna systems in wireless communications. Adaptive implementation of these technologies is important for successful and efficient integration of them into wireless communication systems.

3.2.2 Adaptive System Resource Allocation

In addition to physical link adaptation, system resources can also be allocated adaptively to reduce the interference and to improve the overall system quality. This includes adaptive power control, adaptive channel allocation, adaptive cell assignment, adaptive resource scheduling, adaptive spectrum management, congestion, handoff (mobility), admission, and load control strategies. Adaptive system resource allocation considers the current traffic load as well as the channel and interference conditions. For example, the system could assign more resources to the mobiles that have better link quality to increase the throughput. Alternatively, the system could assign the resources to the user in such a way that the user experiences better quality for the current traffic condition.

Adaptive channel allocation and adaptive cell assignment in hierarchical cellular systems have been studied since the early days of cellular systems. Adaptive channel allocation increases the system capacity through efficient channel utilization and decreased probability of blocked calls [7]. Unlike fixed-channel allocation, in which the channels are assigned to the cells permanently and on the basis of the worst-case scenario, in adaptive channel assignment a common pool of channels is shared by many cells, and the channels are assigned with regard to the interference and traffic conditions.

Adaptive cell assignment can increase capacity without increasing the handoff rate. The cells can be assigned to the users depending on their mobility level. Fast-moving mobiles can be assigned to larger umbrella cells (to reduce the number of handoffs), while slow-moving mobiles are assigned to microcells (to increase capacity) [8].

Research on increasing the average throughput of the system through resource allocation based on water filling has gained significant interest [9–11]. The main idea is to allocate more resources to the users who experience better link quality, resulting in efficient use of the available resources. The high data rate (HDR) system, which is based on a best-effort radio packet protocol, uses an approach based on water filling in allocating system resources. Algorithms that deal with compromising the throughput to achieve fairness have also been studied [10,11].

3.2.3 Receiver Adaptation

Digital wireless communication receiver performance is related to the required value of the signal-to-interference-plus-noise ratio (SINR) so that the BER (or FER) performance can be kept below a certain threshold for reliable communication. For a given complexity, if receiver A requires lower SINR than receiver B to satisfy the same error rate, receiver A is considered to perform better than receiver B.

Receiver adaptation techniques can increase the performance of the receiver, hence reducing the minimum required SINR. As mentioned, this can be used to increase the coverage area for a fixed transmitted power, or it can be used to reduce the transmitted power requirement for a given coverage area. Moreover, receiver adaptation can reduce the average receiver complexity and the power drain from the battery for the same quality of service. To satisfy the desired BER performance, instead of running a computationally complex algorithm for all channel conditions, the receiver can choose the most appropriate algorithm given the system and channel conditions.

Advanced baseband signal-processing techniques play a significant role in receiver adaptation. Baseband algorithms used for time and frequency synchronization, baseband filtering, channel estimation and tracking, demodulation and equalization, interference cancellation, soft information calculation, antenna selection and combining, decoding, and so on can be made adaptive depending on the channel and interference conditions.

Conventional receiver algorithms are designed for the worst-case channel and interferer conditions. For example, the channel estimation and tracking algorithms assume the worst-case mobile speed; the channel equalizers assume the worst-case channel dispersion; the interference cancellation algorithms assume that the interferer is always active and constant; and so on. Adaptive receiver design measures the current channel and interferer conditions and tunes the specific receiver function that is most appropriate for the current conditions. For example, a specific demodulation technique may work well in some channel conditions but might not provide good performance in others. Hence, a receiver might include a variety of demodulators that are individually tuned to a set of channel classes. If the receiver could demodulate the data reliably with a simpler and less-complex receiver algorithm under the given conditions, then it is desired to use that algorithm for demodulation.

3.3 Parameter Measurements

Many adaptation techniques require estimation of various quantities like channel selectivity, link quality, and network load and congestion. Here, we focus more on physical layer measurements from a digital signal-processing perspective.

As discussed, link quality measures have many applications for various adaptation strategies. In addition, information on channel selectivity in time, frequency, and space is useful for adaptation of wireless communication systems. In this section, these important parameters and their estimation techniques are discussed.

3.3.1 Channel Selectivity Estimation

In wireless communications, the transmitted signal reaches the receiver through a number of different paths. Multipath propagation causes the signal to be spread in time, frequency, and angle. These spreads, which are related to the selectivity of the channel, have significant implications on the received signal. A channel is considered to be *selective* if it varies as a function of time, frequency, or space. The information on the variation of the channel in time, frequency, and space is crucial in adaptation of wireless communication systems.

3.3.1.1 Time Selectivity Measure: Doppler Spread

Doppler shift is the frequency shift experienced by the radio signal when either the transmitter or receiver is in motion, and *Doppler spread* is a measure of the spectral broadening caused by the temporal rate of change of the mobile radio channel. Therefore, time-selective fading and Doppler spread are directly related. The coherence time of the channel can be used to characterize the time variation of the time-selective channel. It represents the statistical measure of the time window over which the two signal components have strong correlation, and it is inversely proportional to the Doppler spread. Figure 3.2 shows the effect of mobile speed on channel variation and channel correlation in time as well as the corresponding Doppler spread values in the frequency domain.

In an adaptive receiver, Doppler information can be used to improve performance or reduce complexity. For example, in channel estimation algorithms, whether using channel trackers or channel interpolators, instead of fixing the tracker or interpolation parameters for the worst-case Doppler spread value (as commonly done in practice), the parameters can be optimized adaptively based on Doppler spread information [12,13]. Similarly, Doppler information could be used to control the receiver or transmitter adaptively for different mobile speeds, for example variable coding and interleaving schemes [14]. Also, radio network control algorithms, such as handoff, cell assignment, and channel allocation in cellular systems, can utilize the Doppler information [8]. For example, as will be described, in a hierarchical cell structure, the users are assigned to cells based on their speeds (mobility).

Doppler spread estimation has been studied for several applications in wireless mobile radio systems. Correlation and variation of channel estimates as well as correlation and variation of the signal envelope have been used for Doppler spread estimation [12]. One simple method for Doppler spread estimation is to use

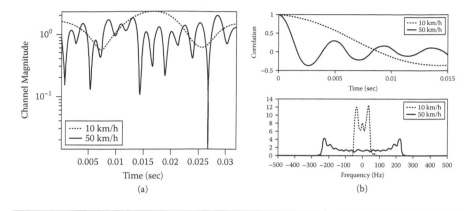

Figure 3.2 **Illustration of the effect of mobile speed on time variation, time correlation, and Doppler spread of radio channel. (a) Channel time variation for different mobile speeds. (b) Time correlation of channel as a function of the time difference (separation in time) between the samples, and the corresponding Doppler spectrum in frequency.**

differentials of the complex channel estimates [15]. The differentials of the channel estimates are very noisy and require low-pass filtering. The bandwidth of the low-pass filter is also a function of the Doppler estimate. Therefore, such approaches require adaptive receivers that continuously change the filter bandwidth depending on the previously obtained Doppler value. A Doppler estimation scheme based on the autocorrelation of complex channel estimates is described in [16]. Also, an approach based on maximum likelihood estimation, given the channel autocorrelation estimate, is utilized for Doppler spread estimation in [17]. Channel autocorrelation is calculated using the channel estimates over the known field of the transmitted data.

Instead of using channel estimates, the received signal can also be used directly in estimating Doppler spread information. In [18], the Doppler frequency is extracted from the samples of the received signal envelope. Doppler information is calculated as a function of the squared deviation of the signal envelope. Similarly, in [19] the mobile speed is estimated as a function of the deviation of the averaged signal envelope in flat fading channels. For dispersive channels, pattern recognition, using the variation of pattern mean, can be used to quantify the deviation of signal envelope. In [20], the filtered received signal is used to calculate the channel autocorrelation values over each slot. Then, the autocorrelation estimate is used for identification of high- and low-speed mobiles. In [21], multiple antennas are exploited; a linear relation between the switching rate of the antenna branches and Doppler frequency is given. Also, the level crossing rate (LCR) of the average signal level has been used in estimating velocity [22,23].

3.3.1.2 Frequency Selectivity Measure: Delay Spread

The multipath signals that reach the receiver have different delays as the paths that the signals travel through have different lengths. When the relative path delays are on the order of a symbol period or more, images of different transmitted symbols arrive at the same time, causing intersymbol interference (ISI).

Delay spread is one of the most commonly used parameters that describe the time dispersiveness of the channel, and it is related to frequency selectivity of the channel. The frequency selectivity can be described in terms of coherence bandwidth, which is a measure of range of frequencies over which the two frequency components have a strong correlation. The coherence bandwidth is inversely proportional to the delay spread [24]. Figure 3.3 shows the effect of time dispersion on channel frequency variation and channel frequency correlation as well as the corresponding power delay profiles (PDPs).

Like time selectivity, the information about the frequency selectivity of the channel can be useful for improving the performance of the adaptive wireless radio systems. For example, in a time-division multiple-access (TDMA)-based Global System for Mobile Communications (GSM; originally Groupe Spécial Mobile), the number of channel taps needed for equalization might vary depending on channel dispersion. Instead of fixing the number of channel taps for the worst-case channel condition, we can change them adaptively [25], allowing simpler receivers with reduced battery consumption and improved performance. Similarly, in [26], a TDMA receiver with adaptive demodulator was proposed, using the measurement about the dispersiveness of the channel. Dispersion estimation can also be used

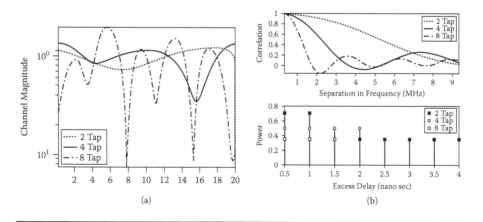

(a)   (b)

Figure 3.3   Illustration of the effect of time dispersion on channel frequency variation, channel frequency correlation, and delay spread. (a) Channel frequency variation for different delay spread values. (b) Channel frequency correlation as a function of separation in frequency and the corresponding power delay profiles.

for other parts of transmitters and receivers. For example, in frequency-domain channel estimation using channel interpolators, instead of fixing the interpolation parameters for the worst-expected channel dispersion, we can change the parameters adaptively depending on the dispersion information [27].

Although dispersion estimation can be useful for many wireless communication systems, it is particularly crucial for orthogonal frequency-division multiplexing (OFDM)-based wireless communication systems. OFDM, which is a multicarrier modulation technique, handles the ISI problem due to high bit rate communication by splitting the high-rate symbol stream into several lower-rate streams and transmitting them on different orthogonal carriers. The OFDM symbols with increased duration might still be affected by the previous OFDM symbols due to multipath dispersion. Cyclic prefix extension of the OFDM symbol avoids ISI from the previous OFDM symbols if the cyclic prefix length is greater than the maximum excess delay of the channel. Because the maximum excess delay depends on the radio environment, the cyclic prefix length needs to be designed for the worst-case channel condition. This makes the cyclic prefix a significant portion of the transmitted data, thereby reducing spectral efficiency. One way to increase spectral efficiency is to adapt the length of the cyclic prefix depending on the radio environment [28].

The adaptation requires estimation of maximum excess delay of the radio channel, which is also related to the frequency selectivity of the channel. In HiperLAN2, which is a wireless local-area network (WLAN) standard, a cyclic prefix duration of 800 ns, which is sufficient to allow good performance for channels with delay spread up to 250 ns, is used. Optionally, a short cyclic prefix with 400-ns duration may be used for short-range indoor applications. Delay spread estimation allows adaptation of these various options to optimize the spectral efficiency. Other OFDM parameters that could be changed adaptively using the knowledge of the dispersion include OFDM symbol duration and OFDM subcarrier bandwidth.

Characterization of the frequency selectivity of the radio channel is reviewed in [29–31] using the LCR of the channel in the frequency domain. Frequency-domain LCR gives the average number of crossings per hertz at which the measured amplitude crosses a threshold level. An analytical expression between LCR and the time-domain parameters corresponding to a specific multipath PDP is given. The LCR is very sensitive to noise, which increases the number of level crossings and severely deteriorates the performance of the LCR measurement [31]. Filtering the channel frequency response reduces the noise effect, but finding the appropriate filter parameters is an issue. If the filter is not designed properly, one might end up smoothing the actual variation of frequency-domain channel response. In [27], instantaneous root mean square (rms) delay spread, which provides information about local (small-scale) channel dispersion, is obtained by estimating the channel impulse response (CIR) in the time domain. The detected symbols in the frequency domain are used to regenerate the time-domain signal through inverse fast Fourier transform (IFFT). This signal is then used to correlate the actual received signal to obtain the CIR, which is then used for delay spread estimation. Because the

detected symbols are random, they might not have good autocorrelation properties, which can be a problem, especially when the number of carriers is low. In
addition, the use of detected symbols for correlating the received samples to obtain
the CIR provides poor results for low SNR values. In [28], the delay spread is also
calculated from the instantaneous time-domain CIR, with the CIR obtained by
taking the IFFT of the frequency-domain channel estimate. Channel frequency
selectivity and delay spread information are calculated using the channel frequency
correlation estimates in [24,32]. An analytical expression between delay spread and
coherence bandwidth is also given.

The level of time dispersion can be obtained by using known training sequences
and an algorithm based on maximum likelihood. The channel can be modeled with
different levels of dispersion. Using these various channel models, the corresponding
channel estimates and the residual error can be calculated. From these residual error
terms, a decision can be made about the level of dispersion. Note that when the channel is overmodeled, the residual error also becomes smaller. Hence, it is not necessarily
true that the model that provides the smallest residual error is the most suitable one.
The most appropriate model can be found by several information criteria algorithms,
like Bayesian information criteria (BIC) or Akaike information criteria (AIC) [33].

3.3.1.3 Spatial Selectivity Measure: Angle Spread

Angle spread is a measure of how multipath signals are arriving (or departing)
with respect to the mean arrival (departure) angle. Therefore, angle spread refers
to the spread of angles of arrival (or departure) of the multipaths at the receiving
(transmitting) antenna array [34]. Angle spread is related to the spatial selectivity
of the channel, which is measured by coherence distance. Like coherence time and
frequency, coherence distance provides the measure of the maximum spatial separation over which the signal amplitudes have strong correlation, and it is inversely
proportional to angular spread; that is, the larger the angle spread is, the shorter
the coherence distance is. Figure 3.4 shows the effect of local scattering on angle
of arrival. The local scattering in the vicinity of Receiver-2 results in larger angular
spreads as the received signals come from many different directions due to a richer
local scattering environment. For a given receiver antenna spacing, this leads to fewer
antenna correlations between the received antenna elements than the correlation of
antennas in Receiver-1. Note that although the angular spread is described as independent of the other channel selectivity values for the sake of simplicity, in reality the
angle of arrival can be related to the path delay. The multipath components that arrive
at the receiver earlier (i.e., with shorter delays) are expected to have similar angles of
arrival (lower angle spread values). Receiver-1 observes less angle spread compared to
Receiver-2. Therefore, receiver antennas in Receiver-1 will have more correlations.

Compared to time and frequency selectivity, spatial selectivity has not been
studied widely in the past. However, recently there has been a significant amount
of work in multiantenna systems. With the widespread application of multiantenna

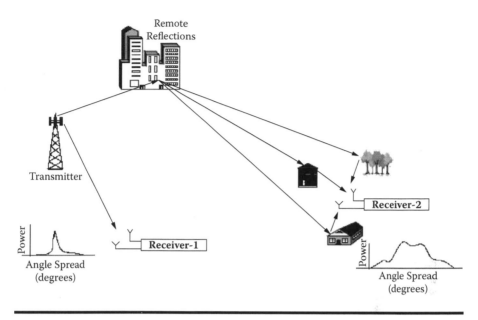

Figure 3.4 Illustration of the effect of different local scattering in angle of arrivals.

systems, it is expected that the need for understanding spatial selectivity and related parameter estimation techniques will gain momentum. Spatial selectivity will be especially useful when the requirement for placing antennas close to each other increases, as in the case of multiple antennas at the mobile units.

Spatial correlation between multiple-antenna elements is related to the spatial selectivity, antenna distance, mutual coupling between antenna elements, antenna patterns, and so on [35,36]. Spatial correlation has significant effects on multiantenna systems. Full capacity and performance gains of multiantenna systems can be achieved only with low antenna correlation values. However, when this is not possible, maximum capacity can be achieved by employing efficient adaptation techniques. Adaptive power allocation is one way to exploit the knowledge of the spatial correlation to improve the performance of multiantenna systems [37]. Similarly, adaptive modulation and coding, which employ different modulation and coding schemes across multiantenna elements depending on the channel correlation, are possible [38,39]. In MIMO systems, adaptive power allocation has been studied using the knowledge of channel matrix estimate and eigenvalue analysis [40,41].

3.3.2 Channel Quality Measurements

Channel quality estimation is by far the most important measurement that can be used in adaptive receivers and transmitters [3]. Different ways of measuring the quality of a radio channel are possible, and many of these measurements are done in the physical layer using baseband signal-processing techniques. In most of the adaptation algorithms, the target quality measure is the FER or BER as these are closely related to higher-level quality-of-service parameters such as speech and video quality. However, reliable measurement of these qualities requires many measurements, and this causes delays in the adaptation as the process could be very long. Therefore, other types of channel quality measurements that are related to these might be preferred. When the received signal is impaired only by white Gaussian noise, analytical expressions can be found relating the BER to other measurements. For other impairment cases, like colored interferers, numerical calculations and computer simulations that relate these measurements to BER can be performed. Therefore, depending on the system, channel quality is related to the BER. Then, for a target BER (or FER), a required signal quality threshold can be calculated to be used with the adaptation algorithm.

The measurements can be performed at various points of a receiver, depending on the complexity, reliability, and delay requirements. There are trade-offs in achieving these requirements at the same time. Figure 3.5 shows a simple example in which some of these measurements can take place. In the following sections, these measurements are discussed briefly.

3.3.2.1 Measures before Demodulation

Received signal strength estimation provides a simple indication of the fading and path loss and provides the information about how strong the signal is at the receiver front end. If the RSS is stronger than the threshold value, then the link is considered to be good. Measuring the signal strength of the available radio channels can be used as part of the scanning and intelligent roaming process in cellular systems. Also, other adaptation algorithms, for example, power control and handoff, can use this

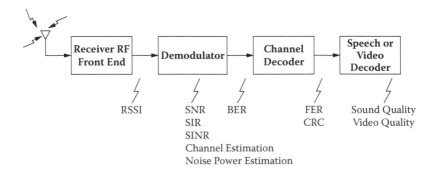

Figure 3.5   A simple wireless receiver that shows the estimation points of commonly used parameters.

information. The RSS measurement is simply reading samples from a channel and averaging them [42]. Compared to other measurements, RSS estimation is simple and computationally less complex as it does not require the processing and demodulation of the received samples. However, the received signal includes noise, interference, and other channel impairments. Therefore, receiving good signal strength does not tell much about the channel and signal quality. Instead, it gives an indication of whether a strong signal is present in the channel of interest. For the measurement of RSS, the transmitter might send a pilot signal continuously, as in the WCDMA cellular system, or a link layer beacon can be transmitted at discrete time intervals, as in IEEE (Institute of Electrical and Electronics Engineers) 802.11 WLANs.

Because the received signal power fluctuates rapidly due to fading, to obtain reliable estimates the signal needs to be averaged over a time window to compensate for short-term fluctuations. The averaging window size depends on the system, application, variation of the channel, and so on. For example, if multiple receiver antennas are involved at the receiver, the window can be shorter than that for a single-antenna receiver.

3.3.2.2 Measures during and after Demodulation

The signal-to-interference ratio (SIR), SNR, and SINR are the most common ways of measuring the channel quality during (or just after) the demodulation of the received signal. SIR (or SNR or SINR) provides information on how strong the desired signal is compared to the interferer (or noise or interference plus noise). Most wireless communication systems are interference limited; therefore, SIR and SINR are more commonly used. Compared to RSS, these measurements provide more accurate and reliable estimates at the expense of computational complexity and with additional delay.

There are many adaptation schemes in which these measurements can be exploited. Link adaptation (adaptive modulation and coding, rate adaptation, etc.), adaptive channel assignment, power control, adaptive channel estimation, and adaptive demodulation are only a few of many applications.

SIR estimation can be employed by estimating signal power and interference power separately and then taking the ratio of these two. In many new-generation wireless communication systems, coherent detection, which requires estimation of channel parameters, is employed. These channel parameter estimates can also be used to calculate the signal power. The training (or pilot) sequences can be used to obtain the estimate of SIR. Instead of the training sequences, the data symbols can also be used for this purpose. For example, in [43], in which SNR information is used as a channel quality indicator for rate adaptation, the cumulative euclidean metric corresponding to the decoded trellis path is exploited for channel quality information.

Another method for channel quality measurement is the use of the difference between the maximum likelihood decoder metrics for the best path and for the second-best path, as described in [44]. In a sense, in this technique some sort of soft information is used for the channel quality indicator. However, this approach does not tell much about the strength of the interferer or the desired signal. There are several other ways of SNR measurement that are based on subspace projection techniques. These approaches can be found in [45] and in the references cited therein.

Often, in obtaining the estimates, the impairment (noise or interference) is assumed to be white and Gaussian distributed to simplify the estimation process. However, in wireless communication systems, the impairment might be caused by a strong interferer, which is colored. In OFDM systems, for example, in which the channel bandwidth is wide and the interference is not constant over the whole band, it is very likely that some part of the spectrum is affected more by the interferer than the other parts. Figure 3.6 shows the OFDM frequency spectrum and two types of noise over this spectrum: colored and white. Hence, when the impairment is colored, estimates that take the color of the impairment into account might be needed [46].

Note that since both the desired signal's channel and interferer conditions change rapidly, depending on the application, both short-term and long-term estimates are desirable. Long-term estimates provide information on long-term fading statistics due to shadowing and lognormal fading as well as average interference conditions. On the other hand, short-term estimates provide measurements of instantaneous channel and interference conditions. Applications such as adaptive channel assignment and handoff prefer long-term statistics, whereas applications such as adaptive demodulation and adaptive interference cancellation prefer short-term statistics.

For some applications, a direct measure of channel quality from channel estimates would be sufficient for adaptation. As mentioned, channel estimates only provide information about the desired signal's power. It is a much more reliable estimate than RSS information as it does not include the other impairments as part

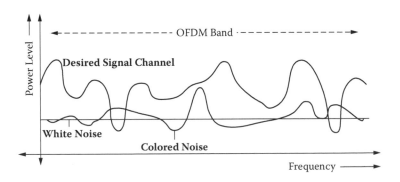

Figure 3.6 Representation of OFDM frequency channel response and noise spectrum. Spectrum for both white and colored noise is shown.

of the desired signal power. However, it is less reliable than SNR (or SINR) estimates since it does not provide information about the noise or interference powers with respect to the power of the desired signal.

Channel estimation for wireless communication systems has a very rich history. A significant amount of work has been done for various systems. In many systems, known information (like pilot symbols, pilot channels, pilot tones, training sequences) is transmitted along with the unknown data to help the channel estimation process. Blind channel estimation techniques that do not require known information transmission have also been studied extensively. For details on channel estimation for wireless communication systems, refer to [47] and [48] and the references listed in them.

3.3.2.3 Measures after Channel Decoding

Channel quality measurements can also be based on postprocessing of the data (after demodulation and decoding). BER, symbol error rate (SER), FER, and CRC information are some of the examples of the measurements in this category. BER (or FER) is the ratio of the bits (or frames) that have errors relative to the total number of bits (or frames) received during the transmission. The CRC indicates the quality of a frame, which can be calculated using parity check bits through a known cyclic generator polynomial.

The FER can be obtained by averaging the CRC information over a number of frames. To calculate the BER, the receiver needs to know the actual transmitted bits, which is not possible in practice. Instead, the BER can be calculated by comparing the bits before and after the decoder. Assuming that the decoder corrects the bit errors that appear before decoding, this difference can be related to the BER. Note that the comparison makes sense only if the frame is error free (good frame), which is obtained from the CRC information.

As mentioned, although these estimates provide excellent link quality measures, reliable estimates of these parameters require observations over a large number of frames. Especially for low BER and FER measurements, extremely long transmission intervals will be needed. Therefore, for some applications these measures might not be appropriate. Note also that these measurements provide information about the actual operating condition of the receiver. For example, for a given RSS or SINR measure, two different receivers that have different performances will have different BER or FER measurements. Therefore, BER and FER measurements also provide information on the receiver capability as well as the link quality.

3.3.2.4 Measures after Speech or Video Decoding

Speech and video quality, delays on data reception, and network congestion are some of the parameters that are related to a user's perception. Essentially, these are the ultimate quality measures that need to be used for adaptive algorithms.

However, these parameters are not easy to measure, and in many cases, real-time measurement might not be possible. On the other hand, these measures are often related to the other measures mentioned. For example, speech quality for a given speech coder can be related to FER of a specific system under certain assumptions [49]. However, as discussed in [49], some frame errors cause more audible damage than others. Therefore, it is still desired to find ways to measure the speech quality more reliably (and timely) and adapt the system parameters accordingly. Speech (or video) quality measures that take the human perception of the speech (or video) into account are highly desirable.

Perceptual speech quality measurements have been studied in the past. Both subjective and objective measurements are available [50]. Subjective measurements are obtained from a group of people who rate the quality of the speech after listening to the original and received speech. Then, a mean opinion score (MOS) is obtained from their feedback. Although these measurements reflect the exact human perception that is desired for adaptation, they are not suitable for adaptation purposes because the measurements are not obtained in real time. On the other hand, the objective measurements can be implemented at the receiver in real time [51]. However, these measurements require a sample of the original speech at the receiver to compare the received voice with the original undistorted voice. Therefore, they are also not applicable for many scenarios.

3.4 Conclusion

Recently, the use of adaptation algorithms for better utilization of the available resources, like power and spectrum, has grown significantly. Several adaptation strategies to increase performance, data rate, capacity, and quality of service of wireless communication systems have been introduced. Many of these adaptation techniques depend on accurate estimation of the various parameters. Therefore, further research on efficient parameter estimation techniques is still needed.

There is a significant amount of work needed for the evolution of the current wireless communication systems to accommodate the future demands. Ultrawideband (UWB), MIMO, and multicarrier wireless communications are some of the technologies that are being studied extensively. All have a common point: their capability to adapt the changing radio channel conditions. Adaptation of multicarrier communications and MIMO schemes has already gained some momentum. Adaptation algorithms for UWB still need to be explored. The flexibility of UWB makes it very attractive for employing successful adaptation schemes.

Acknowledgment

I thank T. Yucek for his help and the anonymous reviewers for their comments.

References

1. Sampei, S., *Applications of Digital Wireless Technologies to Global Wireless Communications*. Englewood Cliffs, NJ: Prentice Hall, 1997.
2. Ikeda, T., Sampei, S., and Morinaga, N., TDMA-based adaptive modulation with dynamic channel assignment for high-capacity communication systems. *IEEE Trans. Commun.*, 49(2), 404–412, 2000.
3. Nanda, S., Balachandran, K., and Kumar, S., Adaptation techniques in wireless packet data services. *IEEE Trans. Commun.*, 38(1), 54–64, 2000.
4. Nobelen, R.V., Seshadri, N., Whitehead, J., and Timiri, S., An adaptive radio link protocol with enhanced data rates for GSM evolution. *IEEE Commun. Mag.*, 54–63, 1999.
5. Furuskar, A., Mazur, S., Muller, F., and Olofsson, H., EDGE: Enhanced data rates for GSM and TDMA/136 evolution. *IEEE Pers. Commun. Mag.*, 6(3), 56–66, 1999.
6. Anderson, S., Dam, H., Forssen, U., Karlsson, J., Kronestedt, F., Mazur, S., and Molnar, K.J., Adaptive antennas for GSM and TDMA systems. *IEEE Pers. Commun. Mag.*, 6, 74–86, 1999.
7. Lee, W.C., *Mobile Cellular Telecommunications*. New York: McGraw-Hill, 1995.
8. Pollini, G., Trends in handover design. *IEEE Commun. Mag.*, 34(3), 82–90, 1996.
9. Bender, P., Black, P., Grob, M., Padavoni, R., Sindhushyana, N., and Viterbi, S., CDMA/HDR: A bandwidth efficient high speed wireless data service for nomadic users. *IEEE Commun. Mag.*, 38, 70–77, 2000.
10. Jalali, A., Padovani, R., and Pankaj, R., Data throughput of CDMA-HDR: A high efficiency-high data rate personal communication wireless system. *Proc. IEEE Veh. Technol. Conf.*, 3, 1854–1858, 2000.
11. Holtzman, J., CDMA forward link waterfilling power control. Proc. IEEE Veh. Technol. Conf., 3, 1663–1667, 2000.
12. Arslan, H., Krasny, L., Koilpillai, D., and Channakeshu, S., Doppler spread estimation for wireless mobile radio systems. *Proc. IEEE WCNC Conf.*, 3, 1075–1079, 2000.
13. Sakamoto, M., Huoponen, J., and Niva, I., Adaptive channel estimation with velocity estimator for W-CDMA receiver. *Proc. IEEE Veh. Technol. Conf.*, 3, 2024–2028, 2000.
14. Mottier, D., and Castelain, D., A Doppler estimation for UMTS-FDD-based on channel power statistics. in *Proc. IEEE Veh. Technol. Conf.*, 5, 3052–3056, 1999.
15. Lindbom, L., Adaptive equalization for fading mobile radio channels. Licentiate thesis, Uppsala University, Uppsala, Sweden, 1992.
16. Morelli, M., Mengali, U., and Vitetta, G., Further results in carrier frequency estimation for transmissions over flat fading channels. *IEEE Commun. Lett.*, 2, 327–330, 1998.
17. Krasny, L., Arslan, H., Koilpillai, D., and Channakeshu, S., Doppler spread estimation in mobile radio systems. *Proc. IEEE WCNC Conf.*, 5(5), 197–199, 2001.
18. Sampath, J.H.A., Estimation of maximum Doppler frequency for handoff decisions. *Proc. IEEE Veh. Technol. Conf.*, 859–862, May 1993.
19. Wang, L., Silventoinen, M., and Honkasalo, Z., A new algorithm for estimating mobile speed at the TDMA-based cellular system. *Proc. IEEE Veh. Technol. Conf.*, 2, 1145–1149, 1996.
20. Xiao, C., Mann, K., and Olivier, J., Mobile speed estimation for TDMA-based hierarchical cellular systems. *Proc. IEEE Veh. Technol. Conf.*, 2, 2456–2460, 1999.

21. Kawabata, K., Nakamura, T., and Fukuda, E., Estimating velocity using diversity reception. *Proc. IEEE Veh. Technol. Conf.*, 1, 371–374, 1994.

22. Lee, W., *Mobile Communications Engineering*. New York: McGraw-Hill, 1998.

23. Austin, M., and Stuber, G., Eigen-based Doppler estimation for differentially coherent CPM. *IEEE Trans. Veh. Technol.*, 43, 781–785, 1994.

24. Arslan, H., and Yucek, T., Delay spread estimation for wireless communication systems. *Proc. 8th IEEE Symp. Computers Commun. (ISCC 2003)*, 282–287, July 2003.

25. Chen, J.-T., Liang, J., Tsai, H.-S., and Chen, Y.-K., Joint MLSE receiver with dynamic channel description. *IEEE J. Sel. Areas Commun.*, 16, 1604–1615, 1998.

26. Husson, L., and Dany, J.-C., A new method for reducing the power consumption of portable handsets in TDMA mobile systems: conditional equalization. *IEEE Trans. Veh. Technol.*, 48(6), 1936–1945, 1999.

27. Schober, H., and Jondral, F., Delay spread estimation for OFDM based mobile communication systems. *Proc. Eur. Wireless Conf.*, 625–628, February 2002.

28. Sanzi, F., and Speidel, J., An adaptive two-dimensional channel estimator for wireless OFDM with application to mobile DVB-T. *IEEE Trans. Broadcasting*, 46(2), 128–133, 2000.

29. Witrisal, K., Kim, Y.-H., and Prasad, R., RMS delay spread estimation technique using non-coherent channel measurements. *IEE Electron. Lett.*, 34(20), 1918–1919, 1998.

30. Witrisal, K., Kim, Y.-H., and Prasad, R., A new method to measure parameters of frequency selective radio channel using power measurements. *IEEE Trans. Commun.*, 49, 1788–1800, 2001.

31. Witrisal, K., and Bohdanowicz, A., Influence of noise on a novel RMS delay spread estimation method. *Proc. IEEE PIMRC Conf.*, 1, 560–566, 2000.

32. Arslan, H., and Yucek, T., Estimation of frequency selectivity for OFDM-based new generation wireless communication systems. *Proc. 2003 World Wireless Congress*, San Francisco, May 2003.

33. Soderstrom, T., and Stoica, P., *Applications of Digital Wireless Technologies to Global Wireless Communications*. Englewood Cliffs, NJ: Prentice Hall, 1997.

34. Paulraj, A., and Ng, B., Space-time modems for wireless personal communications. *IEEE Pers. Commun. Mag.*, 5(1), 36–48, 1998.

35. Ozdemir, M.K., Arslan, H., and Arvas, E., Mutual coupling effect in multi-antenna wireless communication systems. IEEE GlobeCom Conference, San Francisco, CA, December 2003.

36. Ozdemir, M.K., Arslan, H., and Arvas, E., Dynamics of spatial correlation and implications on MIMO systems. *IEEE Commun. Mag.*, June 2004.

37. Shiu, D.S., Foschini, G.J., Gans, M.J., and Kahn, J.M., Fading correlation and its effects on the capacity of multielement antenna systems. *IEEE Trans. Commun.*, 48(3), 502–513, 2000.

38. Ivrlac, M., Kurpjuhn, T., Brunner, C., and Utschick, W., Efficient use of fading correlations in MIMO systems. *Proc. IEEE Conf. Veh. Technol. Conf.*, 4, 2763–2767, 2001.

39. Catreux, S., Erceg, V., Gesbert, D., and Heath, J., Adaptive modulation and MIMO coding for broadband wireless data networks. *IEEE Commun. Mag.*, 40(6), 108–115, 2002.

40. Telatar, E., *Capacity of Multiantenna Gaussian Channels*. Technical Report, AT&T Bell Laboratories, June 1995.

41. Foschini, G.J., and Gans, M.J., On limits of wireless communications in a fading environment when using multiple antennas. *Wireless Pers. Commun.*, 6(3), 311–335, 1998.

42. TIA/EIA 136–131-B, *TDMA Third Generation Wireless: Digital Traffic Channel Layer 1*, March 2000.

43. Balachandran, K., Kabada, S., and Nanda, S., Rate adaptation over mobile radio channels using channel quality information. *Proc. IEEE Globecom' 98 Commun. Theory Mini Conf. Record*, 46–52, 1998.

44. Jacobsmeyer, J., Adaptive data rate modem. U.S. Patent 5541955, July 1996.

45. Turkboylari, M., and Stuber, G.L., An efficient algorithm for estimating the signal-to-interference ratio in TDMA cellular systems. *IEEE Trans. Commun.*, 46(6), 728–731, 1998.

46. Arslan, H., and Reddy, S., Noise variance and SNR estimation for OFDM based wireless communication systems. *Proc. 3rd Int. Conf. Wireless Opt. Commun.*, July 2003.

47. Arslan, G.B.H., Channel estimation in narrowband wireless communication systems. *Wireless Commun. Mobile Comput. (WCMC) J.*, 1(2), 201–219, 2001.

48. Bottomley, G., and Arslan, H., Channel estimation for time-varying channels in wireless communication systems. In *The Wiley Encyclopedia of Telecommunications*. New York: Wiley, 2003.

49. Homayounfar, K., Rate adaptive speech coding for universal multimedia access. *IEEE Signal Proc. Mag.*, 20(2), 30–39, 2003.

50. Wolf, S., Dvorak, C., Kubichek, R., and South, C., How will we rate telecommunications system performance? *IEEE Commun. Mag.*, 29, 23–29, 1991.

51. Voran, S., Objective estimation of perceived speech quality: Part II. *IEEE Trans. Speech Audio Proc.*, 7, 383–390, 1999.

Chapter 4

Coding Techniques

Quazi Mehbubar Rahman and Mohamed Ibnkahla

Contents

4.1 Introduction

A coding technique in general is of two types: source coding and channel coding. The source-coding technique refers to the encoding procedure of the source information signal into digital form. On the other hand, channel coding is applied to ensure adequate transmission quality of the signals. Channel coding is a systematic approach for the replacement of the original information symbol sequence by a sequence of code symbols in such a way that it permits its reconstruction. Here, we focus on the channel-encoding technique.

Channel coding can improve the severe transmission conditions in terrestrial mobile radio communications due to multipath fading. Moreover, it can help to overcome very low signal-to-noise ratios (SNRs) for satellite communications due to limited transmit power in the downlink. The encoding process generally involves mapping every k-bit information sequence into a unique n-bit sequence, in which the latter is called a *code word*.

The amount of redundancy introduced by the encoding process is measured by the ratio k/n, whose reciprocal is known as the *code rate*. The output of the channel encoder is fed to the modulator, which has an output that is transmitted through the channel. At the receiver end, demodulation, decoding, and detection processes are carried out to decide on the transmitted signal information. In the decision process, two different strategies are used: soft decision and hard decision. When the demodulator output consists of discrete elements 0 and 1, the demodulator is said to make a *hard decision*. On the other hand, when the demodulator output consists of a continuous alphabet or its quantized approximation (with greater than two quantization levels), the demodulator is said to make a *soft decision*.

It was theoretically shown by Shannon [1] that the coding technique in general improves the bit error rate (BER) performance of a communications system. Before discussing different channel-coding techniques, we take a look at this theory.

4.2 Shannon's Capacity Theorem

Shannon [1] showed that the system capacity C bits/second of an AWGN (additive white Gaussian noise) channel is a function of the average received signal power S, the average noise power N, and the bandwidth W hertz, which is given by

$$C = W \log_2 \left(1 + \frac{S}{N}\right) \tag{4.1}$$

Theoretically, it is possible to transmit information with arbitrarily small BERs over such a channel at any rate R, where $R \leq C$, using a complex coding scheme. For an information rate of $R > C$, it is not possible to find a code that can achieve an arbitrarily small error probability. Shannon's works show that the values of S, N, and W set a limit on the transmission rate, not on the error probability. In Equation 4.1, the detected noise power N is proportional to the bandwidth, which is given by

$$N = N_0 W \tag{4.2}$$

where N_0 is the noise power spectral density. Substituting Equation 4.2 into Equation 4.1, we get

$$\frac{C}{W} = \log_2 \left(1 + \frac{S}{N_0 W} \right) \tag{4.3}$$

Now, assuming $R = C$, the ratio between the binary signal energy (E_b) and the noise spectral density (N_0) can be written as

$$\frac{E_b}{N_0} = \frac{ST}{N_0} = \frac{S}{N_0 R} = \frac{S}{N_0 C} \tag{4.4}$$

which can be used in Equation 4.3 to yield

$$\frac{C}{W} = \log_2 \left[1 + \frac{E_b}{N_0} \left(\frac{C}{W} \right) \right] \tag{4.5}$$

From Equation 4.5, it can be shown that there exists a limiting value of E_b/N_0 below which there can be no error-free communication at any information rate. This limiting value is known as *Shannon's limit*, which can be calculated by letting

$$x = \frac{E_b}{N_0} \left(\frac{C}{W} \right) \tag{4.6}$$

Substituting this parameter in Equation 4.5, we get

$$\frac{C}{W} = \log_2 [1 + x] \tag{4.7}$$

Equation 4.7 can be rewritten as

$$\frac{C}{W} = x \log_2 [1+x]^{1/x} = \frac{E_b}{N_0} \left(\frac{C}{W} \right) \log_2 [1+x]^{1/x} \tag{4.8}$$

Now, using the identity

$$\lim_{x \to \infty} (1+x)^{2/x} = e$$

in the limit $C/W \geq 0$, Equation 4.8 becomes

$$\frac{E_b}{N_{0\,shannon}} = \frac{1}{\log_2 e} = 0.693 \tag{4.9}$$

In decibel (dB) scale, it is

$$\frac{E_b}{N_{0\,shannon\,dB}} = -1.59\ dB \tag{4.10}$$

Equations 4.9 and 4.10 give the numerical values of Shannon's limit.

4.3 Different Coding Schemes

Channel coding can be classified into two major areas: waveform coding and structured sequences. The objective of waveform coding is to provide an improved waveform set so that the detection process is less subject to errors. Examples of this coding technique include *M*-ary, antipodal, orthogonal, biorthogonal, and transorthogonal signaling. Structured sequences deal with transforming data sequences into better sequences having ordered redundancy in bits. The redundant bits can then be used for the detection and correction of errors. Examples of structured sequence coding include block- and convolutional-coding schemes. This section focuses mainly on the structured sequence-type coding schemes.

Here, the discussion considers only those coding techniques that are being addressed in the recent trend of mobile and satellite communication systems. These schemes include linear block codes (e.g., Hamming codes, BCH [Bose–Chaudhuri–Hocquenghem] codes, and Reed–Solomon [RS] codes) and convolutional codes. Space–time codes and turbo codes are also considered; they are receiving enormous attention in the current developments of both 3G and 4G (third and fourth generations) telecommunications systems. To provide improvement in power efficiency without sacrificing bandwidth efficiency, coded modulation techniques (e.g., trellis-coded modulation, TCM [2]) that combine coding and modulation techniques are good choices. This section also talks briefly about coded modulation techniques.

4.3.1 Block Codes

In the block-coding scheme, each k-bit information symbol block is converted to an n-bit coded symbol block with $(n - k)$ redundancy bits added to the k-bit symbols. These redundancy bits could be parity bits or check bits that do not carry any information. The resulting code is referred to as the (n, k) block code.

Here, the *redundancy* of the code is defined as the ratio between the redundant bits and k-bit symbol, that is, $(n - k)/k$, while the *code rate* is defined as k/n. In block codes, 2^k k-bit message sequences are uniquely mapped into 2^k n-bit codes out of a possible $2n$ n-bit codes.

4.3.1.1 Vector Space and Subspace

A vector space V_n is defined as the set that contains all possible n-bit block codes. On the other hand, a subset S of the vector space V_n is called a subspace if an all-zero vector is in S and the sum (modulo-2) of any two vectors in S is also in S (closure property). The subspace properties are the basis for the algebraic characterization of linear block codes.

4.3.1.2 Linear Block Code

The block codes in which each of the code words can be formed by the modulo-2 sum (EX-OR) of two or more other code words are called *linear block codes*. The code words are said to be linearly dependent on each other.

4.3.1.3 Coding Gain

Coding gain is defined as the improvement in the SNR in decibels at a specified BER performance of an error-correcting coded system over an uncoded one with an identical system scenario.

4.3.1.4 Hamming Codes

Hamming codes are linear block codes characterized by the following (n, k) structure:

$$(n, k) = (2^m - 1, 2^m - 1 - m) \tag{4.11}$$

where $m = 2, 3, \ldots.$

Table 4.1   Hamming Code of Rate 4/7

Block Number	Input Data Block	Output Data Block
0	0000	000 0000
1	1000	110 1000
2	0100	011 0100
3	1100	101 1100
4	0010	111 0010
5	1010	001 1010
6	0110	100 0110
7	1110	010 1110
8	0001	101 0001
9	1001	011 1001
10	0101	110 0101
11	1101	000 1101
12	0011	010 0011
13	1011	100 1011
14	0111	001 0111
15	1111	111 1111

These codes are capable of correcting all single bit errors and detecting all combinations of two or fewer bits in error, which can be seen from the following example: A rate 4/7, that is, $(n, k) = (7, 4)$, Hamming code is shown in Table 4.1, in which each output data block differs from all the other blocks by at least three bits. Hence, if a one- or two-bit error occurs in the transmission of a block, the decoder will detect that error. In the case of a single-bit error, it is also possible for the receiver to match the received block to the closest valid block and thereby correct the single-bit error. If a three-bit error occurs, the original block may be transformed into a new valid block, and all the errors are undetected.

4.3.1.5 Hamming Distance

The difference in the number of bits between two coded blocks is known as the *Hamming distance*. A block code of a Hamming distance d can detect up to $(d-1)$ errors and correct $(d-1)/2$ errors.

4.3.1.6 Implementation Complexity

An increase in the coded block length results in two drawbacks in the block-coding techniques:

- *Transmission delay*: The time taken to collect k bits to form a block increases with increasing block length, introducing delay in the transmission process, which may be unacceptable for real-time applications such as voice transmission.
- *Decoder complexity*: This increases almost exponentially with block length as the decoder searches through 2^k valid code words to find the best match with the incoming 2^n possible coded blocks. In addition to the complexity, the decoding delay can be significant.

4.3.1.7 BCH Codes

The BCH codes are generalizations of Hamming codes that allow multiple error corrections. BCH codes [3] are important because, at a block length of a few hundred, these codes outperform all other block codes with the same block length and code rate. For very high coding overhead with long block length, this coding scheme can be used if reliability of transmission is the key factor and data throughput is less important.

4.3.1.8 Reed–Solomon Codes

The RS codes are a subclass of BCH codes that operate at the block level rather than the bit level. Here, the incoming data stream is first packaged into small blocks, and these blocks are then treated as a new set of k symbols to be packaged into a supercoded block of n symbols. As a result, the decoder is able to detect and correct complete error blocks. This is a nonbinary code set that can achieve the largest possible code minimum distance for any linear code with the same encoder input and output block lengths.

For nonbinary codes, the *distance* between two code words is defined as the number of nonbinary symbols in which the sequences differ. The code minimum distance for the RS codes is given by [4]

$$d_{\min} = n - k + 1 \tag{4.12}$$

These codes are capable of correcting any combination of $(n - k)/2$ or fewer symbol errors. RS codes are particularly useful for burst-type error corrections, so they are effective with the channel with memory. They are also used in error-correcting mechanisms in compact disk (CD) players.

4.3.1.9 Interleaving

The block codes work best when errors are distributed evenly and randomly between incoming blocks. This is usually the case for AWGN channels such as landline telephone links. In a mobile radio environment, however, errors often occur in bursts as the received signal fades in and out due to the multipath propagation and the user's motion. To distribute these errors more evenly between coded blocks, a process known as *interleaving* is used. In general, to accomplish interleaving, the encoded data blocks are read as rows into a matrix. Once the matrix is full, the data can be read out in columns, redistributing the data for transmission.

At the receiver, a deinterleaving process is performed using a similar matrix-filling and -emptying process, reconstructing the original blocks. At the same time, the burst errors are uniformly redistributed across the blocks. The number of rows or columns in the matrix is sometimes referred to as the *interleaving depth*. The greater the interleaving depth is, not only the greater resistance to long fades will be, but also the greater the latency in the decoding process will be as both the transmitter and receiver matrices must be full before encoding or decoding can occur.

4.3.2 Convolutional Codes

A convolutional code is implemented on a bit-by-bit basis from the incoming data source stream. The encoder has memory, and it executes an algorithm using a predefined number of the most recent bits to yield a new coded output sequence. Convolutional codes are linear, with each branch word of the output sequence a function of the input bits and $(k - 1)$ prior bits. Since the encoding procedure is similar to the convolution operation, the coding technique is known as *convolutional coding*. The decoding process is usually a serial process based on present and previous received data bits (or symbols). Figure 4.1 shows an $(n, k) = (2, 1)$ convolutional encoder with a constraint length of $C_{ln} = 3$, which is the length of the shift register.

There are $n = 2$ modulo-2 adders that result in a two-bit coded word for each input bit on EX-OR operation. The output switch samples the output of each modulo-2 adder, thus forming the two-bit code symbol associated with the single input bit. The sampling is repeated for each input bit that results in a two-bit code word. The choice of the connections between the adders and the stages of the register gives rise to the characteristics of the code. The challenge in this case is to find an optimal connection pattern that can provide codes with the best distance properties. Convolutional codes have no particular block size; nonetheless, these are often forced into a block structure by periodic truncation. This requires a number of zero bits to be added at the end of the input data sequence for clearing out the data bits from the encoding shift register. Because the added zeros carry no information, the effective code rate falls below k/n.

The truncation period is generally made as long as practical to keep the code rate close to k/n. Both the encoder and decoder can be implemented using recursive

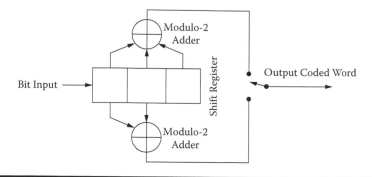

Figure 4.1   Rate convolutional encoder with constraint length of 3.

techniques, with one of the most efficient and well known the Viterbi convolutional decoder [5].

4.3.2.1 Pictorial Representation of a Convolutional Encoder

A convolutional encoder can be represented pictorially in three different ways: state diagram, tree diagram, and trellis diagram.

4.3.2.1.1 State Diagram

In the case of the state diagram, the encoder is characterized by some finite number of states. The state of a rate $1/n$ convolutional encoder is defined as the contents of the rightmost $K-1$ stages (Figure 4.1) of the shift register. The necessary and sufficient condition to determine the next output of a convolutional encoder is to have the knowledge of the current state and the next input. The state diagram for the encoder shown in Figure 4.1 can easily be drawn as shown in Figure 4.2. The states shown in the circles of the diagram represent the possible contents of the rightmost $K-1$ stages of the register, and the paths between the states represent the output branch words resulting from such state transitions. Table 4.2 will help in understanding the state transition mechanism in Figure 4.2. Major characteristics of the state diagram are:

- $2^{k \cdot (K-1)}$ states
- 2^k branches entering each state while the same number of branches leaves each state

4.3.2.1.2 Tree Diagram

The tree diagram in the convolutional encoder incorporates the time dimension in the state transition, which is not provided by the state diagram. Here, the possible code sequences generated by an encoder are represented as branches of a tree.

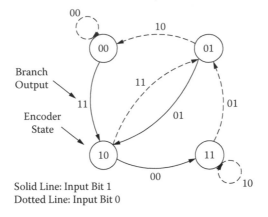

Solid Line: Input Bit 1
Dotted Line: Input Bit 0

Figure 4.2   State diagram for the rate convolutional encoder with constraint length of 3 as shown in Figure 4.1.

With the aid of time dimension, one can easily describe the encoder as a function of a particular input sequence. For examples of tree diagrams, see [6–8]. Since the number of branches in the tree increases as a function of 2^S, S being the sequence length, for a very long sequence this representation is not feasible.

Characteristics of the tree diagram include

- 2^k branches emanating from each node
- The whole tree repeating itself after the Kth stage

Table 4.2   State Transition Mechanism for the State Diagram

Input Bit	Register Content	Present State (Content of the Rightmost K – 1 Stages)	Next State (Content of the Leftmost K – 1 Stages)	Branch Output at Present State
0	000	00	00	00
1	100	00	10	11
1	110	10	11	00
1	111	11	11	10
0	011	11	01	01
0	001	01	00	10
0	010	10	01	11
1	101	01	10	01

4.3.2.1.3 Trellis Diagram

The trellis diagram is an intelligent pictorial representation of the tree diagram in which the repetitive nature of the tree diagram is smartly utilized [6–8]. It is called a trellis diagram because it looks like a garden trellis. Major characteristics of the trellis diagram include

- $2^{k \cdot (K-1)}$ states
- 2^k branches entering each state while the same number of branches leaves each state

For examples of trellis diagrams, see [6–8].

4.3.3 Space–Time Coding

Space–time coding is basically a spatial-type diversity technique in which multiple antennas in the transmitter end are used with either one or more receiving antennas. This technique is known as space–time coding since it involves redundancy by transmitting the same signal using different antennas.

With multiple antennas at the transmitter end, when the receiver end also uses multiple antennas, the system is known as a *multiple-input multiple-output* (MIMO) system. Recently, space–time coding has been receiving increased recognition [9] as a robust coding technique in the field of research due to its application in the 3G scenario.

4.3.4 Turbo Coding

Turbo coding (TC) is a specific decoding technique that was developed from two older concepts: concatenated coding and iterative decoding. These codes are built from parallel concatenation of two recursive systematic block [5] or convolutional codes with nonuniform interleaving. The term *turbo* is used to draw an analogy of this decoding process with a turbo engine in which a part of the output energy is fed back to the input to carry out its operation.

Before discussing the principle of the TC technique, we look at the concept of concatenated coding. In the concatenated-coding method, two or more relatively simple codes are combined to provide much more powerful coding. In its operation, as shown in Figure 4.3, the output of the first encoder (outermost) is fed to the input of the second and so on. In the decoder, the last (or innermost) code is decoded first, and then its output is fed to the next, and so on to the outermost decoder.

The principle of the decoding process of the TC technique can be explained briefly with the aid of Figure 4.4 and in terms of code array. In this case, the decoder first performs row decoding, which generates initial estimates of the data

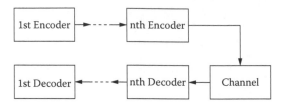

Figure 4.3   Concatenated coding method.

in the array. Here, for each data bit, a tentative decision and a reliability estimate for that decision are provided. The columns in the code array are then decoded by taking both the original input and the previous decoder signals into consideration. In the current decoder, the previous decoded signal information is known as a priori information on the data.

This second decoding further refines the data decision and its reliability estimate. The output of this second decoding stage is fed back to the input of the first decoder. In this case, the information that was missed in the first row decoding is now decoded. The whole procedure continues until the data estimates are converged.

4.3.5 Coded Modulation Techniques

In both block- and convolutional-coding schemes, the coding gain is achieved with the price paid for the bandwidth. Since in these schemes the k-bit information signal is replaced by n-bit coded words ($n > k$), the required bandwidth increases, which is a major bottleneck for the band-limited channels, such as telephone channels. To overcome this problem, combined modulation and coding schemes are considered.

In this case, the coding gain is achieved with the price paid for the decoder complexity. Here, different coded modulation techniques are briefly addressed.

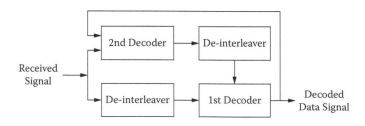

Figure 4.4   Turbo decoder.

4.3.5.1 Trellis-Coded Modulation

TCM is based on the trellis, as used in convolutional coding. In TCM, the trellis branches, instead of being labeled with binary code sequences, are represented as constellation points from the signaling constellation.

4.3.5.2 Block-Coded Modulation

In block-coded modulation (BCM) the incoming data are divided into different levels, and in each level those data streams are block coded at an equal rate.

4.3.5.3 Multilevel Coded Modulation

In multilevel coded modulation (MCM), which is a generalized form of BCM, the incoming data are split in different levels/branches (serial to parallel), and each of these data levels is block coded or convolutionally coded at either equal or unequal rates. Finally, the multiplexed signal results in the MCM.

4.3.5.4 Turbo-Coded Modulation

Based on the combinations of TC and either TCM or MCM, there are many versions of TC modulation techniques available in the research area, for example, turbo trellis-coded modulation (TTCM) [10,11], multilevel TCM (ML-TCM) [12], and so on.

4.4 Coding in Next-Generation Mobile Communications: Some Research Evidence and Challenges

This section mentions some of the current research results on coding techniques that are receiving attention in next-generation mobile communication systems, taking into account both the terrestrial and satellite domains.

In [9], measured MIMO channel data are used to evaluate the performance of the proposed 4G space–time coded orthogonal frequency-division multiplexing (COFDM) system. In the simulation results, it was assumed that the channel responses were constant during the period of two COFDM symbols. BERs for the half-rate convolutional coded quadrature phase-shift keying (QPSK) have been presented for 2-Tx (two transmit antennas) 1-Rx (one receive antenna) and 2-Tx 2-Rx scenarios. Polarization diversity was also considered in the analysis. The results indicated that high gains can be obtained for 2-Tx 2-Rx architecture, with channel correlation coefficients on the order of 0.3 to 0.5.

TC adaptive modulation and channel coding (AMC) is examined in [13] for future 4G mobile systems to observe the throughput gain. Here, a method of generating soft information for higher-order modulations, based on the reuse of the turbo decoding circuitry, is provided. It is shown that 3G-style TC can provide a 0.5- to 4-dB link gain over 256-state convolutional codes, depending on the frame size, modulation, and channel. The link gains from channel coding do not directly translate into throughput gain for AMC, but they are still expected to improve throughput significantly.

In [9], space–time COFDM for the 4G cellular network is proposed in which the individual carriers of the OFDM techniques are modulated using binary phase-shift keying (BPSK), QPSK, and 16-QAM (quadrature amplitude modulation) with coherent detection. The channel encoder consisted of a half-rate convolutional encoder. Here, the channel model considered a wide range of possible delay spreads. The results show that the space–time block codes provide diversity gain and enhance the BER performance.

In [14], to avoid a high degree of complexity in Viterbi decoding, concatenated codes based on MCM are used. In this case, the outer RS code is concatenated with an MCM for high-data-rate application over satellite channels. The results show a significant coding gain in terms of BER with considerably less complexity.

Block turbo codes (BTCs) with trellis-based decoding are proposed in [15] for asynchronous transfer mode (ATM) transmission in digital video broadcasting–return channel via satellite. In [16], an adaptive coding and modulating transmission scheme for 3G mobile satellite systems is proposed. Here, the adaptation mechanism is based on the Rice factor of the channel, which is estimated in real time using an estimation algorithm at the receiver. The transmitter, on receiving the channel information from the receiver, determines the optimal coding and modulation scheme using a lookup table. The coding scheme used convolutional coding of rates 1/2 and 1/3, while for the modulation scheme QPSK and 8PSK modulation formats were used. The simulation results in the satellite UMTS (universal mobile telecommunication systems) environment show that the dynamic range of the transmission power was greatly reduced, which in turn eased the power control requirements.

Besides all this research output, many investigations are in progress considering coding for both the satellite and terrestrial areas. Instead of discussing all these investigation approaches and results, we look at some challenges laid forth in this area.

Now that the 3G system is already in use somewhat successfully in the terrestrial domain, attention in coding challenges is currently focused mainly on the satellite domain for the 4G system. In designing 4G mobile satellite systems, transmitted power is a critical concern. In this case, because of the limited satellite onboard power and the limited life span of the mobile terminal battery, the main challenge is to come up with a power-efficient coding technique. The ACM technique in [16] has already been shown to be a smart solution to this challenge. But, in this case power efficiency can be achieved at the expense of spectral efficiency.

The use of the coded QAM technique with adaptation between different QAM constellations could be a good choice to gain both power and spectral efficiencies. It is difficult to say whether, by taking into account the huge complexity involved in this process, we can still meet our target. This doubt remains strong due to the channel, which plays an important role in the complexity issue. The use of the adaptive OFDM technique with coding can also be explored in this situation on successfully addressing the demerits of the OFDM method discussed.

In general, a multiple-access coding (MAC) scheme offers many users in the communication system the capability to share the same spectrum resource. Different MAC schemes are either in use or in the research domain in both the terrestrial and satellite areas for providing capacity improvement in the system without significantly disturbing the system's performance. Currently, wideband code-division multiple-access (WCDMA) and OFDM/time-division multiple-access (TDMA) techniques are successfully in use in terrestrial mobile multimedia systems. These two MAC schemes are also getting considerable attention [17] in mobile multimedia communications for nongeostationary satellite interfaces.

References

1. Shannon, C.E., A mathematical theory of communication. *Bell Syst. Tech. J.*, 27, 379–423, 623–657, 1948.
2. Ungerboeck, G., Trellis coded modulation with redundant signal sets, part I: Introduction. *IEEE Commun. Mag.*, 25, 5–11, February 1987.
3. Stenbit, J.P., Tables of generators for Bose-Chadhuri codes. *IEEE Trans. Inform. Theory*, IT 10(4), 390–391, October 1964.
4. Fallager, R.G., *Information Theory and Reliable Communication*. New York: Wiley, 1968.
5. Burr, A., *Modulation and Coding for Wireless Communications*. New York: Prentice Hall, 2001.
6. Proakis, J.G., *Digital Communications*, 3rd ed. New York: McGraw-Hill, 1995.
7. Lee, L.H.C, *Convolutional Coding: Fundamentals and Applications*. Norwood, MA: Artech House.
8. Sklar, B., *Digital Communications, Fundamental and Applications*. New York: Prentice Hall, 1988.
9. Doufexi, A., Design considerations and initial physical layer performance results for space time coded OFDM 4G cellular network. PIMRC 2002, Lisbon, Portugal, September 2002.
10. Goff, S. et al., Turbo codes and high spectral efficiency modulation. *Proc. IEEE Int. Conf. Commun. (ICC)*, 645–649, May 1994.
11. Robertson, P., and Worz, T., Bandwidth efficient turbo trellis coded modulation using punctured component codes. *IEEE J. Sel. Areas Commun.*, 16(2), 206–218, 1998.
12. Wachsmann, U., and Huber, J., Power and bandwidth efficient digital communication using the turbo codes in multilevel codes. *Eur. Trans. Telecommun.*, 6(5), 557–567, 1995.
13. Classon, B. et al., Channel coding for 4G systems with adaptive modulation and coding. *IEEE Wireless Commun.*, 8–13, April 2002.
14. Saifuddin, A. et al., HDR codes with concatenated multilevel codes and multiple-symbol differential detection for satellite application. *IEE Electron. Lett.*, 33(16), 1355–1356, July 1997.

15. Vilaipornsawai, U., and Soleymani, M.R., Trellis-based iterative decoding of block codes for satellite ATM. *IEEE Int. Conf. Commun., ICC 2002*, 5, 2947–2951, 2002.
16. Sumanasena, M.A.K., and Evans, B.G., Adaptive modulation and coding for satellite: UMTS. *Proc. 54th IEEE Veh. Technol. Conf.*, 1, 116–120, Fall 2001.
17. Papathanassiou, A. et al., A comparison study of the uplink performance of W-CDMA and OFDM for mobile multimedia communications via LEO satellites. *IEEE Pers. Commun.*, 35–43, June 2001.

Chapter 5

Modulation Methods

Gordon L. Stüber

Contents

5.1 Introduction

Modulation is the process by which the message information is added to the radio carrier. Most first-generation cellular systems, such as the Advanced Mobile Phone Services (AMPS), use analog frequency modulation (FM) because analog technology was mature when these systems were first introduced.

Digital modulation schemes, however, are the obvious choice for future wireless systems, especially if data services such as wireless multimedia are to be supported. Digital modulation can also improve spectral efficiency because digital signals are more robust against channel impairments. Spectral efficiency is a key attribute of wireless systems that must operate in a crowded radio frequency spectrum.

To achieve high spectral efficiency, modulation schemes must be selected that have a high bandwidth efficiency as measured in units of bits per second per hertz of bandwidth. Many wireless communication systems, such as cellular telephones, operate on the principle of frequency reuse, by which the carrier frequencies are reused at geographically separated locations. The link quality in these systems is limited by cochannel interference. Hence, modulation schemes must be identified that are both bandwidth efficient and capable of tolerating high levels of cochannel interference. More specifically, digital modulation techniques are chosen for wireless systems that satisfy the following properties:

Compact Power Density Spectrum: To minimize the effect of adjacent channel interference, it is desirable that the power radiated into the adjacent channel be 60–80 dB below that in the desired channel. Hence, modulation techniques with a narrow main lobe and fast roll-off of side lobes are desirable.

Good Bit Error Rate Performance: A low bit error probability should be achieved in the presence of cochannel interference, adjacent channel interference, thermal noise, and other channel impairments, such as fading and intersymbol interference (ISI).

Envelope Properties: Portable and mobile applications typically employ nonlinear (class C) power amplifiers to minimize battery drain. Nonlinear amplification may degrade the bit error rate performance of modulation schemes that transmit information in the amplitude of the carrier. Also, spectral shaping is usually performed prior to upconversion and nonlinear amplification. To prevent the regrowth of spectral side lobes during nonlinear amplification, the input signal must have a relatively constant envelope.

A variety of digital modulation techniques is currently being used in wireless communication systems. Two of the more widely used digital modulation techniques for cellular mobile radio are $\pi/4$ phase-shifted quadrature phase-shift keying ($\pi/4$-QPSK) and Gaussian minimum-shift keying (GMSK). The former is used in the North American Interim Standard 54 (IS-54) digital cellular system and Japanese personal digital cellular (PDC), whereas the latter is used in the Global System for Mobile Communications (GSM). This chapter provides a discussion of these and other modulation techniques that are employed in wireless communication systems.

5.2 Basic Description of Modulated Signals

With any modulation technique, the bandpass signal can be expressed in the form

$$s(t) = \operatorname{Re}\left\{ v(t)e^{j2\pi f_c t} \right\} \tag{5.1}$$

where $v(t)$ is the complex envelope, f is the carrier frequency, and Re{z} denotes the real part of z. For digital modulation schemes,

$$v(t) = A \sum_k b(t - kT, x_k) \tag{5.2}$$

where A is the amplitude of the carrier $x_k = (x_k, x_{k-1}, \ldots, x_{k-K})$ is the data sequence, T is the symbol or baud duration, and $b(t, x_j)$ is an equivalent shaping function, usually of duration T. The precise form of $b(t, xi)$ and the memory length K depends on the type of modulation that is employed. Several examples are provided in this chapter by which information is transmitted in the amplitude, phase, or frequency of the bandpass signal.

The *power spectral density* of the bandpass signal $S_{ss}(f)$ is related to the power spectral density of the complex envelope $S_{vv}(f)$ by

$$S_{ss}(f) = \frac{1}{2}[S_{vv}(f - f_c) + S_{vv}(f + f_c)] \tag{5.3}$$

The power density spectrum of the complex envelope for a digital modulation scheme has the general form

$$S_{vv}(f) = \frac{A^2}{T} \sum_m S_{b,m}(f) e^{-j2\pi fmT} \tag{5.4}$$

where

$$S_{b,m}(f) = \frac{1}{2} E[B(f, x_m) B^*(f, x_0)] \tag{5.5}$$

$B(f, x_m)$ is the Fourier transform of $b(t, x_m)$, and $E[.]$ denotes the expectation operator. Usually, symmetric signal sets are chosen so that the complex envelope has zero mean; that is, $E[b(t, x_0)] = 0$. This implies that the power density spectrum has no discrete components. If, in addition, x_m and x_0 are independent for $|m| > K$, then

$$S_{vv}(f) = \frac{A^2}{T} \sum_{|m|<k} S_{b,m}(f) e^{-j2\pi fmT} \tag{5.6}$$

5.3 Analog Frequency Modulation

With analog FM, the complex envelope is

$$v(t) = A \exp\left[j2\pi kf \int_0^t [m(t)d\tau] \right] \tag{5.7}$$

where $m(t)$ is the modulating waveform, and kf (in Hz/v) is the frequency sensitivity of the FM modulator. The bandpass signal is

$$s(t) = A \cos\left[2\pi f_c t + 2\pi k_f \int_0^t m(t)dt \right] \tag{5.8}$$

The instantaneous frequency of the carrier $f_i(t) = f_c + k_f m(t)$ varies linearly with the waveform $m(t)$; hence, the term *frequency modulation*. Notice that FM has a constant envelope, making it suitable for nonlinear amplification. However, the complex envelope is a nonlinear function of the modulating waveform $m(t)$; therefore, the spectral characteristics of $v(t)$ cannot be obtained directly from the spectral characteristics of $m(t)$.

With the sinusoidal modulating waveform $m(t) = A_m \cos(2\pi f_m t)$, the instantaneous carrier frequency is

$$f_i(t) = f_c + \Delta_f \cos(2\pi f_m t) \tag{5.9}$$

where $\Delta_f = k_f A_m$ is the peak frequency deviation. The complex envelope becomes

$$v(t) = \exp\left[2\pi \int_0^t [f_i(t)dt] \right] \tag{5.10}$$

$$= \exp[2\pi f_c t + \beta \sin(2\pi f_m t)]$$

where $\beta = \Delta_f / fm$ is called the modulation index. The bandwidth of $v(t)$ depends on the value of β. If $\beta < 1$, then narrowband FM is generated, for which the spectral widths of $v(t)$ and $m(t)$ are about the same, that is, $2f_m$. If $\beta \gg 1$, then wideband FM is generated, for which the spectral occupancy of $v(t)$ is slightly greater than $2\Delta_f$. In general, the approximate bandwidth of an FM signal is

$$W \approx 2\Delta_f + 2f_m = 2\Delta_f\left(1 + \frac{1}{\beta}\right) \tag{5.11}$$

which is a relation known as *Carson's rule*. Unfortunately, typical analog cellular radio systems use a modulation index in the range $1 \leq \beta \leq 3$ where Carson's rule is not accurate. Furthermore, the message waveform $m(t)$ is not a pure sinusoid, so Carson's rule does not directly apply.

In analog cellular systems, the waveform $m(t)$ is obtained by first companding the speech waveform and then hard limiting the resulting signal. The purpose of the limiter is to control the peak frequency deviation Δ_f. The limiter introduces high-frequency components that must be removed with a low-pass filter prior to modulation. To estimate the bandwidth occupancy, we first determine the ratio of the frequency deviation Δ_f corresponding to the maximum amplitude of $m(t)$ and the highest frequency component B that is present in $m(t)$. These two conditions are the most extreme cases, and the resulting ratio, $D = \Delta_f/B$, is called the *deviation ratio*. Then, replace β by D and fm by B in Carson's rule, giving

$$W \approx 2\Delta_f + 2B = 2\Delta_f\left(1+\frac{1}{D}\right) \tag{5.12}$$

This approximation will overestimate the bandwidth requirements. A more accurate estimate of the bandwidth requirements must be obtained from simulation or measurements.

5.4 Phase Shift Keying and π/4 -QPSK

With phase-shift keying (PSK), the equivalent shaping function in Equation 5.2 has the form

$$b(t, x_k) = \varphi_T(t)\exp\left[j\frac{\pi}{M}x_k h_s(t)\right], \quad x_k = x_k \tag{5.13}$$

where $h_s(t)$ is a phase-shaping pulse, $\psi_T(t)$ is an amplitude-shaping pulse, and M is the size of the modulation alphabet. Notice that the phase varies linearly with the symbol sequence $\{x_k\}$, hence, the term *phase-shift keying*. For a modulation alphabet size of M, $x_k - \{\pm1, \pm3, \ldots, \pm(M \vee 1)\}$. Each symbol x_k is mapped onto $\log_2 M$ source bits. A QPSK signal is obtained by using $M = 4$, resulting in a transmission rate of 2 bits/symbol.

Usually, the phase-shaping pulse is chosen to be the rectangular pulse $h_s(t) = u_T(t) - u(t) - u(t-T)$, where $u(t)$ is the unit step function. The amplitude-shaping pulse is very often chosen to be a square root raised cosine pulse, where the Fourier transform of $\psi_T(t)$ is

$$\varphi_T(f) = \begin{cases} \sqrt{T} & 0\le|f|\le\dfrac{(1-\beta)}{2T} \\[4mm] \sqrt{\dfrac{T}{2\left[1-\frac{\sin\pi T}{\beta}\left(f-\frac{1}{2T}\right)\right]}} & \dfrac{1-\beta}{2T}\le|f|\le\dfrac{1+\beta}{2T} \end{cases} \tag{5.14}$$

The receiver implements the same filter $\psi_R(f) = \psi_T(f)$ so that the overall pulse has the raised cosine spectrum $\psi(f) = \psi_R(f)\,\psi_T(f) = |\psi_T(f)|^2$. If the channel is affected by flat fading and additive white Gaussian noise, then this partitioning of the filtering operations between the transmitter and receiver will optimize the signal-to-noise ratio at the output of the receiver filter at the sampling instants. The roll-off factor β usually lies between 0 and 1 and defines the excess bandwidth $100\beta\%$. Using a smaller β results in a more compact power density spectrum, but the link performance becomes more sensitive to errors in the symbol timing. The IS-54 system uses $\beta = 0.35$, while PDC uses $\beta = 0.5$.

The time domain pulse corresponding to Equation 5.14 can be obtained by taking the inverse Fourier transform, resulting in

$$\varphi_T(t) = 4\beta \frac{\cos\left[\frac{(1+\beta)\pi t}{T}\right] + \sin\left[\frac{(1-\beta)\pi t}{T}\right]\left(\frac{4\beta t}{T}\right)^{-1}}{\pi\sqrt{T}\left[1 - \frac{16\beta^2 t^2}{T^2}\right]} \tag{5.15}$$

A typical square root raised cosine pulse with a roll-off factor of $\beta = 0.5$ is shown in Figure 5.1. Strictly speaking, the pulse is noncausal, but in practice a truncated time domain pulse is used. For example, in Figure 5.1 the pulse is truncated to $6T$ and time shifted by $3T$ to yield a causal pulse.

Unlike conventional QPSK, which has four possible transmitted phases, $\pi/4$-QPSK has eight possible transmitted phases. Let $\theta(n)$ be the transmitted carrier phase for the nth epoch, and let $\Delta\theta(n) = \theta(n) - \theta(n - 1)$ be the differential carrier phase between epochs n and $n - 1$. With $\pi/4$-QPSK, the transmission rate is 2 bits/symbol, and the differential phase is related to the symbol sequence $\{x_n\}$

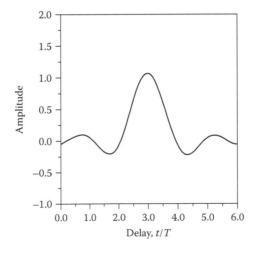

Figure 5.1   Square root raised cosine pulse with roll-off factor $\beta = 0.5$.

through the mapping

$$\Delta\theta(n) = \begin{cases} -\dfrac{3\pi}{4}, & x_n = -3 \\[2mm] -\dfrac{\pi}{4}, & x_n = -1 \\[2mm] \dfrac{\pi}{4}, & x_n = +1 \\[2mm] \dfrac{3\pi}{4}, & x_n = +3 \end{cases} \tag{5.16}$$

Because the symbol sequence $\{x_n\}$ is random, the mapping in Equation 5.16 is arbitrary, except that the phase differences must be $\pm\pi/4$ and $\pm3\pi/4$. The phase difference with the given mapping can be written in the convenient algebraic form

$$\Delta\theta(n) = x_n \frac{\pi}{4} \tag{5.17}$$

which allows us to write the equivalent shaping function of the $\pi/4$-QPSK signal as

$$b(t, -x_1 k) = \varphi(t) \exp\{j[\theta(k-1) + x_1 k\,\pi/4\}$$

$$= \varphi_T(t) \exp\left[j\frac{\pi}{4}\left(\sum_{n=-\infty}^{k-1} x_n + x_k \right) \right] \tag{5.18}$$

The summation in the exponent represents the accumulated carrier phase, whereas the last term is the phase change due to the kth symbol. Observe that the phase-shaping function is the rectangular pulse $uT(t)$. The amplitude-shaping function $JT(t)$ is usually the square root raised cosine pulse in Equation 5.15.

The phase states of QPSK and $\pi/4$-QPSK signals can be summarized by the signal space diagram in Figure 5.2, that shows the phase states and allowable transitions between the phase states. However, it does not describe the actual phase trajectories. A typical diagram showing phase trajectories with square root raised cosine pulse shaping is shown in Figure 5.3. Note that the phase trajectories do not pass through the origin. This reduces the envelope fluctuations of the signal, making it less susceptible to amplifier nonlinearities, and reduces the dynamic range required of the power amplifier.

The power density spectrum of QPSK and $\pi/4$-QPSK depends on both the amplitude- and phase-shaping pulses. For the rectangular phase-shaping pulse $hs(t) = uT(t)$, the power density spectrum of the complex envelope is

$$S_{vv}(f) = \frac{A^2}{T} |\varphi_T(f)|^2 \tag{5.19}$$

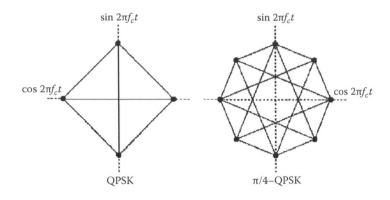

Figure 5.2   Signal-space constellations for QPSK and π/4-QPSK.

With square root raised cosine pulse shaping, $\psi_T(f)$ has the form defined in Equation 5.14. The power density spectrum of a pulse $\tilde{\psi}_T(t)$ that is obtained by truncating $\psi_T(t)$ to length T can be obtained by writing $= \tilde{\psi}_T(t)\,\text{rect}(t/T)$. Then, $\psi_T(f) = T\,\text{sinc}(fT)$, where $*$ denotes the operation of convolution, and the power density spectrum is again obtained by applying Equation 5.19. Truncation of the pulse will regenerate some side lobes, thus causing adjacent channel interference. Figure 5.4 illustrates the power density spectrum of a truncated square root raised cosine pulse for various truncation lengths T.

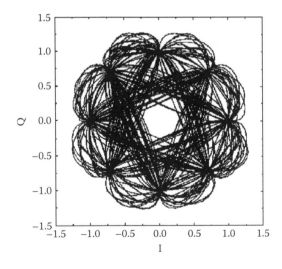

Figure 5.3   Phase diagram of π/4-QPSK with square root raised cosine pulse; $\beta = 0.5$.

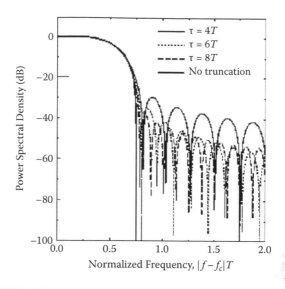

Figure 5.4 **Power density spectrum of truncated square root raised cosine pulse with various truncation lengths; $\beta = 0.5$.**

5.5 Continuous Phase Modulation and Minimum-Shift Keying

Continuous phase modulation (CPM) refers to a broad class of FM techniques by which the carrier phase varies in a continuous manner. A comprehensive treatment of CPM is provided in [1]. CPM schemes are attractive because they have a constant envelope and excellent spectral characteristics. The complex envelope of any CPM signal is

$$v(t) = A \, \exp\left[j2\pi k_f \int_{-\infty}^{t} \sum_n x_n h_s (\tau - nT) d\tau \right] \tag{5.20}$$

The instantaneous frequency deviation from the carrier is

$$f_{dev}(t) = k_f \sum_n x_n h_s (t - nT) \tag{5.21}$$

where *kf* is the peak frequency deviation. If the frequency-shaping pulse $h_s(t)$ has duration *T*, then the equivalent shaping function in Equation 5.2 has the form

$$b(t, x_k) = \exp\left\{ j\left[\beta(T) \sum_{n=-\infty}^{k-1} [x_n + x_k \beta(t)] \right] \right\} u_T(t) \tag{5.22}$$

where

$$\beta T = \begin{cases} 0, & t < 0 \\ \dfrac{\pi h}{\int_0^T h_s(\tau)d\tau} \displaystyle\int_0^t h_s(\tau)d\tau, & 0 \le t \le T \\ \pi h, & t \ge 1 \end{cases} \quad (5.23)$$

is the phase-shaping pulse, and $h = \beta(T)/\pi$ is call the *modulation index*.

Minimum-shift keying (MSK) is a special form of binary CPM ($x_k \in \{-1, +1\}$) that is defined by a rectangular frequency-shaping pulse $h_s(t) = u_T(t)$ and a modulation index $h = 1/2$ so that

$$\beta(T) = \begin{cases} 0, & t < 0 \\ \dfrac{\pi t}{2T}, & 0 \le t \le T \\ \dfrac{\pi}{2}, & t \ge T \end{cases} \quad (5.24)$$

Therefore, the complex envelope is

$$v(t) = A \exp\left(j\frac{\pi}{2}\sum_{n=-\infty}^{k-1} x_n + \frac{\pi}{2}x_k\frac{t - KT}{T} \right) \quad (5.25)$$

An MSK signal can be described by the phase trellis diagram shown in Figure 5.5, which plots the time behavior of the phase

$$\theta(t) = \frac{\pi}{2}\sum_{n=-\infty}^{k-1} x_n + \frac{\pi}{2}x_k\frac{t - kT}{T} \quad (5.26)$$

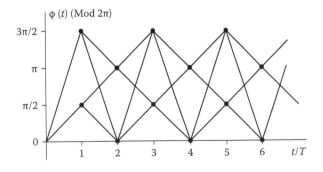

Figure 5.5 Phase trellis diagram for MSK.

The MSK bandpass signal is

$$s(t) = A\cos\left(2\pi f_c t + \frac{\pi}{2}\sum_{n=-\infty}^{k-1} x_n + \frac{\pi}{2}x_k \frac{t-kT}{T}\right)$$

$$(5.27)$$

$$= A\cos\left[2\pi\left(f_c + \frac{x_k}{4T}\right)t - \frac{k\pi}{2}x_k + \frac{\pi}{2}\sum_{n=-\infty}^{k-1} x_n\right] kT \le t \le (k+1)T$$

From Equation 5.27, we observe that the MSK signal has one of two possible frequencies, $f_L = f_c - 1/4T$ or $f_U = f_c + 1/4T$, during each symbol interval. The difference between these frequencies is $f_U - f_L = 1/2T$. This is the minimum frequency difference between two sinusoids of duration T that will ensure orthogonality with coherent demodulation [2], hence, the name *minimum-shift keying*. By applying various trigonometric identities to Equation 5.27, we can write

$$s(t) = A\left[x_k^I \varphi(t - k2T)\cos(2\pi f_c t) - x_k^Q \varphi(t - k2T - T)\sin(2\pi f_c t)\right],$$

$$(5.28)$$

$$kT \le t \le (k+1)T$$

where

$$x_k^I = -x_{k-1}^Q x_{2k-1}$$

$$x_k^Q = x_k^I x_{2k}$$

$$\varphi(t) = \cos\left(\frac{\pi t}{2T}\right), -T \le t \le T$$

Note that x_k^I and x_k^Q are independent binary symbols that take on elements from the set {−1, +1}, and the half-sinusoid amplitude-shaping pulse $\psi(t)$ has duration $2T$ and $\psi(t - T) = \sin(\pi t/2T)$, $0 \le t \le 2T$. Therefore, MSK is equivalent to offset quadrature amplitude shift keying (OQASK) with a half-sinusoid amplitude-shaping pulse.

To obtain the power density spectrum of MSK, we observe from Equation 5.28 that the equivalent shaping function of MSK has the form

$$b(t, x_k) = x_k^I \varphi(t) + jx_k^Q \varphi(t - T)$$

$$(5.29)$$

The Fourier transform of Equation 5.29 is

$$B(f, x_k) = \left(x_k^I + jx_k^Q e^{-j2\pi fT}\right)\psi(f)$$

$$(5.30)$$

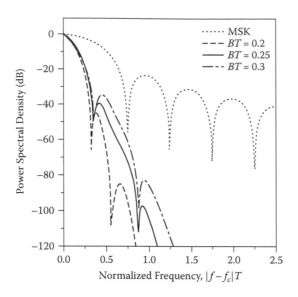

Figure 5.6 Power density spectrum of MSK and GMSK.

Since the symbols and are independent and zero mean, it follows from Equations 5.5 and 5.6 that

$$S_{vv}(f) = \frac{A^2 \, |\varphi(f)|^2}{2T}$$ (5.31)

Therefore, the power density spectrum of MSK is determined solely by the Fourier transform of the half-sinusoid amplitude-shaping pulse $J(t)$, resulting in

$$S_{vv}(f) = \frac{16A^2 T}{\pi^2} \left[\frac{\cos 2\pi fT}{1 - 16 f^2 T^2} \right]^2$$ (5.32)

The power spectral density of MSK is plotted in Figure 5.6. Observe that an MSK signal has fairly large sidelobes compared to $\pi/4$-QPSK with a truncated square root raised cosine pulse (cf. Figure 5.4).

5.6 Gaussian Minimum-Shift Keying

The MSK signals have all of the desirable attributes for mobile radio except for a compact power density spectrum. This can be alleviated by filtering the modulating signal $x(t) = \Sigma_n x_n u_T(t - nT)$ with a low-pass filter prior to FM, as shown in Figure 5.7. Such filtering removes the higher-frequency components in $x(t)$ and therefore yields

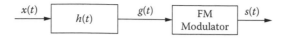

Figure 5.7 Premodulation filtered MSK.

a more compact spectrum. The low-pass filter is chosen to have (1) narrow bandwidth and a sharp transition band, (2) low-overshoot impulse response, and (3) preservation of the output pulse area to ensure a phase shift of $\pi/2$.

GMSK uses a low-pass filter with the following transfer function:

$$H(f) = A \exp\left\{-\left(\frac{f}{B}\right)^2 \frac{\ln^2}{2}\right\} \tag{5.33}$$

where B is the 3-dB bandwidth of the filter, and A is a constant. It is apparent that $H(f)$ is bell shaped about $f = 0$, hence the name Gaussian MSK. A rectangular pulse rect $(t/T) = u_T(t + T/2)$ transmitted through this filter yields the frequency-shaping pulse

$$h_s(t) = A\sqrt{\left(\frac{2\pi}{\ln 2}\right)}(BT)\int_{\frac{t}{T}-\frac{1}{2}}^{\frac{t}{T}+\frac{1}{2}} \exp\left\{-\frac{2\pi^2(BT)^2 x^2}{\ln 2}\right\}dx \tag{5.34}$$

The phase change over the time interval from $-T/2 \le t \le T/2$ is

$$\theta\left(\frac{T}{2}\right) - \theta\left(\frac{-T}{2}\right) = x_0\beta_0(T) + \sum_{\substack{n=-\infty \\ n\neq 0}}^{\infty} x_n\beta_n(T) \tag{5.35}$$

where

$$\beta_n(T) = \frac{\pi h}{\int_{-\infty}^{\infty} h_s(v)dv} \int_{-\frac{T}{2}-nT}^{\frac{T}{2}-nT} h_s(v)dv \tag{5.36}$$

The first term in Equation 5.35 is the desired term, and the second term is the ISI introduced by the premodulation filter, again with GMSK $h = 1/2$, so that a total phase shift of $\pi/2$ is maintained.

Notice that the pulse $h_s(t)$ is noncausal so that a truncated pulse must be used in practice. Figure 5.8 plots a GMSK frequency-shaping pulse that is truncated to $T = 5T$ and time shifted by $2.5T$ for various normalized filter bandwidths BT. Notice that the frequency-shaping pulse has a duration greater than T so that ISI is introduced. As BT decreases, the induced ISI is increased. Thus, whereas a smaller value of BT results in a more compact power density spectrum, the induced ISI will degrade

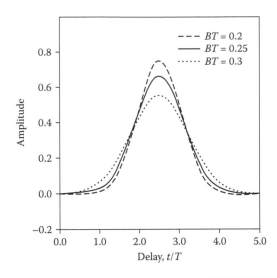

Figure 5.8 **GMSK frequency-shaping pulse for various normalized filter bandwidths BT.**

the bit error rate performance. Hence, there is a trade-off in the choice of *BT*. Some studies have indicated that *BT* = 0.25 is a good choice for cellular radio systems [3].

The power density spectrum of GMSK is quite difficult to obtain but can be computed using published methods [4]. Figure 5.6 plots the power density spectrum for *BT* = 0.2, 0.25, and 0.3, obtained from K. Wesolowski (private communication). Observe that the spectral side lobes are greatly reduced by the Gaussian low-pass filter.

5.7 Orthogonal Frequency-Division Multiplexing

Orthogonal frequency-division multiplexing (OFDM) is a modulation technique that has been suggested for use in cellular radio [5], digital audio broadcasting [6], and digital video broadcasting. The basic idea of OFDM is to transmit blocks of symbols in parallel by employing a (large) number of orthogonal subcarriers. With block transmission, N serial source symbols each with period T_s are converted into a block of N parallel modulated symbols each with period $T = NT_s$. The block length N is chosen so that $NT_s \gg \sigma_T$, where σ_T is the root mean square (rms) delay spread of the channel. Since the symbol rate on each subcarrier is much less than the serial source rate, the effects of delay spread are greatly reduced. This has practical advantages because it may reduce or even eliminate the need for equalization. Although the block length N is chosen so that $NT_s \gg \sigma_T$, the channel dispersion will still cause consecutive blocks to overlap. This results in some residual ISI that will degrade the performance. This residual ISI can be eliminated at the expense of channel capacity by using guard intervals between the blocks that are at least as long as the effective channel impulse response.

The complex envelope of an OFDM signal is described by

$$v(t) = A \sum_{k} \sum_{n=0}^{N-1} x_{k,n} \varphi_n(t - kT) \qquad (5.37)$$

where

$$\varphi_n(t) = \exp\left\{ j \frac{2\pi\left(n - \frac{N-1}{2}\right)}{T} \right\} U_T(t), \quad n = 0.1,\ldots,N-1 \qquad (5.38)$$

are orthogonal waveforms, and $U_T(t)$ is a rectangular shaping function. The frequency separation of the subcarriers, $1/T$, ensures that the subcarriers are orthogonal, and phase continuity is maintained from one symbol to the next but is twice the minimum required for orthogonality with coherent detection.

At epoch k, N-data symbols are transmitted using the N distinct pulses. The data symbols $x_{k,n}$ are often chosen from an M-ary *quadrature amplitude modulation* (M-QAM) constellation, where $x_{k,n} = x^I_{k,n} + jx^Q_{k,n}$ with $x^I_{k,n}, x^Q_{k,n}$ _ {±1, ±3, ... , ± (N − 1)} and $N = \sqrt{M}$.

A key advantage of using OFDM is that the modulation can be achieved in the discrete domain by using either an inverse discrete Fourier transform (IDFT) or the more computationally efficient inverse fast Fourier transform (IFFT). Considering the data block at epoch $k = 0$ and ignoring the frequency offset $\exp\{-j[2\pi(N-1) t/2T]\}$, the complex low-pass OFDM signal has the form

$$v(t) = \sum_{n=0}^{N-1} x_{0,n} \exp\left\{ \frac{j2\pi nt}{NT_s} \right\}, \quad 0 \leq t \leq T \qquad (5.39)$$

If this signal is sampled at epochs $t = kT_s$, then

$$v^k = v(kT_s) = \sum_{n=0}^{N-1} x_{0,n} \exp\left\{ \frac{j2\pi nk}{NT_s} \right\}, \quad k = 0, 1,\ldots,N-1 \qquad (5.40)$$

Observe that the sampled OFDM signal has duration N, and the samples v^0, v^1, ... , v^{N-1} are just the IDFT of the data block $x_{0,0}, x_{0,1}, \ldots , x_{0, N-1}$. A block diagram of an OFDM transmitter is shown in Figure 5.9.

The power spectral density of an OFDM signal can be obtained by treating OFDM as independent modulation on subcarriers that are separated in frequency by $1/T$.

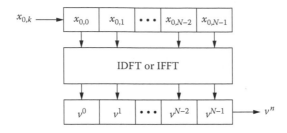

$x_{0,k} \longrightarrow$

| $x_{0,0}$ | $x_{0,1}$ | $\bullet\bullet\bullet$ | $x_{0,N-2}$ | $x_{0,N-1}$ |

IDFT or IFFT

| v^0 | v^1 | $\bullet\bullet\bullet$ | v^{N-2} | v^{N-1} | $\longrightarrow v^n$

Figure 5.9   Block diagram of OFDM transmitter using IDFT or IFFT.

Because the subcarriers are only separated by $1/T$, significant spectral overlap results. Because the subcarriers are orthogonal, however, the overlap improves the spectral efficiency of the scheme. For a signal constellation with zero mean and the waveforms in Equation 5.38, the power density spectrum of the complex envelope is

$$S_{vv}(f) = \frac{A^2}{T} \sigma_x^2 \sum_{n=0}^{N-1} \left| \mathrm{sinc}\left[fT - \left(n - \frac{N-1}{2} \right) \right] \right|^2 \tag{5.41}$$

where $\sigma_x^2 = 1/2\ E[|x_{k,n}|^2]$ is the variance of the signal constellation. For example, the complex envelope power spectrum of OFDM with $N = 32$ subcarriers is shown in Figure 5.10.

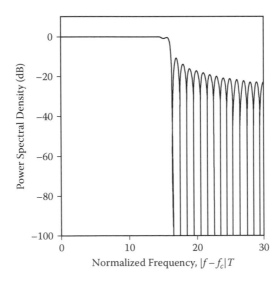

Figure 5.10   Power density spectrum of OFDM with N = 32.

5.8 Conclusions

A variety of modulation schemes are employed in wireless communication systems. Wireless modulation schemes must have a compact power density spectrum while providing a good bit error rate performance in the presence of channel impairments such as cochannel interference and fading.

The most popular digital modulation techniques employed in wireless systems are GMSK in the European GSM system, $\pi/4$-QPSK in the North American IS-54 and Japanese PDC systems, and OFDM in digital audio broadcasting systems.

Glossary

Bandwidth efficiency: Transmission efficiency of a digital modulation scheme measured in units of bits per second per hertz of bandwidth.

Continuous phase modulation: Frequency modulation in which the phase varies in a continuous manner.

Excess bandwidth: Percentage of bandwidth that is in excess of the minimum of *1/2T* (*T* is the baud or symbol duration) required for data communication.

Frequency modulation: Modulation in which the instantaneous frequency of the carrier varies linearly with the data signal.

Gaussian minimum-shift keying: MSK for which the data signal is prefiltered with a Gaussian filter prior to frequency modulation.

Minimum-shift keying: A special form of continuous phase modulation having linear phase trajectories and a modulation index of 1/2.

Orthogonal frequency-division multiplexing: Modulation using a collection of low-bit-rate orthogonal subcarriers.

Phase shift keying: Modulation in which the instantaneous phase of the carrier varies linearly with the data signal.

Power spectral density: Relative power in a modulated signal as a function of frequency.

Quadrature amplitude modulation: Modulation by which information is transmitted in the amplitude of the cosine and sine components of the carrier.

References

1. Anderson, J.B., Aulin, T., and Sundberg, C.-E., *Digital Phase Modulation.* New York: Plenum Press, 1986.
2. Birchler, M.A., and Jasper, S.C., A 64 kbps digital land mobile radio system employing M-16QAM. *Proc. 5th Nordic Sem. Dig. Mobile Radio Commun.,* 237–241, December 1992.
3. Garrison, G.J., A power spectral density analysis for digital FM, *IEEE Trans. Commun.,* COM-23, 1228–1243, November 1975.

4. Le Floch, B., Halbert-Lassalle, R., and Castelain, D., Digital sound broadcasting to mobile receivers, *IEEE Trans. Consum. Elec.,* 35, August 1989.

5. Murota, K., and Hirade, K., GMSK modulation for digital mobile radio telephony, *IEEE Trans. Commun.,* COM-29, 1044–1050, July 1981.

6. Murota, K., Kinoshita, K., and Hirade, K., Spectral efficiency of GMSK land mobile radio. *Proc. ICC'81,* 23.8/1, June 1981.

Chapter 6

Cellular Systems

Lal C. Godara

Contents

6.1 Introduction

The cellular concept was invented by Bell Laboratories, and the first commercial analog voice system was introduced in Chicago in October 1983 [1, 2]. The first-generation analog cordless phone and cellular systems became popular using the design based on a standard known as Advanced Mobile Phone Services (AMPS). Similar standards were developed around the world, including Total Access Communication System (TACS), Nordic Mobile Telephone (NMT) 450, and NMT 900 in Europe; European Total Access Communication System (ETACS) in the United Kingdom; C-450 in Germany; and Nippon Telephone and Telegraph (NTT), Japanese Total Access Communications System (JTACS), and NTACS in Japan.

In contrast to the first-generation analog systems, second-generation systems are designed to use digital transmission. These systems include the Pan-European Global System for Mobile Communications (GSM; originally Groupe Spécial Mobile) and Digital Communications System (DCS) 1800 systems, North American dual-mode cellular system Interim Standard 54 (IS-54) 54, North American IS-95 system, and Japanese personal digital cellular (PDC) system [1,3]. The third-generation mobile communication systems are being studied worldwide under the names Universal Mobile Telecommunications System (UMTS) and International Mobile Telecommunications 2000 (IMT-2000) [4,5]. The aim of these systems is to provide users advance communication services with wideband capabilities using a single standard. Details on various systems can be found in [1,6–9]. In third-generation communication systems, satellites are going to play a major role in providing global coverage [10–16].

The aim of this chapter is to present fundamental concepts of cellular systems by explaining various terminology used to understand the working of these systems. The chapter also provides details on some popular standards. More details on cellular fundamentals may be found in [17–20].

Section 6.2 presents fundamentals of cellular systems for understanding how these systems work. Sections 6.3 and 6.4 are devoted to first-generation and second-generation systems, respectively; a brief description of some popular standards is presented. A discussion of third-generation systems is included in Section 6.5.

6.2 Cellular Fundamentals

The area served by mobile phone systems is divided into small areas known as *cells*. Each cell contains a base station that communicates with mobiles in the cell by transmitting and receiving signals on radio links. The transmission from the base station to a mobile is typically referred to as downstream, forward link, or downlink. The corresponding terms for the transmission from a mobile to the base are upstream, reverse link, and uplink. Each base station is associated with a mobile switching center (MSC) that connects calls to and from the base to mobiles in other cells and the public switched telephone network (PSTN). Figure 6.1 shows a typical setup depicting a group of base stations to a switching center. In this section, terminology associated with cellular systems is introduced with a brief description to facilitate understanding how these systems work [21].

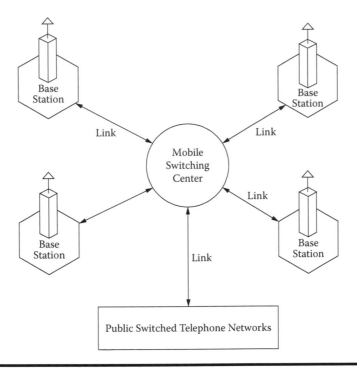

Figure 6.1   A typical cellular system setup.

6.2.1 *Communication Using Base Stations*

A base station communicates with mobiles using two types of radio channels: control channels to carry control information and traffic channels to carry messages. Each base station continuously transmits control information on its control channels. When a mobile is switched on, it scans the control channels and tunes to a channel with the strongest signal. This normally would come from the base station located in the cell in which the mobile is also located. The mobile exchanges identification information with the base station and establishes the authorization to use the network. At this stage, the mobile is ready to initiate and receive a call.

6.2.1.1 *A Call from a Mobile*

When a mobile wants to initiate a call, it sends the required number to the base. The base station sends this information to the switching center, which assigns a traffic channel to this call because the control channels are used only for control information. Once the traffic channel is assigned, this information is relayed to the mobile via the base station. The mobile switches itself to this channel. The switching center then completes the rest of the call.

6.2.1.2 A Call to a Mobile

When someone calls a mobile, the call arrives at the MSC. It then sends a paging message through several base stations. A mobile tuned to a control channel detects its number in the paging message and responds by sending a response signal to the nearby base station. The base station informs the switching center about the location of the desired mobile. The switching center assigns a traffic channel to this call and relays this information to the mobile via the base. The mobile switches itself to the traffic channel, and the call is complete.

6.2.1.3 Registration

A mobile is normally located by transmitting a paging message from various base stations. When a large number of base stations is involved in the paging process, it becomes impractical and costly. This is avoided by a registration procedure by which a roaming phone registers with an MSC closer to itself. This information may be stored with the switching center of the area as well as the home switching center of the phone. The home base of the phone is the one where it is permanently registered. Once a call is received for this phone, its home switching center contacts the switching center where the phone is currently roaming. Paging in the vicinity of the previous known location helps to locate the phone. Once it responds, the call may be connected as discussed.

6.2.2 Channel Characteristics

An understanding of propagation conditions and channel characteristics is important for efficient use of a transmission medium. Attention is being given to understanding the propagation conditions under which a mobile is to operate, and many experiments have been conducted to model the channel characteristics. Many of these results can be found in review articles [22–24] and references therein.

6.2.2.1 Fading Channels

The signal arriving at a receiver is a combination of many components arriving from various directions as a result of multipath propagation. This depends on terrain conditions and local buildings and structures, causing the received signal power to fluctuate randomly as a function of distance. Fluctuations on the order of 20 dB are common within the distance of one wavelength (I λ). This phenomenon is called *fading*. One may think of this signal as a product of two variables.

The first component, also referred to as the *short-term fading component*, changes faster than the second one and has a Rayleigh distribution. The second component is a long-term or slow-varying quantity and has lognormal distribution [17,25]. In other words, the local mean varies slowly with lognormal distribution, and the fast variation around the local mean has Rayleigh distribution.

A movement in a mobile receiver causes it to encounter fluctuations in the received power level. The rate at which this happens is referred to as the *fading rate* in mobile communication literature [26], and it depends on the frequency of transmission and the speed of the mobile. For example, a mobile on foot operating at 900 MHz would cause a fading rate of about 4.5 Hz, whereas a typical vehicle mobile would produce a fading rate of about 70 Hz.

6.2.2.2 Doppler Spread

The movement in a mobile causes the received frequency to differ from the transmitted frequency because of the Doppler shift resulting from its relative motion. As the received signals arrive along many paths, the relative velocity of the mobile with respect to various components of the signal differs, causing the different components to yield a different Doppler shift. This can be viewed as spreading of the transmitted frequency and is referred to as the *Doppler spread*. The width of the Doppler spread in the frequency domain is closely related to the rate of fluctuations in the observed signal [22].

6.2.2.3 Delay Spread

Because of the multipath nature of propagation in the area where a mobile is being used, it receives multiple and delayed copies of the same transmission, resulting in spreading of the signal in time. The root mean square (rms) delay spread may range from a fraction of a microsecond in urban areas to on the order of 100 μs in a hilly area, and this restricts the maximum signal bandwidth to between 40 and 250 kHz. This bandwidth is known as *coherence bandwidth*. The coherence bandwidth is inversely proportional to the rms delay spread. This is the bandwidth over which the channel is flat; that is, it has a constant gain and linear phase.

For a signal bandwidth above the coherence bandwidth, the channel loses its constant gain and linear phase characteristic and becomes frequency selective. Roughly speaking, a channel becomes frequency selective when the rms delay spread is larger than the symbol duration and causes intersymbol interference (ISI) in digital communications. Frequency-selective channels are also known as dispersive channels, whereas the nondispersive channels are referred to as flat-fading channels.

6.2.2.4 Link Budget and Path Loss

Link budget is a name given to the process of estimating the power at the receiver site for a microwave link taking into account the attenuation caused by the distance between the transmitter and the receiver. This reduction is referred to as the *path loss*. In free space, the path loss is proportional to the second power of the distance;

that is, the distance power gradient is two. In other words, by doubling the distance between the transmitter and the receiver, the received power at the receiver reduces to one fourth of the original amount.

For a mobile communication environment utilizing fading channels, the distance power gradient varies and depends on the propagation conditions. Experimental results showed that it ranges from a value lower than two in indoor areas with large corridors to as high as six in metal buildings. For urban areas, the path loss between the base and the cell site is often taken to vary as the fourth power of the distance between the two [22].

Normal calculation of link budget is done by calculating the carrier-to-noise ratio (CNR), where noise consists of background and thermal noise, and the system utility is limited by the amount of this noise. However, in mobile communication systems, the interference resulting from other mobile units is a dominant noise compared with the background and human-made noise. For this reason, these systems are limited by the amount of total interference present instead of the background noise as in the other case. In other words, the signal-to-interference ratio (SIR) is the limiting factor for a mobile communication system instead of the signal-to-noise ratio (SNR), as is the case for other communication systems. The calculation of link budget for such interference-limited systems involves calculating the carrier level above the interference level contributed by all sources [27].

6.2.3 Multiple-Access Schemes

The available spectrum bandwidth is shared in a number of ways by various wireless radio links. The way in which this is done is referred to as a *multiple-access scheme*. There are basically four principle schemes. These are frequency-division multiple access (FDMA), time-division multiple access (TDMA), code-division multiple access (CDMA), and space-division multiple access (SDMA) [28–40].

6.2.3.1 Frequency-Division Multiple-Access Scheme

In an FDMA scheme, the available spectrum is divided into a number of frequency channels of certain bandwidth, and individual calls use different frequency channels. All first-generation cellular systems use this scheme.

6.2.3.2 Time-Division Multiple-Access Scheme

In a TDMA scheme, several calls share a frequency channel [29]. The scheme is useful for digitized speech or other digital data. Each call is allocated a number of time slots based on its data rate within a frame for upstream as well as downstream. Apart from the user data, each time slot also carries other data for synchronization, guard times, and control information.

The transmission from base station to mobile is done in time-division multiplex (TDM) mode, whereas in the upstream direction each mobile transmits in its own time slot. The overlap between different slots resulting from different propagation delays is prevented by using guard times and precise slot synchronization schemes.

The TDMA scheme is used along with the FDMA scheme because several frequency channels are used in a cell. The traffic in two directions is separated by either using two separate frequency channels or alternating in time. The two schemes are referred to as frequency-division duplex (FDD) and time-division duplex (TDD), respectively. The FDD scheme uses less bandwidth than TDD schemes use and does not require as precise synchronization of data flowing in two directions as that in the TDD method. The latter, however, is useful when flexible bandwidth allocation is required for upstream and downstream traffic [29].

6.2.3.3 Code-Division Multiple-Access Scheme

The CDMA scheme is a direct sequence (DS), spread-spectrum method. It uses linear modulation with wide-band pseudonoise (PN) sequences to generate signals. These sequences, also known as *codes*, spread the spectrum of the modulating signal over a large bandwidth, simultaneously reducing the spectral density of the signal. Thus, various CDMA signals occupy the same bandwidth and appear as noise to each other. More details on DS spread spectrum may be found in [36].

In the CDMA scheme, each user is assigned an individual code at the time of call initiation. This code is used for both spreading the signal at the time of transmission and despreading the signal at the time of reception. Cellular systems using CDMA schemes use FDD, thus employing two frequency channels for forward and reverse links.

On forward links, a mobile transmits to all users synchronously, and this preserves the orthogonality of various codes assigned to different users. The orthogonality, however, is not preserved between different components arriving from different paths in multipath situations [34]. On reverse links, each user transmits independently from other users because of their individual locations. Thus, the transmission on reverse link is asynchronous, and the various signals are not necessarily orthogonal.

It should be noted that these PN sequences are designed to be orthogonal to each other. In other words, the cross correlation between different code sequences is zero; thus, the signal modulated with one code appears to be orthogonal to a receiver using a different code if the orthogonality is preserved during the transmission. This is the case on forward link, and in the absence of multipath the signal received by a mobile is not affected by signals transmitted by the base station to other mobiles.

On reverse link, the situation is different. Signals arriving from different mobiles are not orthogonalized because of the asynchronous nature of transmission. This may cause a serious problem when the base station is trying to receive a weak signal from a distant mobile in the presence of a strong signal from a nearby mobile. A strong DS signal from a nearby mobile swamping a weak DS signal from a distant mobile and making its detection difficult is known as the *near–far problem*. It is prevented by controlling the power transmitted from various mobiles such that the received signals at the base station are of almost equal strength. The power control is discussed in Section 6.2.9.

The term *wideband CDMA* (WCDMA) is used when the spread bandwidth is more than the coherence bandwidth of the channel [37]. Thus, over the spread bandwidth of DS-CDMA, the channel is frequency selective. On the other hand, the term *narrowband CDMA* is used when the channel encounters flat fading over the spread bandwidth. When a channel encounters frequency-selective fading over the spread bandwidth, a RAKE receiver may be employed to resolve the multipath components and combine them coherently to combat fading.

A WCDMA signal may be generated using multicarrier (MC) narrowband CDMA signals, each using different frequency channels. This composite MC-WCDMA scheme has a number of advantages over the single-carrier WCDMA scheme. It not only is able to provide diversity enhancement over multipath fading channels but also does not require a contiguous spectrum as is the case for the single-carrier WCDMA scheme. This helps to avoid frequency channels occupied by narrowband CDMA by not transmitting MC-WCDMA signals over these channels. More details on these and other issues may be found in [37] and references therein.

6.2.3.4 Comparison of Different Multiple-Access Schemes

Each multiple-access scheme has its advantages and disadvantages, such as complexities of equipment design, robustness of system parameter variation. For example, a TDMA scheme not only requires complex time synchronization of different user data but also presents a challenge to design portable radio-frequency (RF) units that overcome the problem of a periodically pulsating power envelope caused by short duty cycles of each user terminal. It should be noted that when a TDMA frame consists of N users transmitting equal bit rates, the duty cycles of each user is $1/N$. TDMA also has a number of advantages [29]:

1. A base station communicating with a number of users sharing a frequency channel requires only one set of common radio equipment.
2. The data rate, to and from each user, can easily be varied by changing the number of time slots allocated to the user as per the requirements.
3. It does not require as stringent power control as that of CDMA because its interuser interference is controlled by time slot and frequency channel allocations.
4. Its time slot structure is helpful in measuring the quality of alternative slots and frequency channels that could be used for mobile-assisted handoffs (MAHOs). Handoff is discussed in Section 6.2.7.

It is argued in [34] that, although there does not appear to be a single scheme that is the best for all situations, CDMA possesses characteristics that give it distinct advantages over others:

1. It is able to reject delayed multipath arrivals that fall outside the correlation interval of the PN sequence in use and thus reduces the multipath fading.
2. It has the ability to reduce the multipath fading by coherently combining different multipath components using a RAKE receiver.
3. In TDMA and FDMA systems, a frequency channel used in a cell is not used in adjacent cells to prevent cochannel interference. In a CDMA system, it is possible to use the same frequency channel in adjacent cells and thus increase the system capacity.
4. The speech signal is inherently bursty because of the natural gaps during conversation. In FDMA and TDMA systems, once a channel (frequency or time slot) is allocated to a user, that channel cannot be used during nonactivity periods. However, in CDMA systems the background noise is roughly the average of transmitted signals from all other users; thus, a nonactive period in speech reduces the background noise. Hence, extra users may be accommodated without the loss of signal quality. This in turn increases the system capacity.

6.2.3.5 Space-Division Multiple Access

The SDMA scheme, also referred to as space diversity, uses an array of antennas to enable control of space by providing virtual channels in an angle domain [38]. This scheme exploits the directivity and beam-shaping capability of an array of antennas to reduce cochannel interference. Thus, it is possible using this scheme that simultaneous calls in a cell could be established at the same carrier frequency. This helps to increase the capacity of a cellular system.

The scheme is based on the fact that a signal arriving from a distant source reaches different antennas in an array at different times as a result of their spatial distribution, and this delay is utilized to differentiate one or more users in one area from those in another area. The scheme allows an effective transmission to take place between a base station and a mobile without disturbing the transmission to other mobiles. Thus, it has the potential such that the shape of a cell may be changed dynamically to reflect the user movement instead of currently used fixed-size cells. This arrangement, then, is able to create an extra dimension by providing dynamic control in space [39,40].

6.2.4 Channel Reuse

The generic term *channel* is normally used to denote a frequency in the FDMA system, a time slot in the TDMA system, a code in the CDMA system, or a

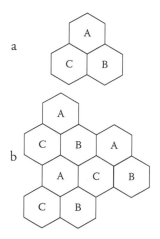

Figure 6.2 **(a) A cluster of three cells. (b) Channel reuse concept using a three-cell cluster.**

combination of these in a mixed system. Two channels are different if they use different combinations of these at the same place. For example, two channels in a FDMA system use two different frequencies. Similarly, in a TDMA system two separate time slots using the same frequency channel are considered two different channels. In that sense, for an allocated spectrum the number of channels in a system is limited. This limits the capacity of the system to sustain simultaneous calls and may be increased only by using each traffic channel to carry many calls simultaneously. Using the same channel again and again is one way of doing it. This is the concept of *channel reuse*.

The concept of channel reuse can be understood from Figure 6.2. Figure 6.2a shows a cluster of three cells. These cells use three separate sets of channels. This set is indicated by a letter. Thus, one cell uses set A, another uses set B, and another set C. In Figure 6.2b, this cluster of three cells is being repeated to indicate that three sets of channels are being reused in different cells. Figure 6.3 shows a similar arrangement with a cluster size of seven cells. Now, let us see how this helps to increase the system capacity.

Assume there are a total of F channels in a system to be used over a given geographic area. Also assume that there are N cells in a cluster that use all the available channels. In the absence of channel reuse, this cluster covers the whole area, and the capacity of the system to sustain simultaneous calls is F. Now, if the cluster of N cells is repeated M times over the same area, then the system capacity increases to MF as each channel is used M times.

The number of cells in a cluster is referred to as the *cluster size*, the parameter $1/N$ is referred to as the *frequency reuse factor*, and a system using a cluster size of N sometimes is also referred to as a system using an N frequency reuse plan. The cluster

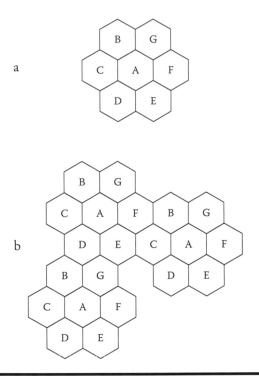

Figure 6.3 **(a) A cluster of seven cells. (b) Channel reuse concept using a seven-cell cluster.**

size is an important parameter. For a given cell size, as the cluster size is decreased, more clusters are required to cover the given area, leading to more reuse of channels; hence, the system capacity increases. Theoretically, the maximum capacity is attained when cluster size is one, that is, when all the available channels are reused in each cell. For hexagonal cell geometry, the cluster size can only have certain values. These are given by $N = i2 + j2 + ij$, where i and j are nonnegative integers.

The cells using the same set of channels are known as *cochannel cells*. For example, in Figure 6.2, the cells using channels A are cochannel cells. The distance between cochannel cells is known as the *cochannel distance*, and the interference caused by the radiation from these cells is referred to as *cochannel interference*. For proper functioning of the system, this needs to be minimized by decreasing the power transmitted by mobiles and base stations in cochannel cells and increasing the cochannel distance. Because the transmitted power normally depends on the cell size, the minimization of cochannel interference requires a minimum cochannel distance; that is, the distance cannot be smaller than this minimum distance.

In a cellular system of equal cell size, the cochannel interference is a function of a dimensionless parameter known as the *cochannel reuse ratio Q*. This is a ratio of the cochannel distance D and the cell radius R, that is,

$$Q = D/R$$

For hexagonal geometry,

$$Q = \sqrt{3N}$$

It follows from these equations that an increase in Q increases the cochannel distance and thus minimizes the cochannel interference. On the other hand, a decrease in Q decreases the cluster size N and hence maximizes the system capacity. Thus, the selection of Q is a trade-off between the two parameters, namely, the system capacity and cochannel interference. It should be noted that for proper functioning of the system, the ratio of signal to cochannel interference should be above a certain minimum value [19].

6.2.5 Cellular Configuration

A cellular system may be referred to as a *macrocell*, a *microcell*, or a *picocell* system depending on the size of cells. Some characteristics of these cellular structures are now described.

6.2.5.1 Macrocell System

A cellular system with a cell size of several kilometers is referred to as a *macrocell system*. Base stations of these systems transmit several watts of power from antennas mounted on high towers. Normally, there is no line of sight (LOS) between the base station and mobiles; thus, a typical received signal is a combination of various signals arriving from different directions. The received signals in these systems experience spreading of several microseconds because of the nature of propagation conditions.

6.2.5.2 Microcell Systems

As cells are split and their boundaries are redefined, their size becomes very small. At a radius less than about a kilometer, the system is referred to as a *microcell system*. In these systems, a typical base station transmits less than 1 W of power from an antenna mounted a few meters above the ground, and normally an LOS exists between the base and a mobile. Cell radius in microcell systems is less than a kilometer, giving rms delay spread on the order of a few tens of nanoseconds, compared with a few micoseconds for macrocell systems. This has an impact on the maximum data rate a channel could sustain.

For microcell systems, the maximum bit rate is about 1 Mbps compared with that of about 300 kbps for macocell systems [27].

Microcell systems are also useful in providing coverage along roads and highways. Because the antenna height is normally lower than the surrounding buildings, the propagation is along the streets, and an LOS exists between the base and a mobile. When a mobile turns a corner, sometimes a sudden drop in received signal strength is experienced because of loss of LOS. Depending on how antennas are mounted on intersections and corners, various cell plans are possible. More details on these aspects may be found in [41] and references therein.

6.2.5.3 Picocell Systems

When cell sizes are reduced below about 100 m covering areas such as large rooms, corridors, underground stations, or large shopping centers, cellular systems are sometimes referred to as *picocell systems,* with antennas mounted below rooftop levels or in buildings. These in-building areas have different propagation conditions from those covered by macrocell and microcell systems and thus require different considerations for developing channel models. Details of various models to predict propagation conditions may be found in [24]. Sometimes, the picocell and microcell systems are also referred to as *cordless communication systems*, with the term *cellular* identifying a macrocell system. Mobiles within these smaller cell systems are called *cordless terminals* or *cordless phones* [1,6,42].

Providing in-building communication services using wireless technology, based on cell shapes dictated by floors and walls, is a feasible alternative and offers many advantages. It is argued in [43] that radio frequencies in the 18-GHz band are ideal for such services because these do not penetrate concrete and steel structures, eliminating the problem of cochannel interferences. These frequencies offer huge bandwidths and require millimeter-size antennas that are easy to manufacture and install.

6.2.5.4 Overlayed System

Small cell systems make very efficient use of the spectrum, allowing large frequency reuse, resulting in increased system capacity. However, these are not suitable for all conditions because of their large handoff requirement. A system of mixed cells with the concept of overlaying is discussed in [41,44–46]. In this system, a hierarchy of cells is assumed to exist. A macrocell system is assumed at the top of the hierarchy with smaller cell systems at the bottom. A mobile with high mobility is assigned to a macrocell system, whereas one with low mobility is assigned to smaller cell systems. A design incorporating various combinations of different multiple-access schemes reflects the ease of handoff and other traffic management strategies. An SDMA scheme has an important role to play in this concept, with various beams placed at the bottom of the hierarchy.

6.2.6 *Channel Allocation and Assignment*

Various multiple-access schemes in the previous discussion are used to divide a given spectrum into a set of disjoint channels. These channels are then allocated to various cells for their use. Channel allocation may be carried out using one of the three basic schemes: fixed channel allocation, dynamic channel allocation, and hybrid channel allocation [47].

6.2.6.1 *Fixed Channel Allocation Schemes*

In fixed channel allocation schemes, a number of channels are allocated to a cell permanently for its use such that these channels satisfy certain channel reuse constraints as discussed. In its simplest form, the same number of channels is allocated to each cell. For a system with uniform traffic distribution across all cells, this uniform channel allocation scheme is efficient in the sense that the average call-blocking probability in each cell is the same as that of the overall system. For systems in which the distribution is not uniform, the call-blocking probability differs from cell to cell, resulting in the call being blocked in some cells when there are spare channels available in other cells.

This situation could be improved by allocating channels nonuniformally as per the expected traffic in each cell or employing one of many prevailing channel-borrowing schemes. One of these is referred to as a *static borrowing scheme*; some channels are borrowed from cells with light traffic and allocated to those with heavy traffic. Rearrangements of channels between cells are performed periodically to meet the variation in traffic load. In this scheme, the borrowed channels stay with the new cell until reallocated. There are other temporary borrowing schemes by which a cell that has used all its channels is allowed to borrow a free channel from a neighbor provided this does not interfere with existing calls.

The borrowed channel is returned to the original cell once the call is complete. Some temporary borrowing schemes allow any channel from a cell to be borrowed, whereas in others only nominated channels are allowed to be borrowed. Many borrowing strategies are available for selecting a channel, ranging from a simple scheme to pick the first-available channel that satisfies the cochannel distance constraints to one that performs an exhaustive search to select a channel that yields maximum SIR and minimizes the future probability of call blocking.

6.2.6.2 *Dynamic Channel Allocation Schemes*

Fixed channel allocation schemes are simple to implement and are generally useful for relatively stable traffic conditions. These schemes are not very efficient for fast-changing user distribution because they are not designed to adapt to short-term variations in traffic conditions. Dynamic channel allocation schemes are most suited for such situations. In these schemes, channels are not allocated to various cells but are kept in a central pool and assigned to calls as they arrive. At the completion of a call, the assigned channel is released and goes back to the pool. The process of channel assignment involves working out a cost of assigning a channel to a call, and a channel with the minimum cost is chosen for the purpose. The various channel assignment schemes differ in the way the cost function is selected using various parameters of interest such as reuse distance, SIR, or probability of call blocking. Some schemes base their assignment only on the current traffic conditions in the service area, whereas others take the past and the present conditions into account.

Dynamic channel assignment schemes may be implemented centrally; a central controller assigns the channels to calls from the pool. The central controller is able to achieve very efficient channel assignment but requires high overhead. The channel assignment may also be implemented in a distributed manner by base stations where calls are originated. The channel implementation by base stations requires less overhead than that required by a central controller and is more suitable for microcell systems. The distributed channel assignment schemes can be divided into two categories. In one case, each base station keeps detailed status information about current available channels in its neighborhood by exchanging status information with other base stations. The schemes in this category may provide near-optimum allocation but pay a heavy price in terms of increased communication with other base stations, particularly in heavy traffic. The other category of distributed channel assignment schemes uses simple algorithms that rely on mobiles to measure signal strength to decide the suitability of a channel.

6.2.6.3 Hybrid Channel Allocation Schemes

The fixed channel allocation schemes are efficient under uniformly distributed heavy traffic. On the other hand, the dynamic channel allocation schemes perform better under low traffic conditions with varying and nonuniformly distributed loads. The hybrid channel allocation schemes maximize advantages of both these schemes by dividing channels into fixed and dynamic sets. The channels in fixed sets are allocated as per fixed channel allocation strategies, and those in the other set are free to be assigned to calls in a cell that has used all its allocated channels. The channels in this set are assigned as per the dynamic channel allocation procedures. Apparently, no optimum ratio of channels is assigned to two sets, and the design parameter is dependent on local traffic conditions. More details on these and related issues may be found in [47] and references therein.

6.2.7 Handoff

It is common for a mobile to move away from its servicing base station while a call is in progress. As the mobile approaches the cell boundary, the strength and quality of the signal it receives start to deteriorate. At some stage, near the cell boundary, it receives a stronger signal from a neighboring base station than it does from its serving base station. At this point, the control of the mobile is handed over to the new base station by assigning a channel belonging to the new cell. This process by which a radio channel used by a mobile is changed is referred to as *handoff* or *handover* [41,44,48–50]. When handoff is between two base stations as described, it is referred to as *intercell handoff*.

On the other hand, when handoff is between two channels belonging to the same base stations, it is referred to as *intracell handoff*. This situation arises when the network, while monitoring its channels, finds a free channel of better quality than that used by a mobile and decides to move the mobile to this new channel to improve the quality of channels in use. Sometimes, the network rearranges channels to avoid congestion and initiates intracell handoff. Handoff is also necessary between different layers of overlaid systems consisting of microcells and macrocells. In these systems, the channels are divided into microcell channels and macrocell channels. When a mobile moves from one microcell to another and there is no available channel for handoff, a macrocell channel is used to meet the handoff request. This avoids the forced termination of a call. Later, if a channel becomes available at an underlaid microcell, then the macrocell channel may be released, and a microcell channel is assigned to the call by initiating a new handoff.

Forced termination of a call in progress is undesirable, and to minimize this a number of strategies are employed. These include reserving channels for handoff, using channel assignment schemes that give priority to a handoff request over new calls, and queuing the handoff request. The channel reservation and handoff priority schemes reduce the probability of forced termination by increasing the probability of blocking new calls. The queuing schemes are effective when handoff requests arrive in groups and there is a reasonable likelihood of channel availability in the near future.

The handoff is initiated when the quality of current channels deteriorates below an acceptable threshold or a better channel is available. The channel quality is measured in terms of bit error rate (BER), received signal strength, or some other signal quality such as eye opening of radio signal that indicates the ratio of signal to interference plus noise.

For handoff initiation, the signal strength is used as an indication of the distance between the base and the mobile. For this reason, a drop in signal strength resulting from Rayleigh fading is normally not used to initiate handoff, and some kind of averaging is used to avoid the problem. In some systems, the round-trip delay between mobile and base is also used as an indication of the distance.

The measurement of various parameters may be carried out either at the mobile or at the base. Depending on where the measurements are made and who initiates the handoff, various handoff implementation schemes are possible, including network-controlled handoff, mobile-controlled handoff, and MAHO.

6.2.7.1 Network-Controlled Handoff

In network-controlled handoff, each base station monitors the signal strength received from mobiles in their cells and makes periodic measurements of the received signal from mobiles in their neighboring cells. The MSC then initiates and completes the handoff of a mobile as and when it decides. The decision is based on the received signal strength at the base station serving the mobiles and base stations in neighboring cells. Because of its centralized nature, the collection of these measurements generates large network traffic. This could be reduced to an extent by making measurements less frequently and by not requiring the neighboring base station to send the measurements continually. However, this reduces accuracy. The execution of handoff by this method takes a few seconds, and for this reason the method is not preferred by microcellular systems where a quick handoff is desirable.

6.2.7.2 Mobile-Controlled Handoff

Mobile-controlled handoff is a highly decentralized method and does not need any assistance from the MSC. In this scheme, a mobile monitors signal strength on its current channel and measures signals received from the neighboring base stations. It receives BER and signal strength information about uplink channels from its serving base stations. Based on all this information, it initiates the handoff process by requesting the neighboring base for allocation of a low-interference channel. The method has a handoff execution time on the order of 100 ms and is suitable for microcell systems.

6.2.7.3 Mobile-Assisted Handoff

In MAHO methods, as the name suggests, a mobile helps the network in the handoff decision making by monitoring the signal strength of its neighboring base stations and passing the results to the MSC via its serving base station. The handoff is initiated and completed by the network. The execution time is on the order of 1 s.

6.2.7.4 Hard Handoff and Soft Handoff

Handoff may be classified into hard handoff and soft handoff. During hard handoff, the mobile can communicate only with one base station. The communication link gets broken with the current base station before the new one is established, and there is normally a small gap in communication during the transition. In the process of soft handoff, the mobile is able to communicate with more than one base station. It receives signals from more than one base station, and the received signals are combined after appropriate delay adjustment. Similarly, more than one station receives signals from mobiles, and the network combines different signals. This scheme is also known as *macroscopic diversity* and is mostly employed by CDMA systems.

Hard handoff, on the other hand, is more appropriate for TDMA and FDMA systems. It is also simple to implement compared with soft handoff. However, it may lead to unnecessary handoff back and forth between two base stations when the signals from two base stations fluctuate. The situation may arise when a mobile, currently being served, for example, by base 1 receives a stronger signal from, say, base 2 and is handed over to base 2. Immediately after that, it receives a stronger signal from base 1 compared to what it receives from base 2, causing a handoff. This phenomenon, known as the *ping-pong effect*, may continue for some time and is undesirable because every handoff has a cost associated with it, requiring network signaling of varying amount for authentication, database updates, circuit switching, and so on. This is avoided by using a hysteresis margin such that the handoff is not initiated until the difference between the signal received from the two base stations is more than the margin. For example, if the margin is ΔdB, then the handoff is initiated when the signal received by the mobile from base 2 is ΔdB more than that from base 1. More details on various handoff implementation issues may be found in [41,48,49] and references therein.

6.2.8 Cell Splitting and Cell Sectorization

Each cell has a limited channel capacity and thus could only serve so many mobiles at a given time. Once the demand in that cell exceeds this limit, the cell is further subdivided into smaller cells, each new cell with its own base station and its frequency allocation. The power of the base station transmitters is adjusted to reflect the new boundaries. The power transmitted by new base stations is less than that of the old one.

The consequence of the cell splitting is that the frequency assignment has to be done again, which affects the neighboring cells. It also increases the handoff rate because the cells are now smaller and a mobile is likely to cross cell boundaries more often compared with the case when the cells are big. Because of altered signaling conditions, this also affects the traffic in control channels.

Cell sectorization refers to when a given cell is subdivided into several sectors, and all sectors are served by the same base station. This is normally done by employing directional antennas such that the energy in each sector is directed by separate antennas. This has the effect of increased channel capacity similar to cell splitting. However, it uses the same base station and thus does not incur the cost of establishing new base stations associated with the cell splitting. This helps in reducing the cochannel interference because the energy is directed toward the sector that does not cause interference in the cochannel cells, particularly in cochannel cells in the opposite direction to the sector. As in the case of cell splitting, this also affects the handoff rate.

6.2.9 Power Control

It is important that a radio receiver receives a power level that is enough for its proper function but not high enough for this level to disturb other receivers. This is achieved by maintaining a constant power level at the receiver by transmitter power control. The receiver controls the power of the transmitter at the other end. For example, a base would control the power transmitted by mobile phones and vice versa. This is accomplished by a receiver monitoring its received power and sending a control signal to the transmitter to control its power transmission as required. Sometimes, a separate pilot signal is used for this purpose.

Power control reduces the near–far problem in CDMA systems and helps to minimize the interference near the cell boundaries when used in a forward link [32,33].

6.3 First-Generation Systems

First-generation systems use analog frequency modulation for speech transmission and frequency shift keying (FSK) for signaling and employ FDMA to share the allocated spectrum. Some of the popular standards developed around the world include AMPS, TACS, NMT, NTT, and C450. These systems use two separate frequency channels, one for base to mobile and the other for mobile to base for full-duplex transmission [1].

6.3.1 Characteristics of Advanced Mobile Phone Service

The AMPS system uses bands of 824 to 849 MHz for uplink and 869 to 894 MHz for downlink transmission. This spectrum is divided into channels of 30-kHz bandwidth. In a two-way connection, two of these channels are used. A pair of channels in a connection is selected such that channels used for uplink and downlink transmission are separated by 45 MHz. This separation was chosen so that inexpensive but highly selective duplexers could be utilized. A typical frequency reuse plan in this system uses either clusters of 12 cells with omnidirectional antennas or 7-cell clusters with three sectors per cell.

There is a total of 832 duplex channels. Of these, 42 are used as control channels, and the remaining 790 channels are used as voice channels. The control channels used for downlink and uplink transmission are referred to as forward control channels (FCCs) and reverse control channels (RCCs), respectively. Similarly, voice channels are referred to as forward voice channels (FVCs) and reverse voice channels (RVCs).

Each base continuously broadcasts FSK data on FCCs and receives on RCCs. A mobile scans all FCCs and locks on an FCC with the strongest signal. Each mobile needs to be locked on an FCC signal to receive and send a call. Base stations monitor their RCCs for transmission from mobiles that are locked on the matching FCCs.

6.3.2 Call Processing

When a mobile places a call, it transmits a message on an RCC consisting of a destination phone number, its mobile identification number (MIN), and other authorization information. The base station monitoring an RCC receives this information and sends it to the MSC. The MSC in turn checks the authentication of the mobile; assigns a pair of FVC and RVC, a supervisory audio tone (SAT), and a voice mobile attenuation code (VMAC); and connects the call to a PSTN. The mobile switches itself to the assigned channels. The SAT is used to ensure the reliable voice communication, and the VMAC is used for power control.

The SAT is an analog tone of 5970, 6000, or 6030 Hz and is transmitted during a call on both FVC and RVC. It is superimposed on the voice signal and is barely audible to the user. It helps the mobile and the base to distinguish each other from cochannel users located in nearby cells. The SAT also serves as a handshake between the base station and the mobile. The base transmits it on an FVC, and the mobile retransmits it on an RVC after detection. If the SAT is not detected within 1 s, both the mobile and the base stop transmission.

A call to a mobile originating at a PSTN is processed by the MSC in a similar fashion. When a call arrives at an MSC, a paging message with MIN is sent out on FCCs of every base station controlled by the MSC. A mobile terminal (MT) recognizes its MIN and responds on the RCC. The MSC assigns a pair of FVC and RVC and a pair of SAT and VMAC. The mobile switches itself to the assigned channel.

While a call is in progress on voice channels, the MSC issues several blank-and-burst commands to transmit signaling information using binary FSK at a rate of 10 kbps. In this mode, voice and SAT are temporarily replaced with this wideband FSK data. The signaling information is used to initiate handoff, to change mobile power level, and to provide other data as required.

The handoff decision is taken by the MSC when the signal strength on the RVC falls below threshold or when the SAT experiences an interference level above a predetermined value. The MSC uses scanning receivers in nearby base stations to determine the signal level of a mobile requiring handoff.

The termination of a call by a mobile is initiated using a signaling tone (ST). ST is a 10-kbps data burst of 1s and 0s. It is sent at the end of a message for 200 ms to indicate "end of call." It is sent along with an SAT and indicates to the base station that the mobile has terminated the call instead of the call dropping out or prematurely terminating. The ST is sent automatically when a mobile is switched off.

6.3.3 Narrowband Advanced Mobile Phone Service, European Total Access Communication System, and Other Systems

A narrowband AMPS (N-AMPS) was developed by Motorola to provide three 10-kHz channels using FDMA in a 30-kHz AMPS channel. By replacing one AMPS channel with three N-AMPS channels at a time, the service providers are able to increase the system capacity threefold. N-AMPS uses SAT, ST, and blank and burst similar to AMPS. Because it uses 10-kHz channels, frequency modulation

Table 6.1 Parameters of Some First-Generation Cellular Standards

Parameters	AMPS	C450	NMT 450	NTT	TACS
Tx frequency (MHz)					
Mobile	824–849	450–455.74	453–457.5	925–940	890–915
Base station	869–894	460–467.74	463–467.5	870–885	935–960
Channel bandwidth (kHz)	30	20	25	25	25
Spacing between forward and reverse channels (MHz)	45	10	10	55	45
Speech signal FM deviation	±12	±5	±5	±5	±9.5
Control signal data rate (kbps)	10	5.28	1.2	0.3	8
Handoff decision is based on	Power received at base	Round-trip delay	Power received at base	Power received at base	Power received at base

Tx, transmission.

(FM) deviation is smaller compared with AMPS; hence, it has a lower signal-to-noise-plus-interference ratio (SNIR), resulting in degradation of audio quality. It has taken measures to compensate this degradation.

The ETACS is identical to AMPS except that it uses 25-kHz channels compared with 30-kHz channels used by AMPS. It also formats its MIN differently from AMPS to accommodate different country codes in Europe. Parameters for these and some other popular analog systems are shown in Table 6.1.

6.4 Second-Generation Systems

In contrast to the first-generation analog systems, second-generation systems are designed to use digital transmission and to employ TDMA or CDMA as a multiple-access scheme. These systems include North American dual-mode cellular system IS-54, North American IS-95 systems, Japanese PDC systems, and European GSM and DCS 1800 systems. GSM, DCS 1800, IS-54, and PDC systems use TDMA and FDD, whereas IS-95 is a CDMA system and also uses FDD for a duplexing technique. Other parameters for these systems are shown in Table 6.2. A brief description of these systems is presented [1].

Table 6.2 Parameters of Some Second-Generation Cellular Standards

Parameter	IS-54	GSM	IS-95	PDC
Tx frequencies (MHz)				
Mobile	824–849	890–915	824–849	940–956 and 1429–1453
Base station	869–894	935–960	869–894	810–826 and 1477–1501
Channel bandwidth (kHz)	30 kHz	200 kHz	1250 kHz	25 kHz
Spacing between forward and reverse channels (MHz)	45	45	45	30/48
Modulation	$\pi/4$ DQPSK	GMSK	BPSK/QPSK	$\pi/4$ DQPSK
Frame duration (ms)	40	4.615	20	20

6.4.1 United States Digital Cellular (IS-54)

United States Digital Cellular (IS-54) is a digital system that uses TDMA as a multiple-access technique compared with AMPS, which is an analog system that uses FDMA. It is referred to as a *dual-mode system* because it was designed to share the same frequency, frequency reuse plan, and base stations with AMPS. It was done so that the mobile and base stations could be equipped with AMPS and IS-54 channels within the same equipment to help migrate from an analog to a digital system and simultaneously to increase system capacity. In this system, each frequency channel of 30 kHz is divided into six time slots in each direction. For full-rate speech, three users equally share six slots, with two slots allocated per user. For half-rate speech, each user uses only one slot. Thus, the system capacity is three times more than that of the AMPS for full-rate speech and double that for the half-rate speech. This system also uses the same signaling (FSK) technique as AMPS for control, whereas it uses $\pi/4$ differential quadrature phase-shift keying (DQPSK) for the voice.

It was standardized as IS-54 by the Electronic Industries Association and Telecommunication Industry Association (EIA/TIA) and was later revised as IS-136. The revised version has digital control channels (DCCs) that provide an increased signaling rate as well as additional features such as transmission of point-to-point short messages, broadcast messages, and group addressing.

As discussed, AMPS has 42 control channels. The IS-54 standard specifies these as primary channels and an additional 42 channels as secondary channels. Thus, it has twice the control channels as AMPS and is able to carry twice the control traffic in a given area. The secondary channels are not monitored by AMPS users and are for the exclusive use of IS-54 users.

Each time slot in each voice channel has one digital traffic channel (DTC) for user data and digitized speech and three supervisory channels to carry control information.

A full-duplex DTC consists of forward DTC to carry data from the base station to the mobile and reverse DTC to carry data from the mobile to the base station. The three supervisory channels are coded digital verification color code (CDVCC), slow associated control channel (SACCH), and fast associated control channel (FACCH).

The CDVCC is a 12-bit message transmitted with every slot containing an 8-bit color code number between 1 and 255. The 12-bit message is generated using shortened Hamming code. It has a similar function to the SAT in AMPS. A station transmits this number on CDVCC channels and expects a handshake from each mobile, which must retransmit this value on a reverse voice channel. If the number is not returned within a specified time, the time slot is relinquished.

The SACCH is a signaling channel and carries control information between base and mobile while a call is in progress. It is sent with every slot carrying information about power level change, handoff, and so on. Mobiles use this channel to send signal strength measurement of neighboring base stations so that the base may implement MAHO.

The FACCH is a second signaling channel to carry control information while the call is in progress. It does not have a dedicated time during each slot as is the case for CDVCC and SACCH. It is similar to blank and burst in AMPS and replaces speech data when used. It carries call release instructions, MAHO, and mobile status requests.

6.4.2 Personal Digital Cellular System

The PDC system, established in Japan, employs the TDMA technique. It uses three time slots per frequency channel and has a frame duration of 20 ms. It can support three users at full-rate speech and six half-rate speech users, similar to IS-54. It has a channel spacing of 25 kHz and uses $\pi/4$ DQPSK modulation. It supports a frequency reuse plan with cluster size four and uses MAHO.

6.4.3 Code-Division Multiple-Access Digital Cellular System (IS-95)

The IS-95 CDMA digital system uses CDMA as a multiple-access technique and occupies the same frequency band occupied by AMPS; that is, the forward-link frequency band is from 869 to 894 MHz, and the reverse-link band is from 824 to 849 MHz. Forward-link and reverse-link carrier frequencies are separated by 45 MHz.

Each channel in IS-95 occupies a 1.25-MHz bandwidth, and this is shared by many users. The users are separated from each other by allocating 1 of 64 orthogonal spreading sequences (Walsh functions). The user data are grouped into 20-ms frames and are transmitted at a basic user rate of 9600 bps. This is spread to a channel chip rate of 1.2288 Mchip/s, giving a spreading factor of 128. RAKE receivers are used at both base station and mobiles to resolve and combine multipath components. During handoff, the standard allows for base station diversity, by which a mobile keeps a link with both the base stations and combines signals from both the stations to improve signal quality as it would combine multipath signals.

In a forward link, a base station transmits simultaneously to all users using 1 of 64 spreading sequences for each user once the user data are encoded using a half-rate convolution code and are interleaved. All signals in a cell are also scrambled using a PN sequence of length 2^{15} to reduce the cochannel interference.

During the scrambling process, the orthogonality between different users is preserved. The forward channel consists of 1 pilot channel, 1 synchronization channel (SCH), up to 7 paging channels, and up to 63 traffic channels. The pilot channel transmits higher power than other channels and is used by mobiles to acquire timing for the forward channel and to compare signal strengths of different base stations.

It also provides phase reference for coherent detection. The SCH operates at 1200 bps and broadcasts a synchronization message to mobiles. The paging channels are used to transmit paging messages from the base station to mobiles and to operate at 9600, 4800, or 2400 bps. The traffic channels support variable data rate operating at 9600, 4800, 2400, and 1200 bps.

On reverse channels, mobiles transmit asynchronously to the base, and orthogonality between different users in a cell is not guaranteed. A strict control is applied to the power of each mobile so that a base station receives constant power from each user, thus eliminating the near–far problem. Power control command is sent by the base to mobiles at a rate of 800 bps. The reverse channels are made up of access channels and reverse traffic channels.

The reverse channels contain up to 32 access channels per paging channel, operate at 4800 bps, and are used by mobiles to initiate communication with the base and to respond to paging messages. The reverse traffic channel is a variable data rate channel and operates similarly to the forward channels at 9600, 4800, 2400, and 1200 bps.

6.4.4 Pan European Global System for Mobile Communications

The Groupe Spécial Mobile was established in 1982 to work toward the evolution of a digital system in Europe, and its work now has become the Global System for Mobile Communications (GSM). Two frequency bands have been allocated for this system. The primary band is at 900 MHz, and the secondary band is at 1800 MHz. The description here mainly concerns the primary band. It has been divided into two subbands of 25 MHz each, separated by 20 MHz. The lower band is used for uplink, and the upper band is used for downlink. Operators are assigned a portion of the spectrum for their use. The carrier frequencies are separated by 200 kHz. This gives the total number of frequency channels over the 25-MHz band as 124. The first carrier is at 890.2 MHz, the second one is at 890.4 MHz, and so on. These carriers are numbered as 0, 1, 2, and so on, respectively. Similarly, 374 different carriers are allocated in the secondary band, which is 75 MHz wide.

6.4.4.1 Multiple-Access Scheme

GSM employs a combination of TDMA and FDMA schemes with slow frequency hopping. GSM transmission takes place by modulating a bundle of about 100 bits known as a *burst*. A burst occupies a finite duration in time and frequency plane. The center frequencies of these bursts are 200 kHz apart, and these are 15/26 ms in duration.

The duration of these bursts is the time unit and is referred to as the *burst period* (BP). Thus, time is measured in BPs. When this burst is combined with slow frequency hopping, a typical transmission appears as shown in Figure 6.4. The hopping sequence is selected randomly using a PN sequence. A channel is defined by specifying which time slot it may use for the transmit burst. That means specifying the time instant and specific frequency. Time slots of a channel are not contiguous in time. All time slots are numbered, and the description of a channel sent to the mobile by the base refers to this numbering scheme. The numbering is cyclic, and each time slot is uniquely identified in this cycle, which is about 3.5 h (3 h, 28 min, 53 s, 760 ms).

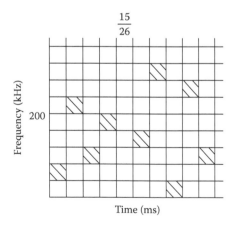

Figure 6.4   GSM multiple access.

Many types of channels are defined in GSM, and each is cyclic. The simplest cycle is of 8 BP. This cycle of eight time slots is also called a *slot*, which is 60/13 ms in duration. The duration of the BP is chosen such that 26 slots are equal to 120 ms, which is a multiple of 20 ms to obtain synchronization with other networks, such as the Integrated Services Digital Network (ISDN). A full dedicated channel is thus cyclic in 120 ms and uses 26 slots. Note that each slot is made up of eight time slots of 15/26 ms known as BPs. Of these 26 slots, 24 slots are used for traffic burst, 1 slot is used for a control burst, and 1 slot is not used.

The transmission between the uplink and the downlink is not independent. Transmission in uplink follows the downlink reception by 3 BP. When the mobile is far from the base, the mobile advances its transmission from the reception to compensate the propagation delay. Hopping sequences in the uplink and the downlink are also related. The hopping sequence in the uplink direction is derived from the one in the downlink direction by adding 45 MHz.

6.4.4.2 Common Channels

Common channels do not carry traffic and are organized based on a cycle of 51 slots (51 × 8 BP). This cycle is deliberately chosen differently from 26 slots of traffic channels so there will not be a common divider between the two. This allows mobiles in a dedicated channel to listen to SCHs and frequency correction channels (FCCHs) of the surrounding cells, which helps mobiles to stay in synchronization. Each SCH and FCCH uses 5 slots in a 51-slot cycle, with SCH following FCCH 8 BP later. This helps a mobile to find the SCH once it has located the FCCH. The other downlink common channels defined include the Broadcast Control Channel (BCCH) for broadcasting and Packet Access Grant Channel (PAGCH) for paging. For uplink, a channel Random Access Channel (RACH) is defined.

6.4.4.3 Burst Format

The quantum of transmission in GSM is 1 BP, which is 7500/13 s in duration and is occupied by about (156 + 1/4) bits. In GSM, several burst formats have been defined, and these are used for different purposes. The access burst is used in the uplink direction from the mobile to the base during the initial phase. This is the first access of the mobile to the base. The burst has a constant amplitude for the period of 87 bits. The structure of the burst consists of 7 bits of tail followed by 41 bits of training sequence, 36 bits of information, and 3 bits of tail on the other side. A single training sequence is specified for this burst. The access burst contains required demodulation information for the base.

The S burst is similar to the access burst, but it is transmitted from the base to the mobile. It is the first burst from the base and has 64 training sequence bits surrounded by 39 information bits and 3 tail bits. The training sequence is unique and chosen so that the mobile knows which sequence the base has chosen. The F burst enables the mobile to find and demodulate the S burst. All of its 148 bits are set to zero, resulting in a pure sine wave of 1625/24 kHz.

The normal burst is used for all other purposes. Its amplitude stays constant, covering 147 bits. It has 26 training sequence bits surrounded by 58 information bits and 3 tail bits. Eight different training sequences have been specified to distinguish cochannel signals. For more details on the GSM system, see, for example, [51].

6.4.5 Cordless Mobiles

The first-generation analog cordless phones were designed to communicate with a single base station, effectively replacing the telephone cord with a wireless link to provide terminal mobility in a small coverage area such as a house or an office. The aim of the second-generation digital cordless system is to use the same terminal in residential as well as public access areas such as offices, shopping malls, and train stations to receive and to originate calls. The cordless systems differ from cellular systems in a number of ways. Their cell size is small, typically less than half a kilometer, and their antenna elevation is low. These are designed for low-speed mobiles, typically on foot, and provide coverage in specific zones instead of the continuous wide-area coverage provided by cellular systems. Cordless handsets transmit very low power. A typical average transmitted power is about 5 to 10 mW, compared with a few hundred milliwatt for cellular handsets [1].

Most of the cordless systems use TDD as duplexing techniques, compared with FDD employed by cellular systems.

Some of the popular digital cordless standards include CT2, a British standard; Digital European Cordless Telecommunication (DECT) standard; personal handyphone system (PHS) of Japan, and personal access communication service (PACS) of the United States. Some of the parameters of these systems are compared in Table 6.3. More details on these may be found in [1] and references therein.

6.5 Third-Generation Systems

The third-generation systems aim to provide a seamless network that can provide users voice, data, multimedia, and video services regardless of their location on the network: fixed, cordless, cellular, satellite, and so on. The networks support global roaming while providing high-speed data and multimedia applications of up to 144 kbps on the move and up to 2 Mbps in a local area.

Table 6.3   Digital Cordless System Parameters

Parameter	CT2	DECT	PACS	PHS
Frequency (MHz)	864–868	1880–1900	1850–1910 and 1930–1990	1895–1918
Channel spacing (kHz)	100	1728	300	300
Duplexing	TDD	TDD	FDD	TDD
Channel rate (kbps)	72	1152	384	384
Transmitted power (mW)				
Average	5	10	25	10
Peak	10	250	200	80
Frame duration (ms)	2	10	2.5	5
Channels per carrier	1	12	8	4

Third-generation systems are currently being defined by both the International Telecommunication Union (ITU) and regional standardization bodies. Globally, the ITU has been defining third-generation systems since the late 1980s through work on the IMT-2000 system, formerly called the Future Public Land Mobile Telecommunications Service (FPLMTS) [52]. The ITU is now in the process of seeking candidate technologies to be evaluated in accordance with agreed guidelines [53]. The European proposal for IMT-2000 is known as the Universal Mobile Telecommunications System (UMTS) (see discussions in [54–57]) and is being defined by the European Telecommunications Standards Institute (ETSI), which has been responsible for UMTS standardization since the 1980s.

Although UMTS will provide significant changes for customers and technologies, systems will be deployed within a short time frame.

IMT-2000 defines systems capable of providing continuous mobile telecommunication coverage for any point on the earth's surface. Access to IMT-2000 is via either a fixed terminal or a small, light, portable MT [58].

Several different radio environments are utilized to provide the required layers of coverage. These range from very small indoor picocells with high capacity, to terrestrial micro- and macrocells, to satellite megacells. IMT-2000 recommendations aim to maximize commonality between the various radio interfaces involved to simplify the task of developing multimode terminals for the various operating environments [59]. In this section, some salient features of IMT-2000 are discussed [4,5].

6.5.1 Key Features and Objectives of IMT-2000

The key features and objectives of IMT-2000 include [60]:

1. Integration of current first- and second-generation terrestrial and satellite-based communications systems into a third-generation system
2. Ensuring a high degree of commonality of design at a global layer
3. Compatibility of services within IMT-2000 and with fixed networks
4. Ensuring high quality and integrity of communications, comparable to the fixed network
5. Accommodation of a variety of types of terminals, including pocket-size terminals
6. Use of terminals worldwide
7. Provision for connection of mobile users to other mobile users or fixed users
8. Provision of services by more than one network in any coverage area
9. Availability to mobile users of a range of voice and nonvoice services
10. Provision of services over a wide range of user densities and coverage areas
11. Efficient use of the radio spectrum consistent with providing service at acceptable cost
12. Provision of a framework for the continuing expansion of mobile network services and for the access to services and facilities of the fixed network
13. Number portability independent of service provider
14. Open architecture that accommodates advances in technology and different applications
15. Modular structure that allows the system to grow as needed

6.5.2 IMT-2000 Services

IMT-2000 supports a wide range of services, including those based on the fixed telecommunication network and those that are specific to mobile users. Services are available in a variety of environments, ranging from dense urban situations, including high-intensity office use, to suburban and rural areas [59]. The actual services obtained by users depend on the capabilities of their terminals, their subscribed set of services, and the services offered by the relevant network operators and service providers [60].

Global roaming users have access to at least a minimum set of services comprising voice telephony, selection of data services, and indication of other services available. IMT-2000 also provides services to fixed users and, if required, can provide rapid and economical implementation of wide-area communications, which is particularly relevant to developing countries [59].

The general service objectives of IMT-2000 are to [61]:

1. Provide a wide range of telecommunication services to mobile or stationary users by means of one or more radio links

2. Make these services available for MTs located anywhere (subject to economic constraints)
3. Provide for flexibility of service provision
4. Promote flexible introduction of services
5. Ensure that a user is provided with an indication of service availability
6. Provide access to voice telephony
7. Provide access to a selection of data services
8. Provide services that depend on terminal type, location, and availability from the network operator
9. Provide a temporary or permanent substitute to fixed networks in rural or urban areas under conditions approved by the appropriate national or regional regulation authority

The general service requirements of IMT-2000 are to provide:

1. Validation and authentication procedures to facilitate billing and accounting based on ITU-T X509
2. An additional layer of security for mobile telecommunication services
3. Privacy of location of a roaming user when desired by the called or calling party
4. Quality of service comparable with that of fixed networks

The general access requirements are:

1. For access to fixed networks, IMT-2000 may be either an adjunct to or an integral part of the PSTN/ISDN.
2. For global use, IMT-2000 should allow international operation and automatic roaming of terminals.
3. For maritime and aeronautical use, operation should be facilitated to the extent permitted by the relevant regulatory body.
4. For satellite operation, IMT-2000 should facilitate direct and indirect satellite operation. The first phase of IMT-2000 provides several telecommunications services, most of which are based on ITU-T E and F series recommendations.

Network services: The following are provided by IMT-2000 [62]:
1. Voice telephony (ITU-T E105)
2. Program sound
3. Message handling (ITU-T F400)
4. Teletex (ITU-T F200)
5. Paging (open loop, closed loop, user acknowledged)
6. Telefax (ITU-T F160 and F180)
7. Point to multipoint

8. Data
9. Video services using MPEG-derived algorithms [63]
10. Short messages
11. Location monitoring
12. Multimedia

Supplementary services: The following services may be provided by IMT-2000:
1. Separation of answering from alerting: Currently, the alerting function resides in the same device used for answering. In IMT-2000, the alerting device may be a pager, and the answering device may be a terminal of the user's choice.
2. Advice of charging: Parties to a call should be able to receive charging information before, during, and after the call.

6.5.3 Planning Considerations

In defining IMT-2000, several factors required consideration: radio access, spectrum requirements, security, network issues, and regulatory environments.

6.5.3.1 Radio Access

IMT-2000 provides access, by means of one or more radio links, to a wide range of services in a wide variety of operating environments. High data rates are required to provide users with the necessary quality of service for multimedia communications, ranging from a few tens of kilobits per second for image transfer, to a couple of hundreds of bits per second for peak Internet transfers, to 2 Mbps for video. The bearers for IMT-2000 are therefore defined as 384 kbps for full-area coverage and 2 Mbps for local-area coverage.

It is essential to optimize third-generation techniques to cater for variable bit rate and packet capabilities because many multimedia applications are packet oriented. Similarly, multimedia support implies flexibility to handle services with different bit rates and E_b/N_0 requirements [64].

The mode of delivery is via either terrestrial or satellite-based radio links, with the possibility of incorporating two or more radio links in tandem. Although it would be desirable for a common radio interface to be provided for the terrestrial and satellite components, this is unlikely to be practical because of spectral and power efficiency design constraints. Therefore, terminals will most likely be required to operate over more than one type of interface, with adaptation controlled by software using digital signal-processing technology. Dual-mode handsets already exist to combine GSM at different frequencies, GSM/DECT, and GSM/satellite. The IMT-2000 design allows for the provision of competitive services to the user in each of these operating conditions [60].

The UMTS radio interface, called UMTS terrestrial radio access (UTRA), will consist of a number of hierarchical layers. The higher layer will use WCDMA, with each user given a special CDMA code and full access to the bandwidth allocated. The macrolayer will provide basic data rates to 144 kbps. The lower layers will provide higher data rates of 384 kbps and 2 Mbps through the use of an FDD. It may also be possible to use TDD through time-division CDMA (TD-CDMA) for higher data rates by dividing the frequency allocation into time slots for the lower layers [64,65].

This compromise between the two competing standards of WCDMA and TD-CDMA means that Europe will have a "family of standards." TD-CDMA provides greater efficiency than GSM and offers reuse of the existing GSM network structure as well as efficient interworking with GSM. TD-CDMA has the same basic frame structure as GSM, each having eight time slots per frame length, but provides higher data rates, up to 2 Mbps indoors. The combination of different access methods is intended to provide flexibility and network efficiency, with the UTMS terminal adopting the access method that best seeks its environment.

6.5.3.2 *Spectrum Requirements*

The work of ITU on the IMT-2000 is aimed at the establishment of advanced global communication services within the frequency bands, 1885 to 2025 MHz and 2110 to 2200 MHz, identified by the World Administrative Radio Convention (WARC-92). Within this bandwidth, the bands 1980 to 2010 MHz and 2170 to 2200 MHz will be utilized by the satellite component [59]. It is important to note that although the WARC-92 frequencies were intended for IMT-2000, their use by other systems such as personal communication services (PCSs) and UMTS is not precluded [66]. WARC-92 resolved that administrations implementing IMT-2000 should make spectrum available in the identified bands for system development and implementation and should use the relevant international technical characteristics that will be developed to facilitate worldwide use and roaming.

Although the intention was to reserve this bank of the spectrum on a worldwide basis for IMT-2000, the Federal Communications Commission (FCC) in the United States engaged in a spectrum auction in late 1994, which resulted in the allocation of large portions of bandwidth in North America to operators providing PCS. The European DECT service and the Japanese PHS service also have spectrum overlaps with the IMT-2000/WARC-92 allocation. The use of this spectrum for other than IMT-2000 services indicates that the allocated spectrum is not enough to meet the growing demand for additional spectrum to provide services such as mobile data services, mobile e-commerce, wireless Internet access, and mobile video services. It should be noted that this spectrum identified in 1992 for global communication services was based on a model in which voice services were assumed to be the major source of traffic, and the services indicated here were not foreseen. In the current climate in which the number of users worldwide is expected to reach 2 billion by 2010 and there is a need to provide common spectrum for global roaming, the World Radio Communication Conference in June 2000 (WRC-2000) decided to increase the available spectrum for IMT-2000 use on a global basis.

This additional spectrum has been identified in three bands: one below 1 GHz (806–960 MHz), another at 1.7 GHz (1710–1885 MHz), and the third band at 2.5 GHz (2500–2690 MHz). Even though these bands are made available on a global basis for countries to implement IMT-2000, a good degree of flexibility has been provided for operators to evolve toward IMT-2000 as per market and other national considerations. The flexibility allows the use of these bands by services other than those for which the spectrum has been made available. Furthermore, it not only enables each country to decide on timing of availability based on national needs, but also permits countries to select those parts of bands where sharing with existing services is most suitable.

6.5.3.3 Security

Because of the radiating nature of wireless communications, IMT-2000 needs to incorporate security measures to prevent easy reception by parties other than the intended recipient. In addition, because of the nature of mobile communications, security measures are required to prevent fraudulent use of services [67]. The security provisions for IMT-2000 are defined with the objective of ensuring interoperability with roaming across international and national network boundaries. Virtually all security requirements and features are related to the radio interface. IMT-2000 security features are categorized as user related or service provider related. Within these categories, they are further categorized as essential or optional [68].

6.5.3.4 Intelligent Network Issues

The standardization of IMT-2000 is considering new and evolving technologies on the telecommunication network side, such as intelligent networks (INs). IMT-2000 network issues are studied in close cooperation by ITU-R/ITU-T and to a great extent as an integral part of ITU-T work on IN concepts and capabilities. It is anticipated that future versions of IN switching and signaling standards will include the management of mobile and radio access as a natural part of the protocols. This includes location registration/updating and paging as well as the various types of handover between radio cells [68].

6.5.3.5 Regulatory Environments

The regulatory considerations for the introduction and use of IMT-2000 include determining the conditions for regulated and nonregulated systems, spectrum sharing, identifying the number of operators and service providers, licensing procedures, and call charging. The provision and establishment of IMT-2000 is subject to the regulatory process in each country's telecommunication authority. It may be necessary to develop new regulatory environments for IMT-2000 that will enable the provision of new services in a variety of ways not anticipated by existing regulations.

6.5.4 Satellite Operation

The satellite component of IMT-2000 enhances the overall coverage and attractiveness of the service and facilitates the development of telecommunications services in developing countries [69]. Satellites are particularly useful in mobile communications because they are able to achieve coverage of very large areas of the earth's surface [70]. To provide service at an acceptable cost, the catchment area must include as many users as possible. In this situation, a globally unique standard formulated by IMT-2000 is preferable to adopting regional solutions. The current version of the recommendation pertaining to the satellite component of IMT-2000 is very generic and does not provide specific details in relation to service, equipment, architecture, or interfaces and protocols [71].

Currently, many satellite PCSs have been proposed based on constellations of orbiting satellites offering continental and worldwide communications, data, tracking, and paging services. The experience gained from these networks in the next few years will provide valuable input into the satellite component of third-generation systems. Depending on the lessons learned, the three possible levels of integration of the satellite component into the terrestrial network are:

1. Network integration at the call level
2. Equipment integration, requiring common service standards and dual-mode terminals
3. System integration, by which the satellite is an integral part of the network and handoff can be supported between terrestrial and satellite megacells [70]

It is anticipated that IMT-2000 will use several satellite constellations, each comprising a number of satellites, radio (*service*) links from the satellite to the IMT-2000 terminal, and radio (*feeder*) links from the satellites to the land earth stations (LESs) [72]. Because the satellite component will have a limited number of LESs, the operation of the network will inherently involve international terrestrial connections, and access to the network may therefore also involve an international connection. A number of non-geostationary earth orbit (non-GEO) satellites based on low-/midearth orbit (LEO/MEO) constellations have been or are being deployed to deliver mobile voice and broadband data services. For the first time at WRC-1997, spectrum was made available for the operation of these satellites, and a provisional power limit was imposed so that they could share the spectrum with GEO satellites. Studies conducted since 1997 on spectrum sharing have found in favor of the concept, and WRC-2000 decided to limit the power of non-GEO satellites to enable their coexistence with GEO satellites, which aim to provide high-speed local access to global broadband services without unacceptable interference [73]. Continuity of coverage will be provided by contiguous footprints of spot beams from one or more satellites in a constellation. For nongeostationary satellites, these footprints will be in motion, and continuity of calls in progress will be achieved by handover between beams, using functionality in both the mobile and satellite components [72].

The following list identifies the key features of the satellite component [72,74,75]:

1. Coverage of any one satellite will be much larger than that of any cluster of terrestrial base stations.
2. Coverage is likely to be by means of a number of spot beams, which will form megacells, with each spot beam larger than any terrestrial macrocell.
3. Satellite coverage can be regional, multiregional, or global.
4. A range of orbit constellations may be used.
5. The number of LESs will be limited.
6. The terrestrial and satellite components should be optimized with respect to each other.
7. The LESs will connect to the satellites using feeder links that operate in frequency bands outside those identified for IMT-2000 operation; the feeder link frequencies may be used by other satellite systems and terrestrial systems, with appropriate sharing criteria.
8. Intersatellite links (ISLs), if used, will operate outside the IMT-2000 band.
9. Provisions must exist to allow multiple service providers to compete in the satellite component.

References

1. Padgett, J.E., Gunther, C.G., and Hattori, T., Overview of wireless personal communications. *IEEE Commun. Mag.*, 33, 28–41, January 1995.
2. Erdman, W.W., Wireless communications: A decade of progress. *IEEE Commun. Mag.*, 31, 48–51, December 1993.
3. Goodman, D.J., Second generation wireless information networks. *IEEE Trans. Veh. Technol.*, 40, 366–374, 1991.
4. Godara, L.C., Ryan, M.J., and Padovan, N., Third generation mobile communication systems: Overview and modelling considerations. *Ann. Telecommun.*, 54(1–2), 114–136, 1999.
5. Padovan, N., Ryan, M., and Godara, L., An overview of third generation mobile communications systems. IEEE Tencon '98: IEEE Region 10 Annual Conference, New Delhi, December 17–19, 1998.
6. Pandya, R., Emerging mobile and personal communication systems. *IEEE Commun. Mag.*, 33, 44–52, June 1995.

7. Baier, P.W., Jung, P., and Klein, A., Taking the challenge of multiple access for third generation cellular mobile radio systems—A European view. *IEEE Commun. Mag.,* 34, 82–89, February 1996.

8. Dasilva, J.S., Arroyo, B., Barni, B., and Ikonomou, D., European third-generation mobile systems. *IEEE Commun. Mag.,* 34, 68–83, October 1996.

9. Re, E.D., A coordinated European effort for the definition of a satellite integrated environment for future mobile communications. *IEEE Commun. Mag.,* 34, 98–104, February 1996.

10. Wu, W.W., Miller, E.F., Pritchard, W.L., and Pickholtz, R.L., Mobile satellite communications. *IEEE Proc.,* 82, 1431–1448, 1994.

11. Abrishamkar, F., and Siveski, Z., PCS global mobile satellites. *IEEE Commun. Mag.,* 34, 132–136, September 1996.

12. Norbury, J.R., Satellite land mobile communication systems. *IEE Electron. Commun. Eng. J.,* 245–253, November/December 1989.

13. Ananasso, F., and Priscoli, F.D., The role of satellite in personal communication services. *IEEE J. Sel. Areas Commun.,* 13, 180–196, 1995.

14. Gaudenzi, R.D., Giannetti, F., and Luise, M., Advances in satellite CDMA transmission for mobile and personal communications. *IEEE Proc.,* 84, 18–39, 1996.

15. Re, E.D., Devieux, C.L., Jr., Kato, S., Raghavan, S., Taylor, D., and Ziemer, R., Eds., Special issue on mobile satellite communications for seamless PCS. *IEEE Trans. Sel. Areas Commun.,* 13, February 1995.

16. Laane, R., Ed.-in-Chief, Special issue on satellite and terrestrial systems and services for travelers. *IEEE Commun. Mag.,* 29, November 1991.

17. Lee, W.C.Y., *Mobile Communication Design Fundamentals.* New York: Wiley, 1993.

18. Lee, W.C.Y., *Mobile Cellular Telecommunications.* New York: McGraw-Hill, 1995.

19. Rappaport, T.S., *Wireless Communications: Principles and Practice.* Englewood Cliffs, NJ: Prentice-Hall, 1996.

20. Garg, V.K., and Wilks, J.E., *Wireless and Personal Communications Systems.* Englewood Cliffs, NJ: Prentice-Hall, 1996.

21. Godara, L.C., Application of antenna arrays to mobile communications—Part I: Performance improvement, feasibility and system considerations. *Proc. IEEE,* 85(7), 1031–1062, July 1997.

22. Pahlavan, K., and Levesque, A.H., Wireless data communications. *IEEE Proc.,* 82, 1398–1430, 1994.

23. Bertoni, H.L., Honcharenko, W., Maceil, L.R., and Xia, H.H., UHF propagation prediction for wireless personal communications. *IEEE Proc.,* 82, 1333–1359, 1994.

24. Fleury, B.H., and Leuthold, P.E., Radiowave propagation in mobile communications: An overview of European research. *IEEE Commun. Mag.,* 34, 70–81, February 1996.

25. French, R.C., The effect of fading and shadowing on channel reuse in mobile radio. *IEEE Trans. Veh. Technol.,* 28, 171–181, 1979.

26. Winters, J.H., Optimum combining for indoor radio systems with multiple users. *IEEE Trans. Commun.,* COM-35, 1222–1230, 1987.

27. Andersen, J.B., Rappaport, T.S., and Yoshida, S., Propagation measurements and models for wireless communications channels. *IEEE Commun. Mag.,* 33, 42–49, January 1995.

28. Adachi, F., Sawahashi, M., and Suda, H., Wideband DS-CDMA for next-generation mobile communications systems. *IEEE Commun. Mag.,* 36, 56–69, September 1998.

29. Falconer, D.D., Adachi, F., and Gudmundson, B., Time division multiple access methods for wireless personal communications. *IEEE Commun. Mag.,* 33, 50–57, January 1995.

30. Raith, K., and Uddenfeldt, J., Capacity of digital cellular TDMA systems. *IEEE Trans. Veh. Technol.,* 40, 323–332, 1991.
31. Lee, W.C.Y., Overview of cellular CDMA. *IEEE Trans. Veh. Technol.,* 40, 291–302, 1991.
32. Gilhousen, K.S. et al., On the capacity of cellular CDMA system. *IEEE Trans. Veh. Technol.,* 40, 303–312, 1991.
33. Pickholtz, R.L., Milstein, L.W., and Schilling, D.L., Spread spectrum for mobile communications. *IEEE Trans. Veh. Technol.,* 40, 313–322, 1991.
34. Kohno, R., Meidan, R., and Milstein, L.B., Spread spectrum access methods for wireless communications. *IEEE Commun. Mag.,* 33, 58–67, January 1995.
35. Abramson, N., Multiple access in wireless digital networks. *IEEE Proc.,* 82, 1360–1370, 1994.
36. Pickholtz, L., Schilling, D.L., and Milstein, L.B., Theory of spread spectrum communications—A tutorial. *IEEE Trans. Commun.,* 855–884, May 1982.
37. Milstein, L.B., Wideband code division multiple access. *IEEE J. Sel. Areas Commun.,* 18(8), 1344–1354, August 2000.
38. Winters, J.H., Salz, J., and Gitlin, R.D., The impact of antenna diversity on the capacity of wireless communication systems. *IEEE Trans. Commun.,* 42, 1740–1751, 1994.
39. Godara, L.C., Application of antenna arrays to mobile communications—Part II: Beamforming and DOA considerations. *Proc. IEEE,* 85(8), 1195–1247, August 1997.
40. Mizuno, M., and Ohgane, T., Application of adaptive array antennas to radio communications. *Electron. Commun. Jpn., Part I: Commun.,* 77, 48–59, 1994.
41. Tripathi, N.D., Reed, J.H., and Van Landingham, H.F., Handoff in cellular systems. *IEEE Pers. Commun.,* 26–37, December 1998.
42. Tuttlebee, W., Cordless personal telecommunications. *IEEE Commun. Mag.,* 30, 42–53, December 1992.
43. Freeburg, T.A., Enabling technologies for wireless in-building network communications—Four technical challenges, four solutions. *IEEE Commun. Mag.,* 29, 58–64, April 1991.
44. Pollini, G.P., Trends in handover design. *IEEE Commun. Mag.,* 34, 82–90, March 1996.
45. Rappaport, S.S., and Hu, L.R., Microcellular communication systems with hierarchical macrocell overlays: Traffic performance models and analysis. *IEEE Proc.,* 82, 1383–1397, 1994.
46. Steel, R., Whitehead, J., and Wong, W.C., System aspects of cellular radio. *IEEE Commun. Mag.,* 33, 80–86, January 1995.
47. Katzola, I., and Naghsineh, M., Channel assignment schemes for cellular mobile telecommunication systems: A comparative survey. *IEEE Pers. Commun.,* 10–31, June 1996.
48. Noerpel, A., and Lin, Y.B., Handover management for a PC network. *IEEE Pers. Commun.,* 18–24, December 1997.
49. Wong, D., and Lim, T.J., Soft handoff in CDMA mobile systems. *IEEE Pers. Commun.,* 6–17, December 1997.
50. Tekinay, S., and Jabbary, B., Handover and channel assignment in mobile cellular networks. *IEEE Commun. Mag.,* 29, 42–46, November 1991.

51. Mouli, M., and Pautet, M.B., The GSM system for mobile communications. *Cell & Sys, Paris* 1992.

52. Van Den Broek, W. et al., Functional models of UMTS and integration into future networks. *Electron. Commun. Eng. J.,* 5(3), 165–172, June 1993.

53. Special issue on IMT-2000. *IEEE Pers. Commun.,* 4(4), August 1997.

54. Shafi, M., Wireless communications in the twenty-first century: A perspective. *Proc. IEEE,* 85(10), 1622–1638, October 1997.

55. Kriaris, I. et al., Third-generation mobile network architectures for the universal mobile telecommunications system (UMTS). *Bell Labs Tech. J.,* 99–117, Summer 1997.

56. Hu, Y., and Sheriff, R., The potential demand for the satellite component of the universal mobile telecommunications system. *Electron. Commun. Eng. J.,* 59–67, April 1997.

57. Johnston, W., Europe's future mobile telephony system. *IEEE Spectrum,* 49–53, October 1998.

58. Chia, S., The Universal Mobile Telecommunications System. *IEEE Commun. Mag.,* 30(12), 54–62, December 1992.

59. Baier, P.W. et al., Taking the challenge of multiple access for third generation cellular mobile radio systems—a European view. *IEEE Commun. Mag.,* 34(2), 82–89, February 1996.

60. International Telecommunications Union, IMT-2000 home page, http://www.itu.int, June 1997.

61. *ITU-M Recommendation M.816, Framework for Services Supported on Future Public Land Mobile Telecommunications Systems,* 22–30, 1994.

62. *ITU-M Recommendation M.817, Future Public Land Mobile Telecommunications Systems Network Architectures,* 1–18, 1994.

63. Da Silva, J.S. et al., European mobile communications on the move. *IEEE Commun. Mag.,* 34(2), February 1996.

64. Dahlman, E. et al., UMTS/IMT-2000 based on wideband CDMA. *IEEE Commun. Mag.,* 70–80, September 1998.

65. Ojanperä, T., An overview of air interface multiple access for IMT-2000/UMTS. *IEEE Commun. Mag.,* 82–95, September 1998.

66. Andermo, P.G., System flexibility and its requirements on a third generation mobile system. In *Third IEEE International Symposium on Personal, Indoor and Mobile Radio Communications, Ottawa,* 1992, 397–401.

67. Beller, M.J., Chang, L., and Yacobi, Y., Security for personal communications services: public key versus private key approaches. In *Third IEEE International Symposium on Personal, Indoor and Mobile Radio Communications, Boston,* 1992, 26–31.

68. *ITU-M Recommendation M.1078, Speech and Voiceband Data Performance Requirements for Future Public Land Mobile Telecommunications Systems,* 110–135, 1994.

69. Ananasso, F. et al., Issues on the evolution towards satellite personal communications networks. In *IEEE Global Telecommunications Conference, Singapore,* November 1995, 541–545.

70. Dondl, P., Standardization of the satellite component of the UMTS. *IEEE Personal Commun.,* 2(5), 68–74, October 1995.

71. *ITU-M Recommendation M.818–1, Satellite Operation within the Future Public Land Mobile Telecommunications Systems,* 49–51, 1994.

72. *ITU-M Recommendation M.1167, Framework for the Satellite Component of Future Public Land Mobile Telecommunications Systems*, 2–10, 1995.
73. WRC-2000. http://www.itu.int./brconf/wrc-2000/docs/index.html.
74. Del Re, E., A coordinated European effort for the definition of a satellite integrated environment for future mobile communications. *IEEE Commun. Mag.*, 34(2), 98–104, February 1996.
75. *ITU-M Recommendation M.1182, Integration of Terrestrial and Satellite Mobile Communications Systems*, 9–15, 1995.

Chapter 7

An Overview of cdma2000, WCDMA, and EDGE

Tero Ojanperä and Steven D. Gray

Contents

7.1 Introduction

In response to the International Telecommunication Union's (ITU's) call for proposals, third-generation cellular technologies are evolving at a rapid pace, with different proposals vying for the future marketplace in digital wireless multimedia communications. While the original intent for the third generation was to have a convergence of cellular-based technologies, this appears to be an unrealistic expectation. As such, three technologies key for the North American and European markets are the third-generation extension of the Telecommunication Industry Association/Electronic Industries Association 95B- (TIA/EIA-95B)-based code-division multiple access (CDMA) called cdma2000, the European third-generation CDMA called wideband CDMA (WCDMA), and the third-generation time-division multiple-access (TDMA) system based on EDGE (Enhanced Data Rates for Global TDMA Evolution). For packet data, EDGE is one case for which second-generation technologies converged to a single third-generation proposal with convergence of the U.S. TDMA system called TIA/EIA-136 and the European system GSM (Global System for Mobile Communications, originally Groupe Spécial Mobile). This chapter provides an overview of the air interfaces of these key technologies. Particular attention is given to the channel structure, modulation, and offered data rates of each technology. A comparison is also made between cdma2000 and WCDMA to help in understanding the similarities and differences of these two CDMA approaches for the third generation.

The promise of a third generation is a world in which the subscriber can access the World Wide Web (WWW) or perform file transfers over packet data connections capable of providing 144 kbps for high mobility, 384 kbps with restricted mobility, and 2 Mbps in an indoor office environment [1]. With these guidelines on rate from the ITU, standards bodies started the task of developing an air interface for their third-generation system. In North America, the TIA evaluated proposals from TIA members pertaining to the evolution of TIA/EIA-95B and TIA/EIA-136. In Europe, the European Telecommunications Standards Institute (ETSI) evaluated proposals from ETSI members pertaining to the evolution of GSM.

While TIA and ETSI were still discussing various targets for third-generation systems, Japan began to roll out its contributions for third-generation technology and develop proof-of-concept prototypes. In the beginning of 1997, the Association for Radio Industry and Business (ARIB), a body responsible for standardization of the Japanese air interface, decided to proceed with the detailed standardization of a WCDMA system. The technology push from Japan accelerated standardization in Europe and the United States. During 1997, joint parameters for Japanese and European WCDMA proposals were agreed upon. The air interface is commonly referred to as WCDMA. In January 1998, the strong support behind WCDMA led to the selection of WCDMA as the UMTS (Universal Mobile Telecommunications System) terrestrial air interface scheme for FDD (frequency division duplex) frequency bands in ETSI. In the United States, third-generation CDMA came through a detailed proposal process from vendors interested in the evolution of TIA/EIA-95B. In February 1998, the TIA committee TR45.5 responsible for TIA/EIA-95B standardization adopted a framework that combined the different vendors' proposals and later became known as cdma2000.

For TDMA, the focus has been to offer Interim Standard 136 (IS-136) and GSM operators a competitive third-generation evolution. WCDMA is targeted toward GSM evolution; however, EDGE allows the operators to supply International Mobile Telecommunications 2000 (IMT-2000) data rates without the spectral allocation requirements of WCDMA. Thus, EDGE will be deployed by those operators who wish to maintain either IS-136 or GSM for voice services and augment these systems with a TDMA-based high-rate packet service. TDMA convergence occurred late in 1997 when ETSI approved standardization of the EDGE concept and in February 1998 when the TIA committee TR45.3 approved the UWC-136 EDGE-based proposal.

The push to the third generation was initially focused on submission of an IMT-2000 radio transmission techniques (RTTs) proposal. The evaluation process has started in the ITU [2]; Figure 7.1 depicts the time schedule of the ITU RTT development. Since at the same time regional standards have started the standards-writing process, the relationship between the ITU and regional standards it is not yet clear. Based on actions in TIA and ETSI, it is reasonable to assume that standards will exist for cdma2000, WCDMA, and EDGE, and all will be deployed based on market demands.

The chapter is organized as follows: Issues affecting third-generation CDMA are discussed, followed by a brief introduction of WCDMA, cdma2000, and EDGE. Table 7.1 compares cdma2000 and WCDMA. For TDMA, an overview of the IS-136-based evolution is given, including the role played by EDGE.

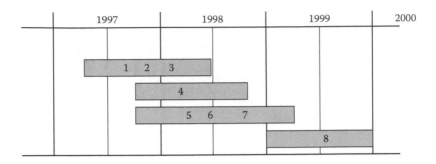

Figure 7.1   ITU timelines: 1, 2, 3—RTTs request, development, and submission; 4—RTT evaluation; 5—review outside evaluation; 6—assess compliance with performance parameters; 7—consideration of evaluation results and consensus on key characteristics; 8—development of detailed radio interface specifications.

Table 7.1   Parameters of WCDMA and cdma2000

	WCDMA	cdma2000
Channel bandwidth	5, 10, 20 MHz	1.25, 5, 10, 15, 20 MHz
Forward-link RF channel structure	Direct spread	Direct spread or multicarrier
Chip rate	4.096/8.192/16.384 Mcps	1.2288/3.6864/7.3728/11.0593/ 14.7456 Mcps for direct spread n_1.2288 Mcps (nD1, 3, 6, 9, 12) for multicarrier
Roll-off factor	0.22	Similar to TIA/EIA-95B
Frame length	10 ms/20 ms (optional)	20 ms for data and control/5 ms for control information on the fundamental and dedicated control channel
Spreading modulation	QPSK (forward link) Balanced dual-channel QPSK (reverse link) Complex spreading circuit	Balanced QPSK (forward link) Dual-channel QPSK (reverse link) Complex spreading circuit
Data modulation	QPSK (forward link) BPSK (reverse link)	QPSK (forward link) BPSK (reverse link)

Table 7.1   Parameters of WCDMA and cdma2000 (*Continued*)

	WCDMA	cdma2000
Coherent detection	User-dedicated time-multiplexed pilot (forward link and reverse link), common pilot in forward link	Pilot time multiplexed with PC and EIB (reverse link) Common continuous pilot channel and auxiliary pilot (forward link)
Channel multiplexing in reverse link	Control and pilot channel time multiplexed I&Q multiplexing for data and control channel	Control, pilot fundamental, and supplemental code multiplexed I&Q multiplexing for data and control channels
Multirate	Variable spreading and multicode	Variable spreading and multicode
Spreading factors	4–256 (4.096 Mcps)	4–256 (3.6864 Mcps)
Power control	Open and fast closed loop (1.6 kHz)	Open loop and fast closed loop (800 Hz)
Spreading (forward link)	Variable-length orthogonal sequences for channel separation; Gold sequences for cell and user separation	Variable-length Walsh sequences for channel separation, M-sequence 3×2^{15} (same sequence with time shift utilized in different cells different sequence in I&Q channel)
Spreading (reverse link)	Variable-length orthogonal sequences for channel separation; Gold sequence 2^{41} for user separation (different time shifts in I&Q channel, cycle 2^{16} 10-ms radio frames)	Variable-length orthogonal sequences for channel separation, M-sequence 2^{15} (same for all users different sequences in I&Q channels), M-sequence 2^{41} for user separation (different time shifts for different users)
Handover	Soft handover Interfrequency handover	Soft handover Interfrequency handover

7.2 CDMA-Based Schemes

Third-generation CDMA system descriptions in TIA and ETSI have similarities and differences. Some of the similarities between cdma2000 and WCDMA are variable spreading, convolutional coding, and quadrature phase-shift keying (QPSK) data modulation. The major differences between cdma2000 and WCDMA occur with the channel structure, including the structure of the pilot used on the forward link. To aid in comparison of the two CDMA techniques, a brief overview is given to some important third-generation CDMA issues, the dedicated channel structure of cdma2000 and WCDMA, and Table 7.1 comparing air interface characteristics.

7.3 CDMA System Design Issues

7.3.1 Bandwidth

An important design goal for all third-generation proposals is to limit spectral emissions to a 5-MHz dual-sided passband. There are several reasons for choosing this bandwidth. First, data rates of 144 and 384 kbps, the main targets of third-generation systems, are achievable within a 5-MHz bandwidth with reasonable coverage. Second, lack of spectrum calls for limited spectrum allocation, especially if the system has to be deployed within the existing frequency bands already occupied by the second-generation systems. Third, the 5-MHz bandwidth improves the receiver's ability to resolve multipath when compared to narrower bandwidths, increasing diversity and improving performance. Larger bandwidths of 10, 15, and 20 MHz have been proposed to support the highest data rates more effectively.

7.3.2 Chip Rate

Given the bandwidth, the choice of chip rate (CR) depends on spectrum deployment scenarios, pulse shaping, desired maximum data rate and dual-mode terminal implementation. Figure 7.2 shows the relation among CR, pulse-shaping filter roll-off factor α, and channel separation Δf. If raised cosine filtering is used, the spectrum is zero (in theory) after $CR/2(1 + \alpha)$. In Figure 7.2, channel separation is selected such that two adjacent channel spectra do not overlap. Channel separation should be selected this way if there can be high power-level differences between the adjacent carriers.

For example, for WCDMA parameters minimum channel separation Δf_{min} for nonoverlapping carriers is $\Delta f_{min} = 4.096 (1 + 0.22) = 4.99712$ MHz. If channel separation is selected in such a way that the spectra of two adjacent channel signals overlap, some power leaks from one carrier to another. Partly overlapping carrier

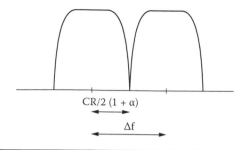

Figure 7.2 Relationship among chip rate (CR), roll-off factor (α), and channel separation (Δf).

spacing can be used, for example, in microcells, in which the same antenna masts are used for both carriers.

A designer of dual-mode terminals needs to consider the relation between the different clock frequencies of different modes. Especially important are the transmitter and receiver sampling rates and the carrier raster. A proper selection of these frequencies for the standard would ease the dual-mode terminal implementation. The different clock frequencies in a terminal are normally derived from a common reference oscillator by either direct division or synthesis by the use of a phase locked loop (PLL). The use of a PLL will add some complexity. The WCDMA CR has been selected based on consideration of backward compatibility with GSM and PDC (personal digital cellular). The cdma2000 CR is a direct derivation of the TIA/EIA-95B CR.

7.3.3 Multirate

Multirate design means multiplexing different connections with different quality-of-service (QoS) requirements in a flexible and spectrum-efficient way. The provision for flexible data rates with different QoS requirements can be divided into three subproblems: how to map different bit rates into the allocated bandwidth, how to provide the desired QoS, and how to inform the receiver about the characteristics of the received signal. The first problem concerns issues such as multicode transmission and variable spreading. The second problem concerns coding schemes. The third problem concerns control channel multiplexing and coding.

Multiple services belonging to the same session can be either time or code multiplexed as depicted in Figure 7.3. The time multiplexing avoids multicode transmissions, thus reducing peak-to-average power of the transmission. A second alternative for service multiplexing is to treat parallel services completely separately with separate channel coding/interleaving. Services are then mapped to separate physical data channels in a multicode fashion as illustrated in the lower part of Figure 7.3. With this alternative scheme, the power, and consequently the quality, of each service can be controlled independently.

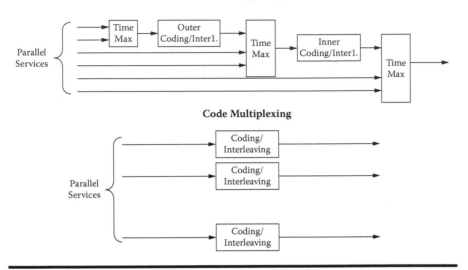

Figure 7.3   Time and code multiplexing principles.

7.3.4 Spreading and Modulation Solutions

A complex spreading circuit as shown in Figure 7.4 helps to reduce the peak-to-average power and thus improves power efficiency.

The spreading modulation can be either balanced- or dual-channel QPSK. In the balanced QPSK spreading, the same data signal is split into I and Q channels. In dual-channel QPSK spreading, the symbol streams on the I and Q channels are

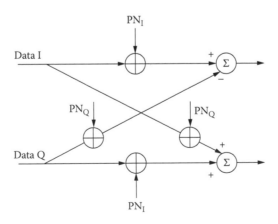

Figure 7.4   Complex spreading of pseudo noise.

independent of each other. In the forward link, QPSK data modulation is used to save code channels and allow the use of the same orthogonal sequence for I and Q channels. In the reverse link, each mobile station uses the same orthogonal codes; this allows for efficient use of binary phase-shift keying (BPSK) data modulation and balanced QPSK spreading.

7.3.5 Coherent Detection in the Reverse Link

Coherent detection can improve the performance of the reverse link up to 3 dB compared to noncoherent reception used by the second-generation CDMA system. To facilitate coherent detection, a pilot signal is required. The actual performance improvement depends on the proportion of the pilot signal power to the data signal power and the fading environment.

7.3.6 Fast Power Control in Forward Link

To improve the forward link performance, fast power control is used. The impact of the fast power control in the forward link is twofold. First, it improves the performance in a fading multipath channel. Second, it increases the multiuser interference variance within the cell since orthogonality between users is not perfect due to the multipath channel. The net effect, however, is improved performance at low speeds.

7.3.7 Additional Pilot Channel in the Forward Link for Beam Forming

An additional pilot channel on the forward link that can be assigned to a single mobile or to a group of mobiles enables deployment of adaptive antennas for beam forming since the pilot signal used for channel estimation needs to go through the same path as the data signal. Therefore, a pilot signal transmitted through an omni-cell antenna cannot be used for the channel estimation of a data signal transmitted through an adaptive antenna.

7.3.8 Seamless Interfrequency Handover

For third-generation systems, hierarchical cell structures (HCSs), constructed by overlaying macrocells on top of smaller micro- or picocells, have been proposed to achieve high capacity. The cells belonging to different cell layers will be in different frequencies, and thus an interfrequency handover is required. A key requirement for the support of seamless interfrequency handover is the ability of the mobile station to carry out cell search on a carrier frequency different from the current one without affecting the ordinary data flow. Different methods have been proposed to obtain multiple carrier frequency measurements. For mobile stations with receiver diversity, there is a possibility for one of the receiver branches to be temporarily reallocated from diversity reception and instead carry out reception on a different carrier. For single-receiver mobile stations, slotted forward-link transmission could allow interfrequency measurements. In the slotted mode, the information normally transmitted during a certain time (e.g., a 10-ms frame) is transmitted in less than that time, leaving an idle time that the mobile can use to measure on other frequencies.

7.3.9 Multiuser Detection

Multiuser detection (MUD) has been the subject of extensive research since 1986, when Verdu formulated an optimum multiuser detector for the additive white Gaussian noise (AWGN) channel, maximum likelihood sequence estimation (MLSE) [3]. In general, it is easier to apply MUD in a system with short spreading codes since cross correlations do not change every symbol as they do with long spreading codes. However, it seems that the proposed CDMA schemes would all use long spreading codes. Therefore, the most feasible approach seems to be interference cancellation algorithms that carry out the interference cancellation at the chip level, thereby avoiding explicit calculation of the cross correlation between spreading codes from different users [4]. Due to complexity, MUD is best suited for the reverse link. In addition, the mobile station is interested in detecting its own signal, in contrast to the base station, which needs to demodulate the signals of all users. Therefore, a simpler interference suppression scheme could be applied in the mobile station. Furthermore, if short spreading codes are used, the receiver could exploit the cyclostationarity (i.e., the periodic properties of the signal) to suppress interference without knowing the interfering codes.

7.3.10 Transmit Diversity

The forward-link performance can be improved in many cases by using transmit diversity. For direct-spread CDMA schemes, this can be performed by splitting the data stream and spreading the two streams using orthogonal sequences or switching the entire data stream between two antennas. For multicarrier CDMA, the different carriers can be mapped into different antennas.

7.4 WCDMA

To aid in the comparison of cdma2000 and WCDMA, the dedicated frame structure of WCDMA is illustrated in Figure 7.5 and Figure 7.6. The approach follows a time multiplex philosophy in which the dedicated physical control channel (DPCCH) provides the pilot, power control, and rate information, and the dedicated physical data channel (DPDCH) is the portion used for data transport. The forward and reverse DPDCH channels have been convolutional encoded and interleaved prior to framing. The major difference between the forward and reverse links is that the reverse channel structure of the DPCCH is a separate code channel from the DPDCH.

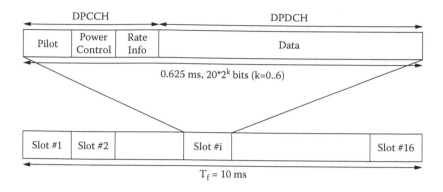

Figure 7.5   Forward-link dedicated channel structure in WCDMA.

After framing, the forward- and reverse-link channels are spread as shown in Figure 7.7 and Figure 7.8, respectively. On the forward-link orthogonal, variable-rate codes c_{ch} are used to separate channels, and pseudorandom scrambling sequences c_{scramb} are used to spread the signal evenly across the spectrum and separate different base stations. On the reverse link, the orthogonal channelization codes are used as in the forward link to separate CDMA channels. The scrambling codes c'_{scramb} and c''_{scramb} are used to identify mobile stations and to spread the signal evenly across the band. The optional scrambling code is used as a means to group mobiles under a common scrambling sequence.

7.4.1 Spreading Codes

WCDMA employs long spreading codes. Different spreading codes are used for cell separation in the forward link and user separation in the reverse link. In the

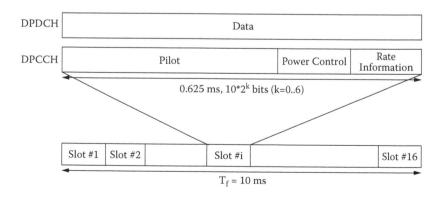

Figure 7.6   Reverse-link dedicated channel structure in WCDMA.

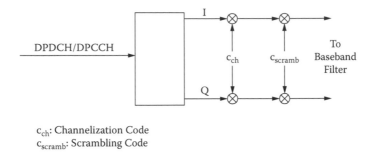

c_{ch}: Channelization Code
c_{scramb}: Scrambling Code

Figure 7.7   Forward-link spreading of DPDCH and DPCCH.

forward link, gold codes of length 2^{18} are truncated to form cycles of 2^{16} times 10-ms frames. To minimize the cell search time, a special short code mask is used. The synchronization channel of WCDMA is masked with an orthogonal short gold code of length 256 chips spanning one symbol. The mask symbols carry information about the base station (BS) long code group. Thus, the mobile station first acquires the short mask code and then searches the corresponding long code. A short very large (VL)–Kasami code has been proposed for the reverse link to ease the implementation of MUD. In this case, code planning would also be negligible because the number of VL-Kasami sequences is more than one million. However, in certain cases, the use of short codes may lead to bad correlation properties, especially with very small spreading factors. If MUD were not used, adaptive code allocation could be used to restore the cross-correlation properties. The use of short codes to ease the implementation of advanced detection techniques is more beneficial in the forward link since the cyclostationarity of the signal could be utilized for adaptive implementation of the receiver.

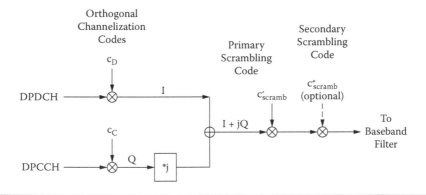

Figure 7.8   Reverse-link spreading for the DPDCH and DPCCH.

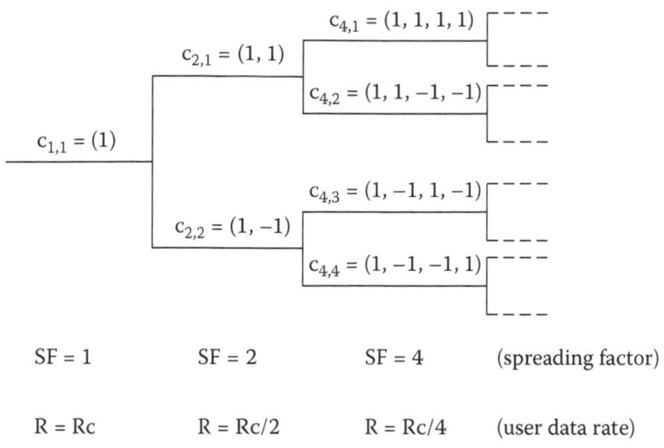

Figure 7.9 Construction of orthogonal spreading codes for different spreading factors.

Orthogonality between the different spreading factors can be achieved by tree-structured orthogonal codes; their construction is illustrated in Figure 7.9 [5]. The tree-structured codes are generated recursively according to the following equation:

$$
c_{2n} = \begin{pmatrix} c_{2n,1} \\ c_{2n,2} \\ \vdots \\ c_{2n,2n} \end{pmatrix} = \begin{pmatrix} \begin{pmatrix} c_{n,1} & c_{n,1} \\ c_{n,1} & -c_{n,1} \end{pmatrix} \\ \vdots \\ \begin{pmatrix} c_{n,n} & c_{n,n} \\ c_{n,n} & -c_{n,n} \end{pmatrix} \end{pmatrix}
$$

where c_{2n} is the orthogonal code set of size $2n$. The generated codes within the same layer constitute a set of orthogonal functions and are thus orthogonal. Furthermore, any two codes of different layers are also orthogonal except for the case when one of the two codes is a mother code of the other. For example, code $c_{4,4}$ is not orthogonal with codes $c_{1,1}$ and $c_{2,2}$.

7.4.2 Coherent Detection and Beam Forming

In the forward link, time-multiplexed pilot symbols are used for coherent detection. Because the pilot symbols are user dedicated, they can be used for channel estimation with adaptive antennas as well. In the reverse link, WCDMA employs pilot symbols multiplexed with power control and rate information for coherent detection.

7.4.3 Multirate

The WCDMA traffic channel structure is based on a single code transmission for low data rates and multicode for higher data rates. Multiple services belonging to the same connection are, in normal cases, time multiplexed as depicted in the upper part of Figure 7.3. After service multiplexing and channel coding, the multiservice data stream is mapped to one or more dedicated physical data channels. In the case of multicode transmission, every other data channel is mapped into Q and every other into I channels. The channel coding of WCDMA is based on convolutional and concatenated codes. For services with a bit error rate (BER) of 10^{-3}, a convolutional code with constraint length of 9 and different code rates (between 1/2 and 1/4) is used. For services with BER = 10^{-6}, a concatenated coding with an outer Reed–Solomon code has been proposed. Typically, block interleaving over one frame is used. WCDMA is also capable of interframe interleaving, which improves the performance for services allowing longer delay. Turbo codes for data services are under study. Rate matching is performed by puncturing or symbol repetition.

7.4.4 Packet Data

WCDMA has two different types of packet data transmission possibilities. Short data packets can be appended directly to a random-access burst. The WCDMA random-access burst is 10 ms long, it is transmitted with fixed power, and the access principle is based on the slotted ALOHA scheme. This method, called *common channel packet transmission*, is used for short infrequent packets, for which the link maintenance needed for a dedicated channel would lead to unacceptable overhead. Larger or more frequent packets are transmitted on a dedicated channel. A large single packet is transmitted using a single-packet scheme in which the dedicated channel is released immediately after the packet has been transmitted. In a multipacket scheme, the dedicated channel is maintained by transmitting power control and synchronization information between subsequent packets.

7.5 cdma2000

The dedicated channels used in the cdma2000 system are the fundamental, supplemental, pilot, and dedicated control channels. Shown for the forward link in Figure 7.10 and for the reverse in Figure 7.11, the fundamental channel provides for the communication of voice, low-rate data, and signaling if power control information for the reverse channels is punctured on the forward fundamental channel. For high-rate data services, the supplemental channel is used where one important difference between the supplemental and the fundamental channel is the addition of parallel-concatenated turbo codes. For different service options, multiple supplemental channels can be used. The code multiplex pilot channel allows for phase coherent

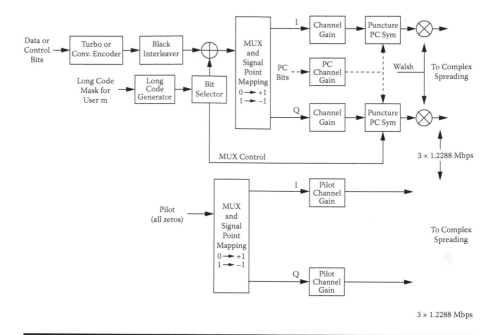

Figure 7.10   Forward-link channel structure in cdma2000 for direct spread. (*Note:* Dashed line indicates using for only the fundamental channel.)

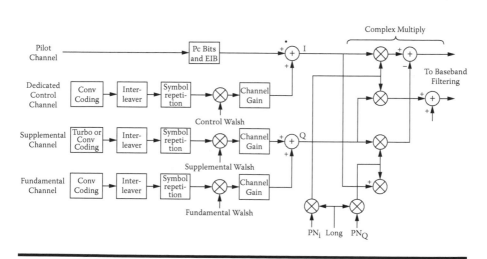

Figure 7.11   Reverse-link channel structure in cdma2000.

detection. In addition, the pilot channel on the forward link is used for determining soft handoff, and the pilot channel on the reverse is used for carrying power control information for the forward channels. Finally, the dedicated control channel, also shown in Figure 7.10 for the forward link and in Figure 7.11, for the reverse, is used primarily for exchange of high-rate media-access control (MAC) layer signaling

7.5.1 Multicarrier

In addition to direct spread, a multicarrier approach has been proposed for the cdma2000 forward link since it would maintain orthogonality between the cdma2000 and TIA/EIA-95B carriers [6]. The multicarrier variant is achieved by using three 1.25-MHz carriers for a 5-MHz bandwidth for which all carriers have separate channel coding and are power controlled in unison.

7.5.2 Spreading Codes

On the forward link, the cell separation for cdma2000 is performed by two M-sequences of length 3×2^{15}, one for the I and one for the Q channel, which are phase shifted by PN (pseudo noise) offset for different cells. Thus, during the cell search process only these sequences are searched. Because there are a limited number of PN offsets, they need to be planned to avoid PN confusion [7]. In the reverse link, user separation is performed by different phase shifts of M-sequence of length 2^{41}. The channel separation is performed using variable spreading factor Walsh sequences, which are orthogonal to each other.

7.5.3 Coherent Detection

In the forward link, cdma2000 has a common pilot channel, which is used as a reference signal for coherent detection when adaptive antennas are not employed. When adaptive antennas are used, an auxiliary pilot is used as a reference signal for coherent detection. Code-multiplexed auxiliary pilots are generated by assigning a different orthogonal code to each auxiliary pilot. This approach reduces the number of orthogonal codes available for the traffic channels. This limitation is alleviated by expanding the size of the orthogonal code set used for the auxiliary pilots. Since a pilot signal is not modulated by data, the pilot orthogonal code length can be extended, thereby yielding an increased number of available codes, which can be used as additional pilots. In the reverse link, the pilot signal is time multiplexed with power control and erasure indicator bit (EIB).

7.5.4 Multirate Scheme

For cdma2000, there are two traffic channel types, the fundamental and the supplemental channel, which are code multiplexed. The fundamental channel is a variable-rate channel that supports basic rates of 9.6 kbps and 14.4 kbps and their corresponding subrates (i.e., Rate Set 1 and Rate Set 2 of TIA/EIA-95B). It conveys voice, signaling, and low-rate data. The supplemental channel provides high data rates. Services with different QoS requirements are code multiplexed into supplemental channels. The user data frame length of cdma2000 is 20 ms. For the transmission of control information, 5- and 20-ms frames can be used on the fundamental channel or dedicated control channel. On the fundamental channel, a convolutional code with constraint length 9 is used. On supplemental channels, convolutional coding is used up to 14.4 kbps. For higher rates, turbo codes with constraint length 4 and rate $1 = 4$ are preferred. Rate matching is performed by puncturing, symbol repetition, and sequence repetition.

7.5.5 Packet Data

Also, cdma2000 allows short data burst using the slotted ALOHA principle. However, instead of fixed transmission power, it increases the transmission power for the random-access burst after an unsuccessful access attempt. When the mobile station has been allocated a traffic channel, it can transmit without scheduling up to a predefined bit rate. If the transmission rate exceeds the defined rate, a new access request has to be made. When the mobile station stops transmitting, it releases the traffic channel but not the dedicated control channel. After a while, it also releases the dedicated control channel as well but maintains the link-layer and network-layer connections to shorten the channel setup time when new data need to be transmitted.

7.5.6 Parametric Comparison

For comparison, Table 7.1 lists the parameters of cdma2000 and WCDMA. cdma2000 uses a CR of 3.6864 Mcps (megachips per second) for the 5-MHz band allocation with the direct spread forward-link option and a 1.2288-Mcps CR with three carriers for the multicarrier option. WCDMA uses direct spread with a CR of 4.096 Mcps. The multicarrier approach is motivated by a spectrum overlay of cdma2000 carriers with existing TIA/EIA-95B carriers [6]. Similar to TIA/EIA-95B, the spreading codes of cdma2000 are generated using different phase shifts of the same M-sequence. This is possible due to the synchronous network operation. Because WCDMA has an asynchronous network, different long codes rather than different phase shifts of the same code are used for the cell and user separation. The code structure determines how code synchronization, cell acquisition, and handover synchronization are performed.

7.6 TDMA-Based Schemes

As discussed, the TIA/EIA-136 and GSM evolutions have had similar paths in the form of EDGE. The UWC-136 IMT-2000 proposal contains, in addition to

Table 7.2   Parameters of 136 HS

	136 HS (Vehicular/ Outdoor)	136 HS (Indoor)
Duplex method	FDD	FDD and TDD
Carrier spacing	200 kHz	1.6 MHz
Modulation	Q-O-QAM B-O-QAM 8PSK GMSK	Q-O-QAM B-O-QAM
Modulation bit rate	722.2 kbps (Q-O-QAM) 361.1 kbps (B-O-QAM) 812.5 kbps (8PSK) 270.8 kbps (GMSK)	5200 kbps (Q-O-QAM) 2600 kbps (B-O-QAM)
Payload	521.6 kbps (Q-O-QAM) 259.2 kbps (B-O-QAM) 547.2 kbps (8PSK) 182.4 kbps (GMSK)	4750 kbps (Q-O-QAM) 2375 kbps (B-O-QAM)
Frame length	4.615 ms	4.615 ms
Number of slots	8	64 (72 µs) 16 (288 µs)
Coding	Convolutional 1/2, 1/4, 1/3, 1/1 ARQ	Convolutional 1/2, 1/4, 1/3, 1/1 Hybrid type II ARQ
Frequency hopping	Optional	Optional
Dynamic channel allocation	Optional	Optional

the TIA/EIA-136 30-kHz carriers, the high-rate capability provided by the 200-kHz and 1.6-MHz carriers shown in Table 7.2. The targets for the IS-136 evolution were to meet IMT-2000 requirements and an initial deployment within the 1-MHz spectrum allocation. UWC-136 meets these targets via modulation enhancement to the existing 30-kHz channel (136C) and by defining complementary wider-band TDMA carriers with bandwidths of 200 kHz for vehicular/

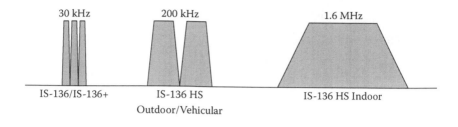

Figure 7.12   UWC-136 carrier types.

outdoor environments and 1.6 MHz for indoor environments. The 200-kHz carrier, 136 HS (vehicular/outdoor) with the same parameters as EDGE, provides medium bit rates up to 384 kbps; the 1.6-MHz carrier, 136 HS (indoor), provides the highest bit rates, up to 2 Mbps. The parameters of the 136 HS proposal submitted to ITU are listed in Table 7.2, and the different carrier types of UWC-136 are shown in Figure 7.12.

7.6.1 Carrier Spacing and Symbol Rate

The motivation for the 200-kHz carrier is twofold. First, the adoption of the same physical layer for 136 HS (vehicular/outdoor) and GSM data carriers provides economics of scale and therefore cheaper equipment and faster time to market. Second, the 200-kHz carrier with higher-order modulation can provide bit rates of 144 and 384 kbps with reasonable range and capacity, fulfilling IMT-2000 requirements for pedestrian and vehicular environments. The 136 HS (indoor) carrier can provide a 2-Mbps user data rate with reasonably strong channel coding.

7.6.2 Modulation

The first proposed modulation methods were quaternary offset quadrature amplitude modulation (Q-O-QAM) and binary offset QAM(B-O-QAM). Q-O-QAM could provide higher data rates and good spectral efficiency. For each symbol, 2 bits are transmitted, and consecutive symbols are shifted by $\pi/2$. An offset modulation was proposed because it causes smaller-amplitude variations than 16-QAM, which can be beneficial when using amplifiers that are not completely linear. The second modulation B-O-QAM has been introduced; it has the same symbol rate of 361.111 ksps, (kilosymbols per second), but only the outer signal points of the Q-O-QAM modulation are used. For each symbol, 1 bit is transmitted, and consecutive symbols are shifted by $\pi/2$. A second modulation scheme, with the characteristic of being a subset of the first modulation scheme and having the same symbol rate as the first modulation, allows seamless switching between the two modulation types between bursts. Both modulation types can be used in the same burst. From a

Table 7.3   Bit Rates for GMSK and 8PSK

Modulation	GMSK	8PSK
Symbol Rate	270 ksymbols/s	270 ksymbols/s
Bits per Symbol	1 bps	3 bps
Modulation Bit Rate	270 kbps	810 kbps

complexity point of view, the addition of a modulation that is a subset of the first modulation adds no new requirements for the transmitter or receiver.

In addition to the originally proposed modulation schemes, Q-O-QAM and B-O-QAM, other modulation schemes (CPM, continuous phase modulation, and 8PSK,) have been evaluated in order to select the modulation best suited for EDGE. The outcome of this evaluation is that 8PSK was considered to have implementation advantages over Q-O-QAM. Parties working on EDGE are in the process of revising the proposals so that 8PSK (phase-shift keying) would replace the Q-O-QAM and GMSK (Gaussian minimum shift keying) can be used as the lower-level modulation instead of B-O-QAM. The symbol rate of the 8PSK is the same as for GMSK, and the detailed bit rate is given in Table 7.3.

7.6.3 Frame Structures

The 136 HS (vehicular/outdoor) data frame length is 4.615 ms, and one frame consists of eight slots. The burst structure is suitable for transmission in a high delay spread environment. The frame and slot structures of the 136 HS (indoor) carrier were selected for cell coverage for high bit rates. The HS-136 indoor supports both FDD and TDD (time-division duplex) methods. Figure 7.13 illustrates the frame and slot structure. The frame length is 4.615 ms, and it can consist of

- 64 1/64 time slots of length 72 μs
- 16 1/16 time slots of length 288 μs

In the TDD mode, the same burst types as defined for the FDD mode are used. The 1/64 slot can be used for every service from low-rate speech and data to high-rate data services. The 1/16 slot is to be used for medium- to high-rate data services. Figure 7.13 also illustrates the dynamic allocation of resources between the reverse link and the forward link in the TDD mode.

The physical contents of the time slots are bursts of corresponding length. Three types of traffic bursts are defined. Each burst consists of a training sequence, two data blocks, and a guard period. The bursts differ in the length of the burst (72 μs and 288 μs) and in the length of the training sequence (27 symbols and 49 symbols), leading to different numbers of payload symbols and different multipath delay performances (Figure 7.14). The number of required reference symbols in

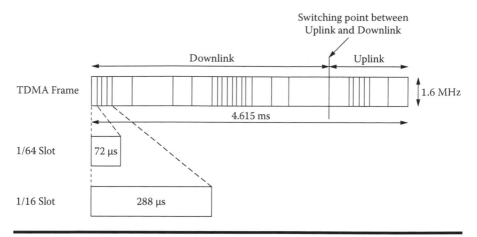

Figure 7.13 Wideband TDMA frame and slot structure.

the training sequence depends on the length of the channel's impulse response, the required signal-to-noise ratio, the expected maximum Doppler frequency shift, and the number of modulation levels. The number of reference symbols should be matched to the channel characteristics, remain practically stable within the correlation window, and have good correlation properties. All 136-based schemes can use interference cancellation as a means to improve performance [8]. For 136 HS (indoor), the longer sequence can handle about 7 μs of time dispersion and the shorter one 2.7 μs. It should be noted that if the time dispersion is larger, the drop in performance is slow and depends on the power delay profile.

7.6.4 Multirate Scheme

The UWC-136 multirate scheme is based on a variable slot, code, and modulation structure. Data rates up to 43.2 kbps can be offered using the 136C 30-kHz carrier

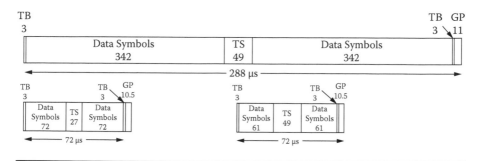

Figure 7.14 Burst structure.

and multislot transmission. Depending on the user requirements and channel conditions, a suitable combination of modulation, coding, and number of data slots is selected. The 136 HS scheme can offer packet-switched and both transparent and nontransparent circuit-switched data services. Asymmetrical data rates are provided by allocating a different number of time slots in the reverse and forward links. For packet-switched services, the radio link control (RLC)/MAC protocol provides fast medium access via a reservation-based medium-access scheme, supplemented by selective ARQ (automatic repeat request) for efficient retransmission.

Similar to 136 HS (outdoor/vehicular), the 136 HS (indoor) uses two modulation schemes and different coding schemes to provide variable data rates. In addition, two different slot sizes can be used. For delay-tolerant packet data services, error control is based on a type II hybrid ARQ scheme [5]. The basic idea is first to send all data blocks using a simple error control coding scheme. If decoding at the receiver fails, a retransmission is requested using a stronger code. After the second retransmission, diversity combining can be performed between the first and second transmissions prior to hard decisions. This kind of ARQ procedure can be used due to the ability of the RLC/MAC protocol to allocate resources quickly and to send transmission requests reliably in the feedback channel [5].

7.6.5 Radio Resource Management

The radio resource management schemes of UWC-136 include link adaptation, frequency hopping, power control, and dynamic channel allocation. Link adaptation offers a mechanism for choosing the best modulation and coding alternative according to channel and interference conditions. Frequency hopping averages interference and improves link performance against fast fading. For 136 HS (indoor), fast power control (frame by frame) could be used to improve the performance when frequency hopping cannot be applied (e.g., when only one carrier is available). Dynamic channel allocation can be used for channel assignments. However, when deployment with minimum spectrum is desired, reuse 1/3 and fractional loading with fixed channel allocation is used.

7.7 Time-Division Duplex

The main discussion about the IMT-2000 air interface has been concerned with technologies for FDD. However, there are several reasons why TDD would be desirable. First, there will likely be dedicated frequency bands for TDD within the identified UMTS frequency bands. Furthermore, FDD requires exclusive paired bands, and spectrum is, therefore, hard to find. With a proper design, including powerful forward error correction (FEC), TDD can be used even in outdoor cells. The second reason for using TDD is flexibility in radio resource allocation; that is, bandwidth can be allocated by changing the number of time slots for the reverse

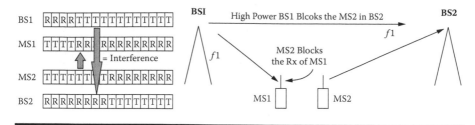

Figure 7.15 TDD interference scenario.

link and forward link. However, the asymmetric allocation of radio resources leads to two interference scenarios that will have an impact on the overall spectrum efficiency of a TDD scheme:

■ Asymmetric usage of TDD slots will have an impact on the radio resource in neighboring cells.
■ Asymmetric usage of TDD slots will lead to blocking of slots in adjacent carriers within their own cells.

Figure 7.15 depicts the first scenario. MS2 (mobile station) is transmitting at full power at the cell border. Since MS1 has a different asymmetric slot allocation than MS2, its forward link slots received at the sensitivity limit are interfered by MS1, which causes blocking. On the other hand, since the BS1 (base station) can have much higher EIRP (effective isotropically radiated power) than MS2, it will interfere with the ability of BS2 to receive MS2. Hence, the radio resource algorithm needs to avoid this situation.

In the second scenario, two mobiles would be connected into the same cell but using different frequencies. The base station receives MS1 on the frequency f1 using the same time slot it uses on the frequency f2 to transmit into MS2. As shown in Table 7.4,

Table 7.4 Adjacent Channel Interference Calculation

BTS transmission power for MS2 in forward link 1W	30 dBm
Received power for MS1	−100 dBm
Adjacent channel attenuation due to irreducible noise floor	50 to 70 dB
Signal to adjacent channel interference ratio	−60 to −80 dB

the transmission will block the reception due to the irreducible noise floor of the transmitter regardless of the frequency separation between f1 and f2.

Both TDMA- and CDMA-based schemes have been proposed for TDD. Most of the TDD aspects are common to TDMA- and CDMA-based air interfaces. However, in CDMA-based TDD systems the slot duration on the forward and reverse links must be equal to enable the use of soft handoff and prevent the interference situation described in the first scenario. Because TDMA systems do not have soft handoff on a common frequency, slot imbalances from one BS to the next are easier to accommodate. Thus, TDMA-based solutions have higher flexibility. The frame structure for the wideband TDMA for the TDD system was briefly discussed in this chapter. WCDMA has been proposed for TDD in Japan and Europe. The frame structure is the same as for the FDD component (i.e., a 10-ms frame split into 16 slots of 0.625 ms each). Each slot can be used for either reverse link or forward link. For cdma2000, the TDD frame structure is based on a 20-ms frame split into 16 slots of 1.25 ms each.

7.8 Conclusions

Third-generation cellular systems are a mechanism for evolving the telecommunications business based primarily on voice telephony to mobile wireless data communications. In light of events in TIA, ETSI, and ARIB, cdma2000, WCDMA, and EDGE are important technologies used to achieve the data communications goal. In comparing CDMA evolution, the European-, U.S.-, and Japanese-based systems have some similarities but differ in the CR and channel structure.

In the best circumstances, some harmonization will occur between cdma2000 and WCDMA, making deployment of hardware capable of supporting both systems easier. In TDMA, the third-generation paths of GSM and TIA/EIA-136 are through a common solution. This alignment will offer TDMA systems an advantage in possible global roaming for data services. In spite of the regional standards differences, the third generation will be the mechanism for achieving wireless multimedia-enabling services beyond the comprehension of second-generation systems.

Acknowledgments

We would like to thank Harri Holma, Pertti Lukander, and Antti Toskala from Nokia Telecommunications; George Fry, Kari Kalliojarvi, Riku Pirhonen, Rauno Ruismaki, and Zhigang Rong from Nokia Research Center; Kari Pehkonen from Nokia Mobile Phones; and Kari Pulli from the University of Stanford for helpful comments. In addition, contributions related to spectrum and modulation aspects from Harri Lilja from Nokia Mobile Phones are acknowledged.

References

1. *ITU-R M.1225, Guidelines for Evaluation of Radio Transmission Technologies for IMT-2000*, 1998.
2. Special issue on IMT-2000: Standards efforts of the ITU. *IEEE Pers. Commun.,* 4(4), August 1997.
3. Verdu, S., Minimum probability of error for asynchronous Gaussian multiple access. *IEEE Trans. IT,* IT-32(1), 85–96, January 1986.
4. Monk, A.M. et al., A noise-whitening approach to multiple access noise rejection—Part I: Theory and background. *IEEE J. Sel. Areas Commun.,* 12(5), 817–827, June 1997.
5. Nikula, E., Toskala, A., Dahlman, E., Girard, L., and Klein, A., FRAMES multiple access for UMTS and IMT-2000. *IEEE Pers. Commun.,* April 1998.
6. Tiedemann, E.G., Jr., Jou, Y-C., and Odenwalder, J.P., The evolution of IS-95 to a third generation system and to the IMT-2000 era. *Proc. ACTS Summit,* 924–929, October 1997.
7. Chang, C.R., Van, J.Z., and Yee, M.F., PN offset planning strategies for non-uniform CDMA networks. *Proc. VTC'97,* 3, 1543–1547, 1997.
8. Ranta, P., Lappeteläinen, A., and Honkasalo, Z-C., Interference cancellation by joint detection in random frequency hopping TDMA networks. *Proc. ICUPC96,* 1, 428–432, September/October 1996.

Chapter 8

Universal Mobile for Telecommunication Systems

Mohamad Assaad and Djamal Zeghlache

Contents

8.1 Introduction

To meet the requirements of the fast-growing demand for wireless services due to the integration of Internet, multimedia, and mobile communications, third-generation (3G) wireless networks are being developed under the International Telecommunication Union (ITU) initiative by the 3GPP and 3GPP2 (Third Generation Pattern Project). The 3GPP is a joint venture of several international standardization organizations from Europe (ETSI, European Telecommunications Standards Institute), Japan (Association for Radio Industry and Business, ARIB// Telecommunications Industry Association (TTC), the United States (T1P1), South Korea (TTA), and China (China Wireless Telecommunication Standard

Group). The 3GPP2 was created in parallel with the 3GGP project, with participation from ARIB, TTC, and CWTS. 3GPP2 focuses more on the Interim Standard 95 (IS-95) evolution.

Third-generation systems developed by 3GPP and 3GPP2 meet the IMT-2000 (International Mobile Telecommunications 2000) requirements, including higher data rate and spectrum efficiency than second-generation (2G) systems, support of both packet-switched (PS) and circuit-switched (CS) data transmission, wide range of services and applications, quality-of-service (QoS) differentiation, and flexible physical layer with variable-bit-rate capabilities to ease the introduction of new services. The most important IMT-2000 proposals are the Universal Mobile Telecommunications System (UMTS) and the CDMA2000 as successors, respectively, to GSM (Global System for Mobile Communications, originally Groupe Spécial Mobile) and IS-95 systems.

The UMTS meets the IMT-2000 requirements by supporting a wide range of symmetric/asymmetric services. Users can access traditional public switched telephone networks (PSTNs)/Integrated Services Digital Network (ISDN) services as well as emerging Internet Protocol (IP) data communications services, including Internet and multimedia applications, with unprecedented efficiency and flexibility in wireless communication systems. The wireless technique adopted for UMTS is code-division multiple access (CDMA).

The UMTS air interface, UMTS terrestrial radio access (UTRA), as introduced in 1998 by ETSI, supports two modes: UTRA FDD, based on wideband CDMA (WCDMA) for frequency-division duplex (FDD) operation, and UTRA TDD, based on time-division CDMA (TD-CDMA) for time-division duplex (TDD) operation. The UTRA FDD mode uses direct sequence CDMA (DSCDMA) technology, with a chip rate of 3.84 Mchips/s (mega chips per second), to operate in paired spectrum bands (5 MHz each for uplink and downlink). The use of WCDMA allows increased multipath diversity and results in signal quality and user bit rate improvements.

The bandwidths used for uplink and downlink are, respectively, 1920, 1980 and 2110, 2170 MHz. The TDD mode uses time-division multiple-access (TDMA) technology and time slot spreading for operation in the unpaired bands: 5 MHz shared dynamically between uplink and downlink. Mode selection depends on spectrum availability and type of coverage: symmetric or asymmetric. The bandwidths reserved for this mode are 1900, 1920 and 2010, 2025.

The evolution of the 3G system inside the 3GPP project has been organized and scheduled in phases and releases (99, 4, 5, and 6). The first release was the UMTS Release 99, introduced at the end of 1999. It supports high-speed services by providing data rates up to 2 Mbps depending on speed and coverage. This release defines and specifies seamless transitions from existing GSM networks to enable transparent intersystem (GSM/WCDMA and vice versa) handovers. This chapter focuses on Release 99 of the WCDMA FDD mode of UMTS.

In Release 4, completed in March 2001, improvements over Release 99 were added to the 3G standards, including the introduction in the TDD mode of a chip rate option of 1.28 Mchips/s to be used in addition to the initial 3.84-Mchips/s rate specified in Release 99. Enhancements in this release also allow the Packet Data Convergence Protocol (PDCP) layer (see Section 8.9) to support new IP header compression algorithms, in particular the protocol described in RFC 3095 [1]. On the core network side, the main improvement is the separation in the mobile switching center (MSC) of the user and control planes, respectively, into the media gateway (MGW) and MSC.

In Release 5, the main improvement is the development of the high-speed down-link packet access (HSDPA) to achieve higher aggregate bit rates on the downlink. HSDPA is based on the introduction of new enhancement techniques, such as fast scheduling, adaptive modulation and coding, and hybrid automatic repeat request (HARQ). HSDPA relies on a distributed architecture in which more intelligence is introduced in the node B (base station) to handle packet data processing thereby allowing faster scheduling and retransmission mechanisms.

In addition to HSDPA, Release 5 introduces the IP Multimedia Subsystem (IMS) to support IP-based transport and service creation.

To support multiple services at higher bit rates, Release 6 and beyond focus on the introduction of new features and enhancements, including [2,3]

- New transport channel in the uplink called the enhanced dedicated channel (E-DCH) to improve coverage and capacity to provide higher-bit-rate services
- Advanced antennas technologies, such as beam-forming and multiple-input multiple-output (MIMO)
- Introduction of new higher-bit-rate broadcast and multicast services (see Section 8.10)
- Addition of new features to the IMS
- Possibility of new frequency variant use (for WCDMA), such as use of the 2.5- and 1.7/2.1-GHz spectrum
- Improvements of GSM/EDGE Radio Access Network (GERAN) radio flexibility to support new services

The UMTS services and their QoS requirements are described in Section 8.2. The general architecture of a UMTS network is introduced in Section 8.3. The rest of this chapter focuses essentially on the Universal Terrestrial Radio Access Network (UTRAN) entities and protocols. Section 8.4 presents the radio interface protocol architecture used to handle data and signaling transport among the user, UTRAN, and the core network. Sections 8.5 to 8.11 describe these protocol layers. Section 8.12 provides a description of the automatic repeat request (ARQ) protocol used in the UMTS at the radio link control (RLC) layer. Power control and handover for the UMTS air interface are respectively described in Sections 8.13 and 8.14.

8.2 UMTS Services

UMTS is required to support a wide range of applications with different QoS requirements (bit rate, error rate, delay, jitter, etc.). These applications can be classified into person-to-person services, content-to-person services, and business connectivity [3]. Person-to-person services consist of peer-to-peer services between two or more subscribers. Depending on the QoS requirements and the type of application, these services can be offered either in the CS domain (e.g., speech, video telephony) or the PS domain (e.g., multimedia messages [MMS], push-to-talk, voice-over IP [VoIP], multiplayer games) of the UMTS network. Content-to-person services consist of content downloading, access to information, as well as broadcast/multicast at high rate. These services include Wireless Access Protocol (WAP), browsing, audio and video streaming, and multimedia broadcast/multicast services (MBMSs). Note that MBMS consists of sending audio, streaming, or file downloading services to all users or a group of users using specific UMTS radio channels (see Section 8.10).

Business connectivity consists of access to Internet or intranet using laptops via the UMTS radio interface. The effect of severe wireless conditions represents a key aspect to address and to assess overall performance. To guarantee QoS requirements of UMTS applications and services (using a remote host), several studies have analyzed interactions between layers and explored solutions to mitigate their negative effects.

In the 3GPP, services are classified into groups according to their QoS requirements, which define priorities between services and enable allocation of appropriate radio resources to each service. When cells are heavily loaded, the network blocks or does not accept (if new service arrives) services with low priority or services requiring more resources than those available. The UMTS system can also delay the transmission of data services that have low-delay sensitivity and allocate the resources to the services that should be transferred quickly (e.g., real-time services).

Four distinct QoS classes are defined in the 3GPP specifications [4]: (1) conversational, (2) streaming, (3) interactive, and (4) background. The conversational class has the most stringent QoS requirements, whereas the background class has the most flexible QoS requirements in terms of delay and throughput.

8.2.1 Conversational Class Applications

Conversational applications refer essentially to multimedia calls such as voice and video telephony. Applications and services such as NetMeeting, Intel VideoPhone, and multiplayer games are good examples of applications that map onto this QoS class [5].

Conversational applications are the most delay-sensitive applications since they carry real-time traffic flows. An insufficiently low transfer delay may result in service QoS degradations. The QoS requirements of this class depend essentially on the subjective human perception of received applications traffic (audio and video), as well as the performance of the used codec (i.e., audio/video source coding). Studies of the human perception of the audio/video traffic have shown that an end-to-end delay of 150 ms is acceptable for most user applications [6]. In the 3GGP specifications [7], it is stated that 400 ms is an acceptable delay limit in some cases, but it is preferable to use 150 ms as an end-to-end delay limit.

Concerning the time relation (variation) between information entities of the conversational stream (i.e., jitter), the audio/video codec does not tolerate any jitter. According to [7], the jitter should be less than 1 ms for audio traffic. Note that audio traffic is bursty, with silent periods between bursts depending on the codec, user behavior, and application nature (radio planning). Video traffic is transmitted regularly with variable packet lengths, and this results in burstiness, in particular when variable-bit-rate codecs are used. The bit error rate (BER) target (or limit) should vary between $5*10^{-2}$ and 10^{-6} depending on the application [4]. This corresponds to frame error rates (FERs) less than 3% for voice and 1% for video as advocated in [7].

These stringent delay requirements prevent the data link layer from using retransmission protocols. ARQ consists of retransmitting erroneous packets, but this results in additional delays that are unacceptable for conversational applications. Conversational traffic is consequently carried over the User Datagram Protocol (UDP) instead of the Transmission Control Protocol (TCP). The reliability of TCP is achieved via retransmissions and congestion control that consists of retransmitting packets (subject to errors or congestion). A dynamic transmission window is also used to regulate TCP traffic. These mechanisms result in additional delays. TCP is mostly used for non-real-time services, while UDP, which does not use any flow congestion control, is used for real-time and conversational traffic classes.

8.2.2 Streaming Class Applications

Streaming applications have fewer delay requirements than conversational services. The fundamental characteristic of this applications class is to maintain traffic jitter under a specific threshold. Jitter relates to the time relation between received packets. This threshold depends on the application, the bit rate, and the buffering capabilities at the receiver. The use of a buffer at the receiver smoothes traffic jitter and reduces the delay sensitivity of the application. Video sequences are managed by the client at the receiver, which plays the sequences back at a constant rate. A typical buffer capacity at the receiver is 5 s. This means that the application streams could not be delayed in the network more than 5 s. In the 3GPP specifications [7], it is indicated that the startup delay of streaming applications should be less than 10 s and the jitter less than 2 s. If the buffer is empty, due to jitter and delay latency, the application will pause until enough packets to play are stored in the buffer. Typical examples of streaming application software that is able to play audio and video streaming are RealPlayer and Windows Media Player.

The most suitable protocol stack to handle streaming services is RTP (Real-Time Protocol)/RTCP (Real-Time Control Protocol)/UDP since it can achieve low delays and low jitter. However, streaming is carried in certain cases over TCPs (e.g., network containing firewalls requiring use of TCP, nonlive streaming applications that are completely downloaded before being played).

The use of retransmission mechanisms is acceptable as long as the number of retransmissions and the overall delay are limited. The use of the HARQ mechanism in the medium-access control–high speed (MAC-hs) sublayer in HSDPA is acceptable. The ARQ protocol of the RLC sublayer, however, is not suitable for this class of applications due to the important delays introduced in the received information. The MAC-hs and RLC sublayers, as well as the ARQ and HARQ mechanisms, are described later in this chapter. Concerning the tolerated error rates at the receiver, the target or limit BER can vary between $5*10^{-2}$ and 10^{-6} depending on the application, as indicated in the 3GPP specifications [4].

8.2.3 *Interactive Class Applications*

The interactive class applications have lower delay requirements than conversational and streaming classes. This class consists of applications with request response patterns such as those of the Web and WAP browsing. In this class, applications are essentially of the server-to-person type; the user requests information and waits for a response from the server in a reasonably short delay. The quantity of transfer information on the downlink is more important than on the uplink in these cases. Because more delay latency can be tolerated, retransmission mechanisms such as ARQ protocol at MAC-hs or RLC sublayers can be used to achieve reliability (error-free reception) at the link layer with less radio resource consumption (more details on the use of ARQ in wireless system can be found in Section 8.12).

Web-browsing applications are basically conveyed over TCP, whereas WAP services are carried over UDP or a WAP-specific protocol called Wireless Data Protocol (WDP). Finally, interactive services can tolerate large delays and jitter, and a high-reliability, low-residual error rate is easily achieved. The residual error rate afforded via ARQ should be less than $6*10^{-8}$ and the delay latency less than 4 s per page [6,7].

8.2.4 *Background Class Applications*

The background class presents the most delay latency tolerance since the destination does not expect the data within a certain time. This application data can be sent in the background of other application classes. Typical examples of this class are e-mail, File Transfer Protocol (FTP), short messages (short message service, SMS), and MMS. The link carrying these services should present high reliability since these applications need to be received correctly (e.g., payload content

Table 8.1 Overview of Services Classes QoS Requirements [6,7]

Service Class	Conversational	Streaming	Interactive	Background
One-way delay	<150 ms (preferred) and < 400 ms (limit)	< 5 sec	< 4 sec/page	No limit
Bit error rate	Between $5*10^{-2}$ and 10^{-6}	Between $5*10^{-2}$ and 10^{-6}	Between $4*10^{-3}$ and $6*10^{-8}$	Between $4*10^{-3}$ and $6*10^{-8}$
Delay variation	< 1 ms	< 2 sec	N.A.	No limit
Use of ratransmision mechanism	not used	MAC-hs	MAC-hs, RLC	MAC-hs, RLC
Transport Layer	UDP	UDP and sometimes TCP	TCP (WDP for WAP)	TCP

must be preserved). The residual BER should be less than $6*10^{-8}$ [4]. Therefore, ARQ mechanisms at the MAC-hs and RLC sublayers can be used, as well as a reliable transport protocol such as TCP. An overview of QoS requirements of conversational, streaming, interactive, and background services is presented in Table 8.1 [6,7].

8.2.5 Quality of Service Parameters

In addition to the traffic classes, other QoS parameters have been defined in [5,8] to differentiate between services and to achieve the appropriate service quality of each application.

- *Maximum bit rate.*
- *Guaranteed bit rate.*
- *Maximum service data unit (SDU) size* (at the radio link control layer; see Section 8.8 for more details). This parameter is used in admission control and policing.
- *SDU error ratio* indicates the fraction of erroneous SDUs.
- *SDU format information* indicates all possible sizes of SDU. This information is used essentially in RLC transparent mode.
- *Delivery order* indicates whether the SDUs are delivered in sequence to the upper layers.

- *Residual bit error ratio* indicates the undetected BER in the delivered SDUs to the upper layers (i.e., beyond the RLC layer).
- *Delivery of erroneous SDUs,* which indicates whether erroneous SDUs at the RLC layer are delivered to the upper layer depending on the RLC mode used.
- *Discard timer* indicates the time after which an erroneous SDU is not retransmitted (see Section 8.8).
- *Transfer delay* defined in [5,8] as *the maximum delay for the 95th percentile of the distribution of delay of all delivered SDUs during the life time of the connection.*
- *Allocation/retention priority* indicates the priority or the relative importance of a UMTS connection compared to other connections. This information is used during admission control and resource allocation procedures.
- *Source traffic characteristics.*

8.3 General Architecture

The general architecture of a UMTS network is shown in Figure 8.1 [9–11]. Network elements in this architecture can be grouped into three domains: user equipment (UE) domain, UTRAN domain, and core network domain. Domains and entities in each domain are separated by reference points serving as interfaces [3].

This architecture has been conceived to vehicle and manage CS and PS traffic. Thus, the external networks can be divided into two groups: (1) CS networks, such as PSTN and ISDN; and (2) PS networks, such as Internet and X25.

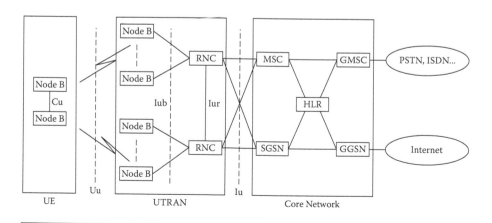

Figure 8.1   General architecture of UMTS network.

8.3.1 User Equipment Domain

User equipment is a device allowing a user access to network services [10]. This domain can be divided into two parts or subdomains called the mobile equipment domain and user services identity module domain (USIM), separated by the Cu interface. The UE domain is connected to the UTRAN by the Uu interface [9].

The mobile equipment is the terminal that performs radio transmissions and related functions. The USIM is the smart card that contains the user identity and subscription information regardless of the terminal used. This device holds user profile data, procedures, and authentication algorithms that allow secure user identification.

8.3.2. UTRAN Domain

The access network domain provides to the UE the radio resources and the mechanisms necessary to access the core network domain. The UTRAN contains entities that control the functions related to mobility and network access. They also allocate or release connections (radio bearers). The UTRAN consists of radio subsystems radio network subscriber (RNS) connected to the core network by the Iu interface. Each RNS includes one radio network controller (RNC) and one or more node Bs [9,12].

The node B is in reality the base station. It is the entity that allocates and releases radio channels, partially manages the radio resources, controls the transmission power in the downlink, and converts the data flow between Iu and Uu interfaces. In Release 99, the node B contains procedures that manage only the physical layer, such as coding, spreading, transmission and detection, and physical layer signaling [12]. In Releases 5 and 6, intelligence is introduced in the node B, which is then able to perform MAC functions, especially scheduling and HARQ.

The RNC is the main element and the intelligent part of the RNS. The RNC controls the use and the reliability of the radio resources. It performs the functions of the MAC/RLC layer and terminates the Radio Resource Control (RRC) Protocol. In 3GPP [12], three types of RNC have been specified: SRNC (serving RNC), DRNC (drift RNC), and CRNC (controlling RNC). The SRNC is the entity that holds the RRC connections with the UE. This entity is the point of connection between the UTRAN and the core network and is involved in the user mobility management within the UTRAN.

When a UE moves in the connection state from a cell managed by the SRNC to another associated with a different RNS, the RNC of the new cell is called DRNC. The RRC connection is still handled by the SRNC. In this case, the DRNC serves as a simple relay to forward information between the SRNC and the UE.

The CRNC is the RNC performing the control and the configuration of a node B. This entity holds the responsibility of load control in its own cells as well as the admission control of and code allocation to new users accessing the system.

8.3.3 Core Network Domain

The core network domain is basically inherited from the General Pocket Radio Services (GPRS) network architecture according to a transition phase, from GPRS to UMTS networks, specified in 3GPP. The core network consists of physical network entities integrating both CS and PS domains. The CN domain provides various support functions for services traffic conveyed over the UMTS system. The services correspond to management of user location formation, control of network features, and transfer mechanisms for signaling [9]. The core network includes switching functions for CS services via the MSC and the gateway MSC. The home location register (HLR) and the visitor location register (VLR) are the databases responsible for handling, respectively, user subscriptions and terminals visiting various locations. To manage packet data services, the packet domain relies on the serving GPRS support node (SGSN) and gateway GPRS support node (GGSN), which serve, respectively, as routers and gateways.

The SGSN and the GGSN are involved in the management of session establishment (i.e., packet data protocol contexts) and in the mobility of data services. In certain cases, mobility management is achieved jointly by CS and PS domain via the cooperation of the SGSN, GGSN, MSC, HLR, and VLR via dedicated interfaces fully described in 3GPP specifications.

The HLR is a database that handles maintenance of user subscription data and profiles. This information is transferred to the adequate VLR or the SGSN to achieve location and mobility management. In addition, the HLR provides routing information for mobile calls and SMS.

The VLR is involved in user location updates in the CS domain (functions inherited from the GSM architecture). It contains subscriber information required for call and mobility management of a subscriber visiting the VLR area. The MSC is a switch used mainly for voice and SMS. It is involved, with the PSTN, in the establishment of end-to-end CS connections via Signaling System 7 (SS7). In addition, it is coupled to the VLR to achieve mobility management.

The gateway MSC provides switching for CS services between the core network and external CS networks and is involved in international calls. The SGSN plays in the packet domain a similar role to the MSC/VLR in the CS domain. It handles location and mobility management, by updating routing area, and performs security functions and access control over the packet domain.

The GGSN serves as an edge router in the core network to convey data between the UMTS network and external packet networks (e.g., Internet). In other words, it has in the packet domain the same role that the Gateway Mobile Switching Center (GMSC) does in the CS domain. The GGSN is involved in packet data management, including session establishment, mobility management, and billing (accounting). In addition, the GGSN includes firewall and filtering of data entering the core network to protect the UMTS network from external packet data networks (e.g., Internet).

8.3.4 Interfaces

The interfaces between the logical network elements of the UMTS architecture are defined in the 3GPP specifications [9,10,13–15]. The main interfaces are Cu, Uu, Iu, Iur, and Iub.

The Cu interface is the electrical interface between the USIM and the mobile equipment. Uu is the WCDMA air interface described separately in this chapter.

8.3.4.1 Iu Interface

The interface between the UTRAN and the core network, called Iu and described in [13], is involved in several functions handling control and signaling of mobile calls. These functions include establishment, management, and release of radio-access bearers (connections). The Iu interface achieves the connection between the UTRAN and the CS and PS core network domains. This interface allows the transfer of signaling messages between users and core network and supports cell broadcast service via the Iu broadcast. In addition, the Iu interface supports location services by controlling location reporting in the RNC and by allowing transfer of the location report (including geographical area identifier or global coordinates) to the core network.

The Iu interface is also involved in relocation of the serving radio network subsystem (SRNS), as well as intra- and intersystem handover management. The SRNS relocation is caused in general by a hard handover at the Uu interface that generates a change of radio resources at this interface. Another target for the relocation is to keep all radio links in the same DRNC. In this case, radio resources are not altered, and data flows remain uninterrupted.

8.3.4.2 Iur Interface

The Iur interface, used between RNCs and described in [14], provides the capability to support essentially four distinct functions: (1) inter-RNC mobility, (2) radio resource management, (3) dedicated channel traffic (mobility and management), and (4) common channel traffic (mobility and management). Note that the definitions of dedicated and common channels are given in Section 8.5.

The inter-RNC mobility function allows the support of radio interface mobility between RNSs, including SRNS relocation, packet error reporting, and inter-RNC registration area update. This function does not support any user data traffic exchange between RNCs. The radio resource management function, introduced in the subsequent Releases 5 and 6 for Iur optimization purposes [3], provides transfer of signaling information between RNCs, including node B timing, positioning parameters, and cell information and measurements.

The Iur interface is involved in the mobility management of the dedicated and shared channel traffic. In the case of handover, this interface provides the capability of supporting establishment and release of the dedicated channel in the DRNC of the new cell area. It provides support of the radio links management in the DRNC by allowing the exchange of measurement reports and power control settings. In addition, it achieves transfer of the dedicated channel transport blocks between serving and drift RNCs.

In addition to the dedicated and shared channels, the Iur interface supports the mobility management of the common channel traffic, including setup and release of the transport connection, as well as flow control handling between RNCs. This function has the drawback of introducing more complexity in the Iur interface and generating inefficiency in resource utilization [3] by splitting the MAC layer into two entities, MAC-d and MAC-c located, respectively, in the SRNC and DRNC.

8.3.4.3 Iub Interface

The Iub interface [15], used between the RNC and node B, supports location-handling service and mobility management. It is involved in the admission control of mobile users, resource allocation (e.g., outer-loop power control), connection establishment, and release as well as handover management. This interface also allows control of the radio equipment and radio-frequency (RF) allocation in node B.

Other interfaces are described in the 3GPP specifications [10]. These interfaces include those between MSCs, VLRs, HLR and MSC; HLR and VLR; MSC and GMSC. Interfaces within the packet domain and between the packet and circuit domains are also readily available.

8.4 UTRAN Protocol Architecture

To manage and handle mobile calls, data transfer, and signaling information between users and the core network (across the different interfaces described), the UTRAN is comprised of nodes that handle the information (data and control) at different protocol layers. Figure 8.2 depicts the UTRAN protocol architecture, which consists of several protocol layers visible in UTRAN [16]:

1. Physical layer
2. Data link layer, which contains
   - MAC
   - RLC
   - PDCP
   - Broadcast/multicast control (BMC)
3. RRC at the network layer

The UTRAN distinguishes two message planes: (1) user plane to convey user messages and (2) control plane to manage signaling and control messages.

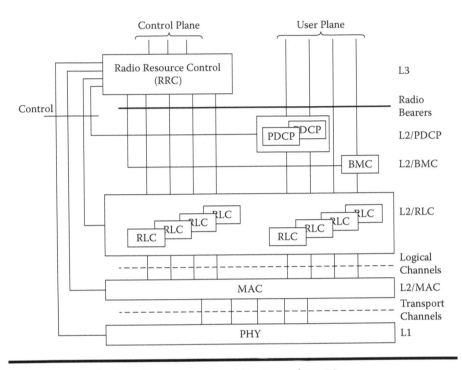

Figure 8.2   Radio interface protocol architecture of UMTS.

In the control plane, the data link layer (layer 2) is split into two sublayers: MAC and RLC. In the user plane, layer 2 contains, in addition to MAC and RLC, two sublayers: PDCP and BMC. Layer 3 (network layer), visible in the UTRAN, contains only one sublayer, called the RRC, located in the control plane. Note that call control, session management (SM), and mobility management are transparent to UTRAN and thus not included in the radio interface protocol architecture.

An overview of the UTRAN protocol sublayers is presented in Sections 8.5 to 8.11. Particular attention is given to MAC and RLC layers, which interact with the TCP/IP stack and affect system efficiency and performance.

8.5 UMTS Channels

To carry and manage several traffic types over the air interface, the 3GPP specifications define several channels, each having a specific role in establishing and maintaining sessions in the UMTS access network. These channels can be divided into three groups: logical channels, transport channels, and physical channels.

8.5.1 Logical Channels

A logical channel is defined according to the type of information it transports. One can distinguish two classes of logical channels: those for control and those for traffic.

8.5.1.1 Logical Control Channels

Logical control channels are used for the transfer of information in the user plane. The various logical control channels are [16,17]:

- Broadcast control channel (BCCH), used on the downlink to broadcast system and network information in all the cells.
- Paging control channel (PCCH), used on the downlink to carry paging information for mobile-terminated calls and sessions.
- Common control channel (CCCH), used on both uplink and downlink to transport signaling information to all users.
- Dedicated control channel (DCCH), used on both uplink and downlink to transport signaling information between the UTRAN. This channel is a spreading code dedicated to a user.
- Shared control channel (SHCCH), used on both uplink and downlink in TDD mode only to transmit control information between UTRAN and mobile stations.
- MBMS point-to-multipoint control channel (MCCH) used on the downlink to carry MBMS control information from the UTRAN to the UE (user equipment). (MBMS is described in Section 8.10.)
- MBMS point-to-multipoint scheduling channel (MSCH) used on the downlink to carry MBMS scheduling control information for one or more MTCHs (MBMS point-to-multipoint traffic channels). Logical traffic channels are used for the transfer of information in the user plane. The three types of specified traffic channels are [16,17]
  - Dedicated traffic channel (DTCH), used on both uplink and downlink for the data transmission between the UTRAN and a dedicated user
  - Common traffic channel (CTCH), used for the transport of messages to all cell users
  - MTCH, used on the downlink to carry MBMS traffic data from the network to the UE

8.5.2 Transport Channels

Transport channels are services offered by layer 1 to the higher layers [18,19]. The transport channel is unidirectional (i.e., uplink or downlink) and consists of the characteristics required for the data transfer over the radio interface. For example, the size of the transport block (to transport a data unit of the MAC layer) is one of the transport channel characteristics. Note that the corresponding period to transmit a transport block is known as the transmit time interval (TTI).

In Release 99, the TTI can take the values of 10, 20, 40, or 80 ms. For voice services, the TTI is fixed at 10 ms, whereas for data services it changes according to the service used. In Release 5, the TTI of the high-speed downlink shared channel (HS-DSCH) has been reduced to 2 ms (for data services) to introduce finer-grain control and scheduling in the system.

Transport channels are classified in three groups: dedicated channels, common channels, and shared channels. The mapping of the logical channels onto the transport channels is depicted in Figure 8.3 [16,17].

The dedicated channel (DCH) is a point-to-point channel used on both uplink and downlink to carry data and control information (from higher layer) between the UTRAN and a specific (dedicated) user. The nature of transmitted information is transparent to the physical layer, which carries user data and control information in the same way. However, the physical layer parameters established by the UTRAN

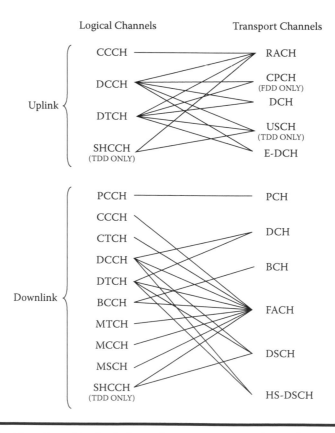

Figure 8.3 Mapping of logical channels onto transport channels.

depend on the nature of the transmitted information on the channel (i.e., data or control). Note that the dedicated channel can be transmitted over the entire cell or over only a part of the cell. The DCH can support soft handover and fast power control.

In Release 6, enhancements are introduced in the UMTS standard to achieve a higher data rate on the uplink. A new transport channel is specified for this purpose to support higher-order modulation and HARQ techniques. This channel, called the enhanced dedicated channel (E-DCH), is subject to fast power control and scheduling controlled by node B. More details on this channel can be found in [16].

The specified shared channels in UMTS UTRA FDD mode are DSCH and HS-DSCH. The downlink shared channel (DSCH) is a downlink transport channel shared by several users. The DSCH is consequently associated with one or several downlink dedicated channels when shared by multiple users. The DSCH is transmitted over the entire cell or over only a part of the cell using, for example, beam-forming antennas. The HS-DSCH is a downlink transport channel shared by several UE and is associated with one downlink dedicated physical channel (DPCH) and one or several HS-SCCH.

In the TDD mode, the 3GPP specifies [16] another shared transport channel to be used on the uplink. This channel, called the uplink shared channel (USCH), is shared by UE carrying dedicated control or traffic data.

A common channel is a unidirectional point-to-multipoint channel used on either uplink or downlink to transfer information between the UTRAN and one or more users. The main common channels specified in [16] are:

- The broadcast channel (BCH) is a downlink transport channel used to broadcast system and cell-specific information over the entire cell. This channel always has a single transport format independent of the radio environment and type of information.
- The random-access channel (RACH) is an uplink transport channel received from users in the entire cell. It contains control information that allows the mobile to access the network via connection or channel requests. The transmission power of this channel is estimated using open-loop power control. Users attempting initial access to network resources via the RACH rely on received downlink channels (especially pilot channels or the BCH) to estimate the required amount of power needed for RACH transmission. The RACH is shared by users using a carrier sense multiple access/collision avoidance (CSMA/CA) technique to manage collisions and contention over the radio interface. In addition to the random access, the RACH could also be used for transmitting very short packets in RLC unacknowledged mode (UM).
- The forward access channel (FACH) is a downlink point-to-point transport channel. The FACH is transmitted over the entire cell and is used to carry signaling information to the user. For example, the FACH is used to carry

access grant messages in response to channel requests or random-access messages received from users (that want to access the network) on the RACH.

■ The paging channel (PCH) is a downlink transport channel transmitting paging control information toward mobile stations when the system does not know the precise user location. It is typically transmitted in several cells to locate the user. The transmission of the PCH is associated with the transmission of the physical paging indicator channel (PICH), allowing the support efficient sleep-mode procedures. The paging messages awaken terminals from sleep mode.

■ The common packet channel (CPCH) is an uplink random-access transport channel similar to the RACH. The main difference between these two channels is that the CPCH is used only in connected mode. The CPCH was conceived to transfer larger-size packets, whereas the RACH is mostly used for random access or short packets in RLC UM mode. The CPCH is associated with a dedicated channel on the downlink that provides power control and CPCH control commands (e.g., emergency stop) for the uplink CPCH. The CPCH is also characterized by initial collision and contention. It supports inner-loop power control (fast power control) as opposed to the RACH, which relies only on the open-loop power control to transmit the random-access request.

8.5.3 *Physical Channels*

To carry the information contained in the transport channels, the physical channels adapt these transport channels (in terms of coding and flows) to the physical layer procedures.

The physical channels consist of frames and slots with a basic radio frame period of 10 ms consisting of 15 slots. The number of bits per slot depends on the physical channel used (spreading factor [SF], modulation, coding rate, etc.). Each physical channel has a specific structure and specific SF according to the number of bits transmitted on the channel and importance of the information (degree of protection, coding, spreading, etc.).

The main physical channels, specified in Release 99 [18], are the synchronization channel (SCH), the common pilot channel (CPICH), the acquisition indication channel (AICH), the CPCH status indication channel (CSICH), the PICH, the collision detection/channel assignment indication channel (CD/CA-ICH), the dedicated physical data channel (DPDCH), the dedicated physical control channel (DPCCH), the physical random-access channel (PRACH), the primary common control physical channel (PCCPCH), the secondary common control physical channel (SCCPCH), the physical downlink shared channel (PDSCH), and the physical common packet channel (PCPCH).

In Release 5, three physical channels have been added to the 3GPP specifications: the high-speed physical downlink shared channel (HS-PDSCH) intended to carry the HS-DSCH transport channel and two other associated channels intended to carry the related physical control information, the high-speed shared control channel (HS-SCCH) and the high-speed dedicated physical control channel (HS-DPCCH).

In Release 6, the E-DCH dedicated physical data channel (E-DPDCH) has been introduced to carry the E-DCH transport channel. The physical-related control information of this channel is carried on the E-DCH dedicated physical control channel (E-DPCCH), the E-DCH absolute grant channel (E-AGCH), the E-DCH relative grant channel (E-RGCH), and the E-DCH hybrid ARQ indicator channel (E-HICH). In addition, a new physical channel, called MBMS notification indicator channel (MICH), has been specified to support the introduction of new high-rate MBMS. The MICH is always associated with the S-CCPCH to which the FACH transport channel is mapped.

Figure 8.4 illustrates the mapping of the transport channels onto the physical channels [18]. This figure shows that only P-CCPCH, S-CCPCH, DPDCH,

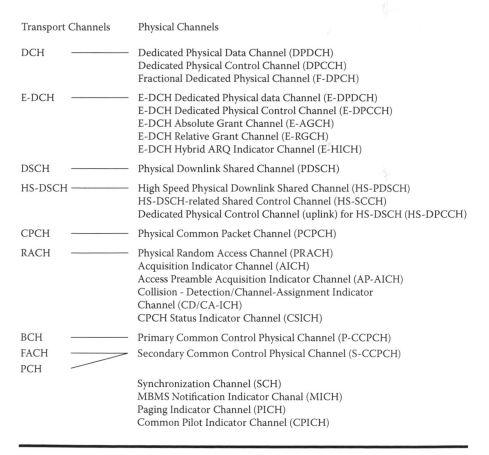

Transport Channels   Physical Channels

DCH ——————— Dedicated Physical Data Channel (DPDCH)
            Dedicated Physical Control Channel (DPCCH)
            Fractional Dedicated Physical Channel (F-DPCH)

E-DCH ——————— E-DCH Dedicated Physical data Channel (E-DPDCH)
             E-DCH Dedicated Physical Control Channel (E-DPCCH)
             E-DCH Absolute Grant Channel (E-AGCH)
             E-DCH Relative Grant Channel (E-RGCH)
             E-DCH Hybrid ARQ Indicator Channel (E-HICH)

DSCH ——————— Physical Downlink Shared Channel (PDSCH)

HS-DSCH ——————— High Speed Physical Downlink Shared Channel (HS-PDSCH)
              HS-DSCH-related Shared Control Channel (HS-SCCH)
              Dedicated Physical Control Channel (uplink) for HS-DSCH (HS-DPCCH)

CPCH ——————— Physical Common Packet Channel (PCPCH)

RACH ——————— Physical Random Access Channel (PRACH)
            Acquisition Indicator Channel (AICH)
            Access Preamble Acquisition Indicator Channel (AP-AICH)
            Collision - Detection/Channel-Assignment Indicator
            Channel (CD/CA-ICH)
            CPCH Status Indicator Channel (CSICH)

BCH ——————— Primary Common Control Physical Channel (P-CCPCH)
FACH ——————— Secondary Common Control Physical Channel (S-CCPCH)
PCH

            Synchronization Channel (SCH)
            MBMS Notification Indicator Chanal (MICH)
            Paging Indicator Channel (PICH)
            Common Pilot Indicator Channel (CPICH)

Figure 8.4   Mapping of transport channels onto physical channels.

PDSCH, HS-PDSCH, E-DPDCH, PCPCH, and PRACH can carry higher-layer information (data or signaling) since they correspond to transport channels. The other physical channels do not correspond to any transport channel. These channels are intended to carry control information related to the physical layer only. They are unknown and completely transparent to higher layers. They are exclusively used by the physical layer to harness and control the radio link.

The channels used usually for data transmission are the DPDCH, the HS-PDSCH, and the E-DPDCH. The PCPCH and the S-CCPCH can transmit data in certain cases (e.g., broadcast/multicast services on the S-CCPCH). Detailed structures of all UMTS channels are presented in [18].

8.5.3.1 Dedicated Physical Channel

The DPCH consists of two channels: the DPDCH and the DPCCH. The DPDCH is used to carry the dedicated channel transport channel.

The DPCCH is used to carry control information generated at layer 1. The layer 1 control information consists of known pilot bits to support channel estimation for coherent detection, transmit power control (TPC) commands, feedback information (FBI), and an optional transport-format combination indicator (TFCI). The TFCI informs the receiver about the instantaneous transport format combination of the transport channels mapped to the simultaneously transmitted DPDCH radio frame.

In the uplink, the DPCCH and the DPDCH are transmitted in parallel using one spreading code for each. However, in the downlink these two channels are multiplexed in time and are transmitted using the same spreading code.

Figures 8.5 and 8.6 show the frame structure of the DPDCH and the DPCCH, in the uplink and the downlink respectively, [18]. Each radio frame of

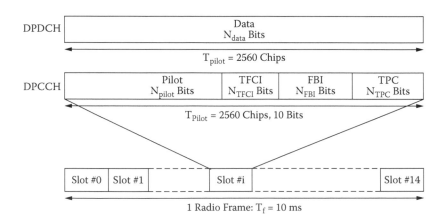

Figure 8.5 Frame structure of the uplink DPCH channel.

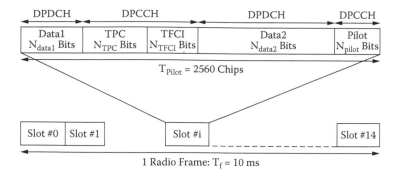

Figure 8.6 Frame structure of the downlink DPCH channel.

10 ms is split into 15 slots, each of length $T_{slot} = 2560$ chips because the spreading chip rate is 3.84 Mcps. This physical channel supports inner-loop power control at a frequency of 1.5 KHz that corresponds to a fast power control period of one slot. The DPDCH spreading factor varies from 256 to 4 according to the application data rate. The spreading factor of the DPCCH (control) is always equal to 256, resulting in 2560 chips per DPCCH slot (i.e., there are 10 bits per DPCCH slot). More details on these two channels and other physical channels can be found in [18].

8.6 Physical Layer

The physical layer of the UTRAN, in particular the Uu interface, relies on WCDMA. The physical layer offers services to the upper layer (MAC layer) via the transport channels. The services offered by the physical layer include multiplexing/demultiplexing of transport channels and mapping of coded composite transport channels onto physical channels [19,20]. The physical layer handles in addition encoding/decoding of transport channels using a forward error-correcting (FEC) code to protect the information transmitted over the radio interface against errors. Even though FEC is used, the received signals still contain errors due to severe radio conditions in general. The erroneous packets are retransmitted at the RLC layer using the ARQ protocol described in Section 8.12.

To achieve a target QoS, that is, target BER or target signal-to-noise ratio (SNR), the physical layer supports measurements of received signal quality via a pure physical channel called the CPICH. The channel quality is measured in terms of FER, signal-to-interference ratio (SIR), interference power, transmission power, and so on. Based on these measurements, the physical layer supports closed-loop power control that adapts the transmission power of physical channels to the short-term radio channel variations to achieve a given target QoS for the transmitted information over the radio interface (more details are provided in Section 8.13).

The physical layer also supports soft handover and macrodiversity combining over dedicated physical channels. This is not applied on shared channels. By combining two or more signals, containing the same information and transmitted via two or more node Bs, channel diversity is achieved and signal quality improved at the receiver. The soft handover allows uninterrupted transmission of mobile calls and data sessions, which in turn improves the QoS during handover (see Section 8.14).

In addition to services described, the physical layer performs other basic operations, including [19,20]:

■ RF processing
■ Rate matching
■ Modulation and spreading of physical channels
■ Demodulation and despreading of physical channels
■ Frequency and time synchronization (chip, bit, slot, frame)
■ Power weighting and combining of physical channels
■ Synchronization and timing advance on uplink channels (TDD only)

8.7 Medium-Access Control

The MAC sublayer provides services to the RLC sublayer via logical channels (control and traffic) and coordinates access to the physical layer by mapping these logical channels onto the transport channels. Information at the RLC sublayer, bundled into packet data units (PDUs), is multiplexed by the MAC sublayer into transport blocks and delivered to the physical layer. This multiplexing function allows the mapping of several RLC instances onto the same transport channel; in other words, it supports the multiplexing of several logical channels into the same transport channel as shown in Section 8.5. Note that on the receiver side, the MAC transport blocks delivered from the physical layer are demultiplexed into RLC PDUs, using the multiplexing identification contained in the MAC protocol control information.

During this mapping, the MAC sublayer performs control of transport formats by assigning the appropriate format for each transport channel depending on the instantaneous source rate to achieve efficient use of transport resources. When several logical channels belonging to different users are transported by the same common channel (e.g., FACH, RACH), the UE identification (i.e., cell radio network temporary identity [C-RNTI] or UTRAN radio network temporary identity [U-RNTI]) present in the MAC header is used to identify the logical channels of each user on the receiver side.

Priorities between different data flows of one user or between different users sent over common, shared, and E-DCH transport channels can be handled also by the MAC sublayer. Priority between data flows of the same user can be performed by assigning adequate transport formats to each flow so that high-priority flows can be transferred over layer 1 with high bit rate and low priority flows with low bit rate. On the uplink, RACH and CPCH resources (i.e., access slots and pre-amble signatures for UMTS FDD) are organized in different access service classes (ASCs) (up to eight ASCs are specified in [16,17]). The MAC sublayer is respon-sible for applying and indicating to the physical layer the ASC partition associated to a given MAC PDU. This function provides different priorities on RACH and CPCH [16].

The MAC sublayer is also involved in traffic measurement and monitoring. The amount of RLC PDUs corresponding to a given transport channel is compared to a threshold specified by the RRC layer. The traffic volume measurement is reported to the RRC layer to handle reconfiguration or transport channel-switching deci-sions. Ciphering for transparent RLC mode (see the next section for TM definition) and execution of the switching between common and dedicated transport channels (decided by RRC) are also performed by the MAC sublayer.

In Release 5, a new MAC entity, called the MAC-hs and located in node B, has been introduced in the 3GPP specifications. This entity is responsible for handling and managing the HARQ mechanism introduced in HSDPA. The MAC-hs entity is responsible for assembling, disassembling, and reordering higher-layer PDUs. The PDUs are delivered in sequence to higher layers.

8.7.1 MAC Architecture

To handle the functions described previously, the MAC layer is divided into the following domains or entities [17]:

- MAC-b is the entity that handles the BCH channel. There is only one MAC-b in each UE and one in the UTRAN (node B) as specified by the 3GPP.
- MAC-d is the entity that handles the DCH channel. This entity is specific to each user. In the UTRAN, this entity is located in the SRNC. Note that ciphering is performed by this entity.
- MAC-c/sh/m is the entity that handles the FACH, PCH, RACH, CPCH, DSCH (TDD only), and USCH (TDD only), including ASC selection, transport formats selection, scheduling/priority handling, and so on. There is one MAC-c/sh/m located in the UE and one in the UTRAN (located i the CRNC). Note that in Release 99 this entity was named MAC-c/sh Release 6, it is named MAC-c/sh/m since it is involved in the MBMS se (see Section 8.10) by multiplexing and reading the MBMS Id (which to distinguish between MBMS services).
- MAC-hs is the entity that handles the HS-DSCH channel Release 5. This entity, located in node B and in the UE, is HARQ functionality, transport format selection, and sched

■ MAC-m is the entity that controls access to the FACH channel when it is used to carry MTCH and MSCH logical channels. This entity is added to the 3GPP specifications in Release 6. It exists only in the UE side of the MAC architecture in the case of selective combining of MTCH channels from multiple cells.

■ MAC-e/es are the entities that handle the E-DCH channel. These entities are introduced by the 3GPP in Release 6.

The general MAC architectures on the UE and the UTRAN sides are, respectively, depicted in Figures 8.7 and 8.8 [17].

8.7.2 Protocol Data Unit

Peer-to-peer communication is achieved by the exchange of PDUs. The MAC PDU presented in Figure 8.9 [17], consists of a MAC header and a MAC SDU both of variable size. The content and the size of the MAC header depend on the type of the logical channel, and in some cases none of the parameters in the MAC header is needed. The size of the MAC SDU depends on the size of the RLC PDU, which is defined during the setup procedure.

In the 3GPP specifications [17], the MAC PDU structure is described in detail for all transport channels. The MAC-d PDUs of HS-DSCH and E-DCH are similar to the MAC PDU described in this section.

Only, the MAC-hs and MAC-e/es PDUs are distinct when HS-DSCH and E-DCH are used. The MAC PDU header optionally contains the following fields [17]:

■ The target channel type field (TCTF) is a flag that identifies logical channels such as BCCH, CCCH, CTCH, SHCCH, MCCH, MTCH, MSCH carried on FACH, USCH (TDD only), DSCH (TDD only), and RACH transport channels. Note that the size of the TCTF field of FACH for FDD is 2, 4, or 8 bits.

■ The C/T field is used to identify the logical channel instance and type carried on dedicated transport channels and potentially on the FACH and RACH channels (only when they are used for user data transmission). The identification of the logical channel instances is mandatory when multiple logical channels are carried on the same transport channel. The size of the C/T field is fixed to four bits for both common transport channels and dedicated transport channels.

■ The UE identity type field provides identification of the UE on common transport channels. Two types of UE identities are defined by the 3GPP: First, U-RNTI is used only in the downlink direction (never in the uplink) when DCCH logical channel of RLC UM is mapped onto common transport channels. The RLC UM mode is described in Section 8.8. Second, the C-RNTI is used for DTCH and DCCH in uplink, DTCH (and maybe DCCH) in downlink, when mapped onto common transport channels.

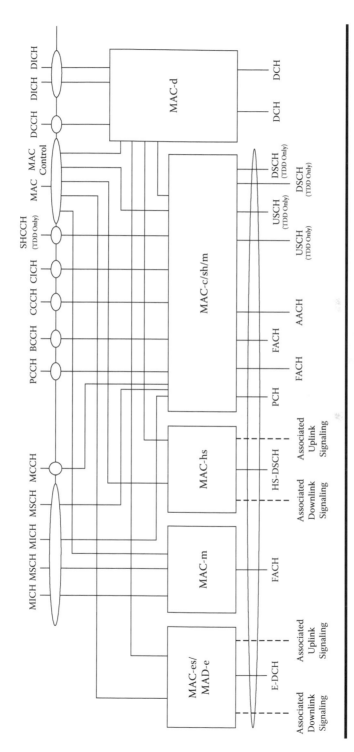

Figure 8.7 General MAC architecture of the UMTS Release 6 on the UE side.

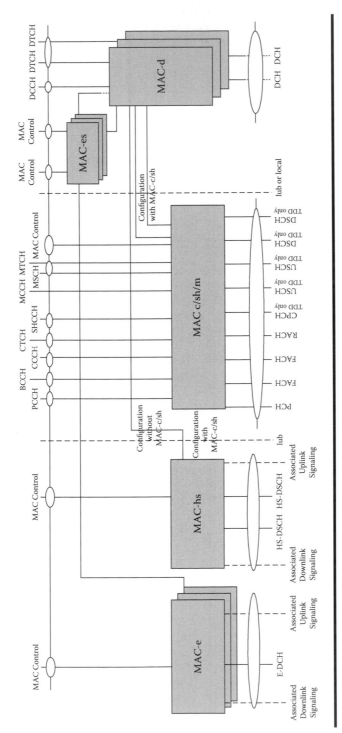

Figure 8.8 General MAC architecture of the UMTS Release 6 on the UTRAN side.

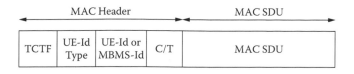

Figure 8.9   Structure of a MAC protocol data unit (PDU).

- The UE identity type field helps the receiver to correctly decode the UE identity in MAC headers.
- The MBMS identity, added to the specifications in Release 6, is used only in the downlink to provide identification of MTCH for an MBMS service carried on the FACH transport channel.

As we have indicated, these MAC header fields are used optionally, that is, depending on the logical channel and the transport channel on which this logical channel is mapped. To explain this further, Figure 8.10 provides an example of a MAC PDU header when DTCH or DCCH is mapped onto dedicated, common, or shared transport channels (except E-DCH and HS-DSCH) [17]. In this figure, five cases are considered:

1. DTCH or DCCH is mapped onto DCH. In this case, none of the MAC header field is used since no multiplexing of dedicated channels at the MAC sublayer is considered.
2. DTCH or DCCH is mapped onto DCH. In this case, the C/T field is included in the MAC header since multiplexing of dedicated channels on MAC is considered.
3. DTCH or DCCH is mapped onto RACH/FACH. In this case, the fields TCTF, C/T, UE identity type, and UE identity are included in the MAC header. For FACH, the UE identity field used can be either the C-RNTI or the U-RNTI, whereas for RACH only the C-RNTI is used.

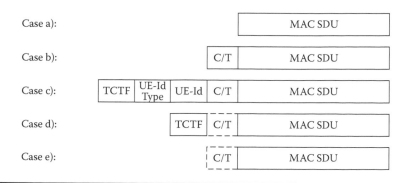

Figure 8.10   Example of a MAC PDU header structure.

4. DTCH or DCCH is mapped onto DSCH or USCH for UMTS TDD only. In this case, the TCTF field is included in the MAC header. In addition, the C/T field may be used if multiplexing on MAC is applied. Note that this case exists in TDD mode only (since DSCH and USCH are used for TDD mode).

5. DTCH or DCCH is mapped onto DSCH or USCH where DTCH or DCCH are the only logical channels. Only the C/T field may be included in the MAC header if multiplexing on MAC is applied. Note that this case exists in TDD mode only (since DSCH and USCH are used for TDD mode).

8.8 Radio Link Control

The RLC [21] sublayer provides radio link services for use between the UE and the network. The RLC sublayer contains discrete RLC entities required to create radio bearers. For each radio bearer, the RLC instance is configured by RRC [22] to operate in one of three modes: TM, UM, or acknowledged mode (AM). The modes are used according to the application characteristics.

The upper-layer PDUs are delivered to the RLC sublayer and fit into RLC SDUs. Each RLC SDU is then segmented in one or more RLC PDUs, which are mapped onto MAC PDUs. Examples of data flow mapping between RLC and MAC layers for transparent and nontransparent (UM or AM) modes are depicted, respectively, in Figures 8.11 and 8.12. The two types of RLC PDUs are defined in the 3GPP specifications [21]. The first corresponds to PDUs or data PDUs used to carry the upper-layer information (data or signaling). The second is the status PDU

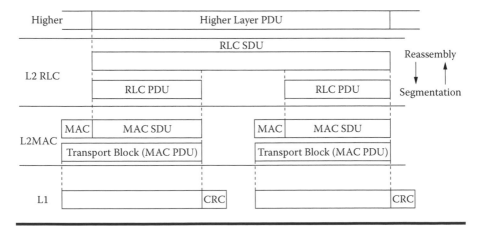

Figure 8.11 Example of data flow mapping between RLC transparent mode and MAC.

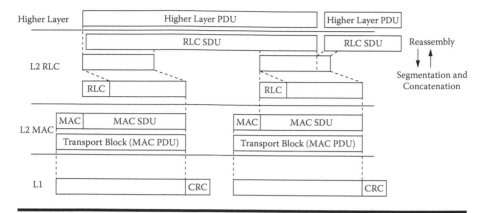

Figure 8.12 **Example of data flow mapping between RLC nontransparent mode (UM or AM) and MAC.**

used to carry the related RLC control information. The detailed structures of these PDUs are given in [21].

8.8.1 Transparent Mode

The TM is used in general for time-critical data such as speech services. Because no signaling between RLC entities and RLC PDUs is specified by the 3GPP for this mode, the TM is not capable of detecting PDUs lost in transmission from the peer RLC entity. In this mode, the RLC layer performs segmentation of upper-layer PDUs into RLC PDUs and vice versa, that is, assembly of RLC PDUs into variable-length higher-layer PDUs. In addition, this layer contains a discard timer allowing the RLC transmitter to discharge a given RLC SDU from the buffer if the timer has expired. The SDU discard function is described in more detail in this section. To summarize, the RLC layer performs the following functions [21]:

- Segmentation/reassembly of the upper-layer PDUs
- Transfer of user data
- SDU discard

8.8.2 Unacknowledged Mode

In the UM, RLC signaling is added to the data. A sequence number is assigned to each PDU so that the UM RLC can indicate the PDUs lost in transmission from the peer UM entity. A lost PDU is detected by receiving a PDU with an out-of-order sequence number. This error detection function is called a *sequence number check.* At the receiver side, the RLC entity can deliver the recovered RLC SDUs to the higher layer using two strategies:

1. Immediate delivery of the recovered SDUs even if the previous SDUs are not recovered correctly, which generates out-of-sequence data at the higher layer. This strategy is suitable for time-critical data services.
2. In-sequence delivery of the SDUs to the higher layer by reordering the received SDUs at the receiving entity of the RLC sublayer. This strategy results in additional delays and is more suitable for non-time-critical services.

Note that in both strategies, erroneous SDUs are discarded, and this results in missed SDUs at the higher layer. This mode is only suitable for services with a low residual SDU error ratio requirement. Note that missed SDUs can be generated by discharging SDUs at the RLC transmitter buffer due to timer expiration. The UM is essentially used for services carried over the UDP protocol and not TCP. Missing data at the transport layer (due to missed erroneous SDUs) may be misinterpreted by the TCP as congestion. The TCP would trigger unwanted retransmissions and congestion window control. To take maximum advantage of the variable radio channel capacity (which requires variable MAC PDU sizes), the UM provides the possibility to segment or concatenate data from SDUs to fit MAC PDUs of various sizes. Padding bits can be used to fill an RLC PDU when concatenation is not applicable and the remaining data to be transmitted do not fill an entire RLC PDU. The RLC UM mode is characterized by the following RLC functions [21]:

- Segmentation and reassembly
- Concatenation
- Padding
- Transfer of user data
- Ciphering to avoid unauthorized data acquisition
- Sequence number check
- SDU discard
- Out-of-sequence SDU delivery
- Duplicate avoidance and reordering

8.8.3 Acknowledged Mode

In AM, additional PDUs are defined to allow bidirectional signaling between peer RLC entities. These additional PDUs are used to request the retransmission of missing or erroneous data at the receiver. In TM and UM, the erroneous PDUs are discarded. In the case of AM, a STATUS–REPORT is sent to the peer RLC entity asking it to retransmit the erroneous PDUs using the ARQ protocol (see Section 8.12 for more details on ARQ). Consequently, this mode is suitable for non-time-critical services with high-quality/high-integrity requirements carried over TCP or UDP transport protocols (but essentially TCP). The recovered SDUs are delivered to the upper layer in sequence or out of order.

The sequence delivery of SDUs results in better performance at the TCP layer since out-of-sequence reception of packets may cause a triple duplicate phenomenon (misinterpreted as congestion by TCP), which results in packet retransmissions and congestion window downsizing at the TCP layer. The functions handled by the RLC sublayer in the AM mode can be summarized as [21]:

- Segmentation and reassembly.
- Concatenation.
- Padding.
- Transfer of user data.
- Ciphering to avoid unauthorized data acquisition.
- Error correction (using ARQ).
- SDU discard.
- In-sequence SDU delivery.
- Duplicate detection used when the RLC receiver detects the reception of the same RLC PDU more than one time. The RLC entity delivers the resultant upper-layer PDU only once to the upper layer.
- Flow control to control the transmission rate between the peer RLC entities.
- Protocol error detection and recovery.

8.8.4 SDU Discard at the RLC Sender

In the 3GPP specifications [21], the RLC sender entity can be configured by upper layers (in particular, the RRC layer) to discard SDUs from the sender buffer in certain cases. The objective of this function is to manage the QoS requirements, in terms of delays and error ratio, of the service application carried over the UMTS system.

The reasons to discard SDUs at the RLC sender change from one RLC mode to another. In all RLC modes (TM, UM, and AM), once the RLC layer receives an SDU from the upper layer (PDCP sublayer), a timer specific to this SDU is started. If the timer expires before delivering the corresponding SDU to the MAC layer (i.e., before transmitting the SDU from the sender), the corresponding SDU is discarded. The presence of the timer in the RLC sender entity delimits the transfer delay of the SDUs to the receiver at the expense of increased SDU loss ratio. The RRC layer configures the RLC timer as well as the radio bearer between the UE and UTRAN to meet the QoS requirements (i.e., SDU loss ratio and SDU transfer delay).

Note that, depending on the RLC modes used, two timer discard modes can be distinguished: (1) timer-based discard without explicit signaling used in UM and (2) TM modes and timer-based discard with explicit signaling used in the AM mode.

In the RLC AM mode, in addition to timer-based discard, SDU discard may be generated when the maximum number of a given PDU retransmission is reached. In certain cases, severe wireless channel conditions may result in successive retransmission failure (i.e., erroneous) of the same PDU. The RLC discards the corresponding SDU when the number of retransmission reaches a maximum, noted as MaxDAT, to limit the SDU transfer delay and meet QoS requirements. For services with no delay restrictions (e.g., background applications class), the SDU is not discarded even if the maximum number of the corresponding PDU retransmissions (MaxDAT) is reached. Four SDU discard modes have been specified by the 3GPP in [21]:

- Timer-based discard with explicit signaling applicable to RLC AM mode
- Timer-based discard without explicit signaling applicable to RLC TM and UM modes
- SDU discard after MaxDAT transmissions applicable to RLC AM mode
- No discard after MaxDAT transmissions applicable to RLC AM mode

8.8.4.1 Timer-Based Discard with Explicit Signaling

The timer-based discard with explicit signaling mode is only applicable for RLC AM mode. For every delivered SDU from the upper layer (PDCP sublayer), a timer is started to monitor the lifetime of the SDU in the RLC sender buffer. The values of the timer can range from 100 ms to 7.5 s. If the timer of a given SDU expires, the corresponding SDU is then discarded from the buffer. The RLC sender explicitly notifies the receiver about this SDU discard event using the field super field more receiving window (SuFi MRW), contained in the so-called status PDU. The receiver acknowledges the receipt of this super field via the SuFi MRW-ACK sent to the sender and advances its receiving window to skip the missing PDU.

When the SuFi MRW field is transmitted, a timer is started in the RLC sender entity. If the timer expires before receiving the SuFi MRW-ACK from the receiver (e.g., the SuFi MRW gets lost), the SuFi MRW is retransmitted, and the timer is reinitialized. Note that the timer expires after a time ranging between 50 and 900 ms according to the configuration signaled by the RRC layer. If the number of transmissions of SuFi MRW reaches the maximum number MaxMRW (configured by RRC in a range from 1 to 32), the explicit SDU discard procedure is stopped, and a reset procedure is initiated by the RLC sender.

The reset procedure consists of transmitting a reset PDU, initializing a specific timer, and expecting reception of a reset PDU-ACK from the receiver before the timer expires. In case of timer time-out, the reset PDU is retransmitted, and the timer is reinitialized. The timer time-out can range from 50 ms to 1 s, and the maximum number of reset PDU transmissions lies in the range 1–32. If the maximum number of reset PDU transmissions is reached, the RLC signals to the RRC an unrecoverable error and a failure of the reset procedure. The RRC layer can decide then to reconfigure or to release the corresponding radio bearer.

8.8.4.2 Timer-Based Discard without Explicit Signaling

The timer-based discard of SDUs without explicit signaling is applicable for RLC TM and UM modes. In this mode, timer expiration results in discarding the corresponding SDU like the previous discard mode. However, the sender does not explicitly inform the receiver about this SDU discard event. The sender continues the transmission of the subsequent SDUs. In the UM mode, the receiver deduces implicitly the discard event by receiving out-of-sequence SDUs (since the sequence number check function is used in this mode). In the TM mode, no error detection in the received PDUs is used. The RLC receiver entity delivers merely the received data to the upper layer (transfer of user data function mentioned here).

8.8.4.3 SDU Discard after MaxDAT Transmissions

In the RLC AM mode, the use of ARQ results in retransmitting the erroneous PDUs to deliver error-free data to the upper layer. If the number of transmissions of a given PDU reaches the maximum MaxDAT, configured by the RRC layer, this PDU and the other PDUs of the same SDU (since an SDU may contain more than one PDU) are discarded. As in the timer-based discard with explicit signaling, the RLC sender explicitly notifies the receiver about this discard event using the field SuFi MRW contained in the corresponding status-PDU. The same explicit signaling and subsequent reset procedures, described for the timer-based discard, are used in this SDU discard mode. Note that the parameter MaxDAT, configured by RRC, can range from 1 to 40.

8.8.4.4 No Discard after MaxDAT Transmissions

The mode of no discard after MaxDAT transmissions is similar to the SDU discard after MaxDAT transmissions mode described, except for when the number of transmissions MaxDAT is reached. The explicit signaling procedure is skipped, and the reset procedure is initiated immediately.

8.9 Packet Data Convergence Protocol

The PDCP sublayer [23] belongs to the data link layer in the user plane and for the switched domain only. The PDCP allows the use of network, transport, and upper-layer protocols like TCP/IP or UDP/IP over the UTRAN by converting the network packets into RLC SDUs and vice versa.

In addition, this sublayer provides header compression and decompression of the IP packets conveyed in the UTRAN. Since only a few TCP/IP header fields change from one IP packet to another and the majority of header fields remain intact, the PDCP can support the compression of subsequent IP packets, which results in more efficient use of the radio resources. In the 3GPP specifications [23], several header compression algorithms are presented (in particular, those proposed in [1,24]).

The PDCP also supports reliability of data transfer in the UTRAN. In certain cases (e.g., SRNS relocation), the RLC does not provide reliability for data transmission. In addition, when upper layers (e.g., transport layer) are not reliable (e.g., UDP) and that reliability is achieved at the application layer (e.g., streaming), errors caused by SRNS relocation cannot be recovered by retransmission of erroneous packets at the transport layer. In this case, by maintaining the same PDCP sequence number, the PDCP sublayer avoids losses and provides reliability for these data services.

8.10 Broadcast/Multicast Control and Multimedia Broadcast/Multicast Service

The BMC, described in [25] and active only in the user plane, supports and controls cell broadcast service (CBS) over the UMTS radio interface. The SMS cell broadcast service is the only service, used in Release 99, that utilizes this protocol sublayer. The SMS CBS service relies on the UM of the RLC and uses the logical CTCH, which is mapped into the transport FACH.

The BMC is the entity that stores the cell broadcast messages (associated with scheduling information) received over the interface between the cell broadcast center (CBC) and the RNC. On the UTRAN side, the BMC estimates the appropriate transmission rate of the broadcast service over the FACH channel and requests the required resources from the RRC. In addition, the BMC generates and transmits, to the cell users, schedule messages that indicate the radio frame that should contain the transmitted CBS messages. In the UE, the BMC sublayer receives the schedule messages and determines the appropriate radio frame containing the CBS messages. This information is then transferred to the RRC responsible for management and configuration of the physical layer for discontinuous reception. Consequently, only the radio frame containing the CBS messages is received by the cell users. In the UE, the BMC delivers the error-free CBS messages to the upper layers.

Basically, the CBS is used for low-data-rate services (SMS cell broadcast). In Release 6 [26], the MBMS is introduced to convey higher-rate broadcast/multicast information over the radio interface (e.g., 64 kbps). According to the number of cell users, the system could decide to transmit the messages via MBMS using point-to-point or point-to-multipoint transmission.

When point-to-point transmission is used, the MBMS content is transmitted over the dedicated transport channel (DCH) of each user (that should receive this service). In this case, the logical channel can be the DCCH (for related control information) or the DTCH (for user data), and the DCH transport channel is mapped into the DPDCH physical channel.

When the MBMS content is transmitted using point-to-multipoint transmission, two new logical channels are being specified in Release 6 to carry the MBMS information: MCCH to carry control information and MTCH to transport user data. These two logical channels are mapped onto the FACH transport channel, which in turn is mapped onto the SCCPCH.

8.11 Radio Resource Control

The RRC [22] layer is the most complex layer in the UTRAN. It is involved in the management of connection between the UEs and RNC by handling the control plane signaling of layer 3 between the UEs and UTRAN.

The principal function of RRC is the establishment, reconfiguration, and release of radio bearers, transport channels, and physical channels on request from higher layers. Establishment and reconfiguration consists of performing admission control and selection of parameters allowing description of the processing of radio bearers in layers 1 and 2. Note that a radio bearer is an association of functions at various levels to handle information transmission (signaling and nonsignaling) between the UE and UTRAN. On request from higher layers on the UE side to establish or release the UE signaling connection, the RRC performs establishment, management, and release of RRC connections. Note that RRC connection release can be generated by a request from a higher layer or by the RRC layer itself in case of RRC connection failure.

In addition, the RRC layer is involved in the UE measurement reporting, including control of the parameters to be measured (e.g., air interface quality, traffic) and the period of measurements as well as the report format. Based on these measurements, the RRC performs control of radio resources in both uplink and downlink, including coordination of radio resources allocation between the different radio bearers of the same RRC connection, open-loop power control to set the target of the closed-loop power control, allocation of sufficient radio resources to achieve the radio bearer's QoS, and so on. The RRC keeps track of the UE location and handles mobility functions for the RRC connection, allowing the handover mechanism, including inter- and intrafrequency hard handover, intersystem handover, and intersystem cell reselection (intersystem means UTRAN and another system, e.g., GPRS).

The RRC layer also handles system information broadcasting from the network to all users or to a specific group of users. It is involved in the broadcasting of paging information initiated by higher layers or the RRC layer itself during the establishment of the RRC connection. This layer controls the BMC sublayer and MBMS services (Section 8.10). Depending on the traffic requirements of the BMC sublayer, the RRC performs initial configuration and radio resource allocation of the CBS (e.g., mapping schedule of CTCH onto FACH) and configures the physical layer at the UE for discontinuous reception. Consequently, only the radio frame containing the CBS messages is received by the cell users.

In addition to the functions described, the RRC performs other procedures, such as ciphering control (i.e., on/off) between the UE and UTRAN, routing of higher-layer PDUs to the appropriate higher-layer entity, and initial cell selection based on measurements in idle mode and selection procedures.

8.12 Automatic Repeat Request Protocol

The information transmitted over the air interface is protected against errors by the use of the FEC code called the channel code. In addition, to counteract the fast-fading effect, some averaging techniques, such as long interleaving, wide-band spread spectrum, and frequency hopping, are used to average the effect of the fading over all the transmission time so that temporary bad channel conditions are compensated for by short-term good channel conditions. These techniques use fixed parameter values to deal with the worst channel conditions. In other words, channel coding rate, interleaving, and spreading bandwidth are all fixed during the system conception phase. In UMTS, as in the majority of wireless systems, the adaptation to the short-term channel variations is performed using fast power control.

When services, either voice or data, are transmitted over wireless systems, required QoS constraints must be satisfied. The service QoS is generally characterized by a delay constraint and a tolerance to errors. For voice service, the BER should not exceed 10^{-3}, but no delay is tolerated.

For non-real-time data services (i.e., interactive and background classes), the BER should be less than 10^{-8}. Some delay in receiving NRT (now real time) services is usually acceptable. To achieve a target BER of 10^{-8}, two possibilities can be envisaged: increasing the transmitted power at node B or increasing the redundancy in channel coding. The first solution results in a drastic decrease of cell capacity since more power from the total and limited available node B power of 43 dBm should be allocated to each user. In this case, the interference induced by a user on other users would also increase, thereby decreasing the cell capacity even more. The second solution lowers the achievable user bit rate since more header or redundancy is added to the information sequences. Consequently, both solutions induce a capacity loss. To alleviate this degradation, the ARQ protocol has been proposed and widely used in current wireless systems to achieve error-free data delivery over the air interface for NRT applications.

The ARQ protocol consists of retransmitting erroneously received information until error-free reception of the information packets occurs. Since NRT data tolerate certain delays, the idea operates at much higher BER (e.g., 10^{-3} instead of 10^{-8}) and compensates for the increased error rates by retransmitting erroneously received packets until error-free transmission is achieved. Solutions are developed in a wireless system: The target BER is increased to a higher value (e.g., 10^{-3}), which causes an increase of errors in the received information. This solution results in savings of resources through operation at higher BER at the expense of increased delays in packet reception.

A trade-off should be found between capacity improvement and increased delays. The incurred delays over the radio interface can also have a drastic effect on overall end-to-end system performance since the TCP is often used for NRT applications in the fixed networks and the Internet to manage congestion control. The ARQ protocols and TCP can negatively interact and can lead to effective system throughput and capacity degradation. This would defeat the original purpose of improving cell capacity in the radio-access networks and the end-to-end applications throughput as well.

For most of the data network systems, three ARQ protocols can be used: the SW (stop and wait) protocol, the go-back-*n* protocol, and the SR (selective repeat) protocol [28]. In the UMTS R99 [21], the SR-ARQ is used at the RLC level. In fact, the cyclic redundancy check (CRC) decoder detects the presence of errors in each RLC PDU at the UE. In case the RLC PDU is in error, the UE informs the RNC via node B by transmitting a nonacknowledgment of the PDU over the uplink DPCCH. The RNC retransmits the erroneous RLC PDU until it is received without errors.

8.12.1 SW Protocol

The SW protocol is the simplest ARQ protocol. The sender, or the RNC in the case of UMTS R99, classes the packets to transmit in a first input, first output (FIFO) buffer, transmits the first packet in the buffer to the receiver, starts a timer, and waits for an acknowledgment from the receiver. A nonacknowledgment or a timer expiration causes a retransmission of the same packet by the sender. Once a positive acknowledgment is received before timer expiration at the sender, the next packet in the buffer is transmitted to the receiver. This strategy causes high delays since the packets in the buffer cannot be transmitted before receiving the acknowledgment of the previous packet. The time of inactivity elapsing between the transmission of a packet and the reception of the acknowledgment makes this protocol inefficient [28]. Note that in UMTS R99 this protocol is not used.

8.12.2 Sliding Window Protocol

To deal with the inefficiency problem of the SW strategy, a sliding window protocol was developed. Instead of transmitting a packet and waiting for the acknowledgment before sending another packet, the sender transmits W packets, where W is a transmission window size, before receiving the acknowledgment of the first one. Once the acknowledgment of the first packet is received, another packet is transmitted, so that the total number of the transmitted packets that await acknowledgment is maintained equal to the window size W. This strategy increases the efficiency of the system but requires larger headers; for example, a sequence number should be attributed to each packet.

If the received packet contains errors, a negative acknowledgment is sent, for example, on the uplink DPCCH to the sender (e.g., the RNC). In this case, two control strategies can be applied: go-back-*n* and SR. The go-back-*n* control protocol manages several blocks at a time. When a received packet is erroneous, a negative acknowledgment is transmitted to the sender. All transmitted packets starting from the erroneous one have to be retransmitted. At the receiver, all packets received past the erroneous one are discarded.

This protocol requires more header and more complex receivers. An out-of-sequence problem can arise since some packets can be correctly received and decoded before others. This can have a drastic effect on higher-layer protocols such as TCP. Therefore, in-sequencing of these packets should be performed before their transmission to higher layers.

8.13 Power Control

Power control is one of the key techniques used in wireless networks. The radio channel changes instantaneously according to such things as mobile position, environment, scatters, and shadowing. To achieve a required QoS per user, the transmission power must be adapted instantaneously to the channel variations to maintain a given signal level at the receiver. This would maintain a given BER at the receiver.

As QoS requirements depend on service type, different target BERs are expected. For speech service, the target BER is 10^{-3}. For data, the target BER is about 10^{-8} to achieve the high integrity requirements of data. For a given mobile scenario (e.g., environment, channel type, frequency, mobile speed), the target BER corresponds to a specific value of SIR called the target SIR. If the scenario changes, the target SIR must be changed to maintain the same BER. Consequently, power control must provide an instantaneous adaptation of the data transmission to the QoS requirements and the mobile scenario by adapting the target SIR and the instantaneous transmitted power to achieve this target SIR.

In UMTS, this adaptation can be conducted continuously by using the radio link quality measurement when connectivity is already established and before the connection to the RRC that regulates access over the air interface. Consequently, two forms of power control exist in UMTS: open loop and closed loop [29,30]. The first is used on the uplink when users attempt access for the first time, and the second is used during active connections or sessions.

8.13.1 Open-Loop Power Control

Open-loop power control consists of an estimation of the initial uplink and downlink transmission powers. By receiving the CPICH and the control parameters on the BCCH, the UE sets the initial output power, for uplink channels (such as PRACH or the first transmission of the DPCCH before the start of the inner loop), to a specific value to achieve a target SIR at the receiver without generating a high interference level on the other users' signals. In the downlink, node B on UE measurements sets the transmission power (e.g., DPCH first transmission before the inner-loop start, SCCPCH carrying FACH and PCH) at an adequate level to achieve as much as possible an acceptable signal level without generating much interference. Note that the RNC is not involved in the open-loop power control since it is a one-way power control between node B and the UE only.

8.13.2 Closed-Loop Power Control

Closed-loop power control is performed continuously during active sessions known as RRC connections and involves three entities: the UE, the node B, and the RNC. Two mechanisms, inner loop and outer loop, are running among these three entities to adjust the instantaneous power and the target SIR.

The inner-loop power control is the first part of the closed-loop power control performed between the UE and node B. The inner-loop power control, based on SIR measurements at receivers, is used to combat fast channel variations and interference fluctuations. The power control algorithm aims at keeping the user SIR at an appropriate level by adjusting the transmission powers up or down. This power control operates on each slot—that is, at a frequency of 1500 Hz. In case of uplink, each base station in the active set evaluates the received SIR by estimating the received DPCH power after RAKE processing of the concerned mobile signal and the total uplink received interference. In case of downlink, the mobile station computes the received SIR by estimating the DPCH power after RAKE combining of the received signal and the total downlink interference power received at that mobile station.

In both uplink and downlink, TPC commands are generated and should be transmitted over each slot according to the following rule: If $SIR_{estimated} > SIR_{target}$ then the TPC command to transmit is 0, whereas if $SIR_{estimated} < SIR_{target}$ then the TPC command to transmit is 1. Note that in the downlink the TPC command is unique on each slot or can be repeated over three slots according to the mode used. In addition, the power step size can be equal to 0 in the downlink if a certain "limited power increase used" parameter is used and the so-called power-raise-limit is reached (see [29] for more details). In the uplink, if the TPC command is 1 (or 0) the UE might increase (or decrease) the output power of the DPCCH channel by a step size of 1 dB. A step size of 2 dB is also specified for use in certain cases.

Because the DPCCH and DPDCH channels are transmitted in parallel using two different codes, they are transmitted at two different power levels depending on the transport format combination (TFC) of these channels. The UE estimates a gain factor between the powers of these channels and adjusts the power of the DPDCH according to the DPCCH power. In the downlink, the DPCCH and DPDCH channels are transmitted on the same channel (i.e., same code). In this

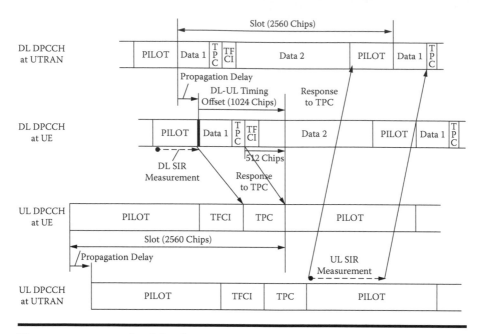

Figure 8.13   Example of the fast power control timing for DPCH channel [29].

case, node B adapts the transmission power by a step size of 0.5, 1, 1.5, or 2 dB according to the received TPC command. Note that the support of a 1-dB step size by node B is mandatory, while the other step sizes are optional. An example of the fast power control timing in uplink and downlink is depicted in Figure 8.13. Note that the power control dynamic range is 25 dBm. The maximum allowed output power per channel is 36 dBm, and the maximum base station output power is limited to 43 dBm.

The outer-loop power control maintains the quality of communication for each service by using power as low as possible. The RNC performs outer-loop control by adjusting the SIR target dynamically to keep the block error rate (BLER) in a given range. The uplink outer loop is located in the RNC. The RNC adjusts the target SIR by receiving the measured SIR from node B and by measuring the BLER. The downlink outer loop is located in the UE. This procedure is slow and can be conducted 10 to 100 times per second (i.e., at a frequency ranging from 10 to 100 Hz).

8.14 Handover

The UMTS system supports the use of hard, soft, and softer handover [30] between cells. The objective of the handover use in the cellular system is to achieve an optimum fast closed power control, in other words, to allow mobile connection to the strongest cells. The strongest cell is the cell that has the best wireless link conditions to transmit the data to a specific user, that is, that allocates the lowest power resources to this user achieving the required service QoS (target BLER).

■ The hard handover is applied when the link quality of a user to an alternative cell becomes better than that to the serving cell. The active connection in this case changes from the serving cell to the other cell. This kind of handover is applied in the following four cases: handover between GSM and UMTS, intermode handover (FDD/TDD and vice versa), intramode handover across different frequencies (the serving cell and the alternative cell do not have the same frequency), and intrafrequency handover for services using a shared channel.

■ In the 3GPP specifications, soft/softer handover can be used for services using a dedicated channel and between cells having the same radio access technology (i.e., CDMA) and the same frequency. Note that softer handover consists of a soft handover between different sectors of the same cell. The basic idea of a soft handover is to allow mobile station connection to more than one node B at the same time. This results in a seamless handover with no disconnections of radio bearers. In addition, the mobile station combines, using maximal ratio combining (MRC), the signals received from different node Bs (called *macrodiversity*). This increases the SIR level and ensures good signal quality received by users at the cell border (where in general insufficient signal level is obtained from a single cell). In the uplink, the macrodiversity can also be applied by combining the signals transmitted by the same mobile and received by different node Bs (different sectors in the case of softer handover). These received signals are combined in the RNC in the case of soft handover or in the base station in the case of softer handover, which improves the uplink received SIR and allows reduction of the mobile transmission power.

A mobile station in CDMA systems is connected to more than one base station (these base stations are called the *active set*) if the path losses are within a certain handover margin or threshold denoted by *hm* or "reporting range." Handovers due to mobility are performed using a hysteresis mechanism defined by the handover hysteresis *hhyst* and replace hysteresis (*rephyst*) parameters. Figure 8.14 [30] depicts an example of soft handover in which three cells are considered, and ΔT is the time to trigger. A node B is added into the active set if the received pilot signal level (i.e., SIR over the CPICH that reflects the path loss) is greater than the best pilot level (i.e., the best path loss) by more than *hhyst* − *hm* (*pilotlevel* > *best pilot level* −*hm* + *hhyst*). A base station is removed from the active set if *pilot level* > *best pilot level* − *hm* − *hhyst*. Similarly, a base station outside the active set can replace an active base station if its path loss becomes more than *rephyst* of that base station. For more details on handover procedures, consult [30].

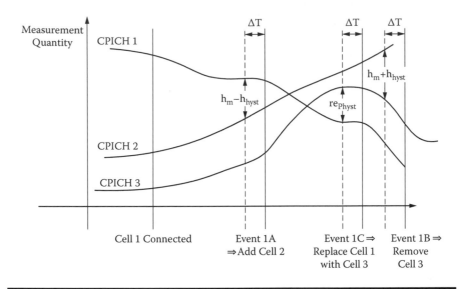

Figure 8.14 Example of soft handover algorithm [30].

References

1. Bormann, C., Burmeister, C., Fukushima, H., Hannu, H., Jonsson, L.E., Hakenberg, R., Koren, T., Le, K., Liu, Z., Martensson, A., Miyazaki, A., Svanbro, K., Wiebke, T., Yoshimura, T., and Zheng, H., *RObust Header Compression (ROHC): Framework and Four Profiles: RTP, UDP, ESP, and Uncompressed RFC 3095*. July 2001.
2. Tanner, R., and Woodard, R., *WCDMA Requirements and Practical Design*. Chichester, UK: Atrium, 2004.
3. Holma, H., and Toskala, A., *WCDMA for UMTS. Radio Access for Third Generation Mobile Communications*, 3rd ed. Chichester, UK: Wiley, 2004.
4. 3GPP TS 23.107 V6.2.0, *Quality of Service (QoS) Concept and Architecture*. Release 6, December 2004.
5. Laiho, J., Wacker, A., and Novosad, T., *Radio Network Planning and Optimisation for UMTS*. London: Wiley, 2002.
6. ITU-T, *Recommendation G114. One-Way Transmission Time*. 1996.
7. 3GPP TS 22.105 V6.4.0, *Services and Service Capabilities*. Release 6, September 2005.
8. Ahvonen, K., IP telephony signalling in a UMTS all IP network. Master's thesis, Helsinki University of Technology, November 2000.
9. 3GPP TS 23.101 V6.0.0, *Universal Mobile Telecommunications System (UMTS) Architecture*. Release 6, December 2004.
10. 3GPP TS 23.002 V6.7.0, *Network Architecture*. Release 6, March 2005.
11. 3GPP TS 23.121 V3.6.0, *Architectural Requirements for Release 1999*. Release 1999, June 2002.
12. 3GPP TS 25.401 V6.5.0, *UTRAN Overall Description*. Release 6, December 2004.

13. 3GPP TS 25.410 V6.2.0, *UTRAN Iu Interface: General Aspects and Principles*. Release 6, December 2004.
14. 3GPP TS 25.420 V6.3.0, *UTRAN Iur Interface General Aspects and Principles*. Release 6, March 2005.
15. 3GPP TS 25.430 V6.5.0, *UTRAN Iub Interface: General Aspects and Principles*. Release 6, June 2005.
16. 3GPP TS 25.301 V6.2.0, *Radio Interface Protocol Architecture*. Release 6, March 2005.
17. 3GPP TS 25.321 V6.5.0, *Medium Access Control (MAC) Protocol Specification*. Release 6, June 2005.
18. 3GPP TS 25.211 V6.6.0, *Physical Channels and Mapping of Transport Channels onto Physical Channels (FDD)*. Release 6, September 2005.
19. 3GPP TS 25.302 V6.3.0, *Services Provided by the Physical Layer*. Release 6, March 2005.
20. 3GPP TS 25.201 V6.2.0, *Physical Layer—General Description*. Release 6, June 2005.
21. 3GPP TS 25.322 V6.3.0, *Radio Link Control (RLC) Protocol Specification*. Release 6, March 2005.
22. 3GPP TS 25.331 V6.7.0, *Radio Resource Control (RRC) Protocol Specification*. Release 6, September 2005.
23. 3GPP TS 25.323 V6.3.0, *Packet Data Convergence Protocol (PDCP) Specification*. Release 6, September 2005.
24. Degermark, M., Nordgren, B., and Pink, S., *IP Header Compression*. RFC 2507, February 1999.
25. 3GPP TS 25.324 V6.4.0, *Broadcast/Multicast Control (BMC)*. Release 6, September 2005.
26. 3GPP TS 25.346 V6.6.0, *Introduction of the Multimedia Broadcast Multicast Service (MBMS) in the Radio Access Network (RAN)*. Release 6, September 2005.
27. Reed, I.S., and Chen, X., *Error-Control Coding for Data Networks*. Boston: Kluwer Academic, 1999.
28. Bertsekas, D., and Gallager, R., *Data Networks*, 2nd ed. Englewood Cliffs, NJ: Prentice Hall, 1992.
29. 3GPP TS 25.214 V6.5.0, *Physical Layer Procedures (FDD)*. Release 6, March 2005.
30. 3GPP TR 25.922 V6.0.1, *Radio Resource Management Strategies*. Release 6, April 2004.

Chapter 9

High-Speed Downlink Packet Access

Mohamad Assaad and Djamal Zeghlache

Contents

9.1 Introduction

The Universal Mobile Telecommunications System (UMTS) proposed for third-generation (3G) cellular networks in Europe is meant to provide enhanced spectral efficiency and data rates over the air interface. The objective for UMTS, known as wideband code-division multiple access (WCDMA) in Europe and Japan, is to support data rates up to 2 Mbps in indoor or small-cell outdoor environments and up to 384 Kbps in wide-area coverage for both packet- and circuit-switched data. The 3GPP (Third Generation Partnership Project), responsible for standardizing the UMTS system, realized early that the first releases for UMTS would be unable to fulfill this objective. This was evidenced by the limited achievable bit rates and aggregate cell capacity in Release 99. The original agenda and schedule for UMTS evolution has been modified to meet these goals by gradual introduction of advanced radio, access, and core network technologies through multiple releases of the standard. This phased roll-out of UMTS networks and services also would ease the transition from second- to third-generation cellular for manufacturers as well as network and service providers. In addition, to meet the rapidly growing needs in wireless Internet applications, studies initiated by the 3GPP since 2000 not only anticipated this needed evolution but also focused on enhancements of the WCDMA air interface beyond the perceived third-generation requirements.

The high-speed downlink packet access (HSDPA) system [1–5] has been proposed as one of the possible long-term enhancements of the UMTS standard for downlink transmission. It has been adopted by the 3GPP and is used in Europe. HSDPA introduces first-adaptive modulation and coding, retransmission mechanisms over the radio link and fast packet scheduling, and later, multiple transmit-and-receive antennas. This chapter describes the HSDPA system and some of these related advanced radio techniques. Similar enhancements are envisaged for the UMTS uplink but are not covered here. Refer to the 3GPP standard for UMTS, especially the high-speed uplink packet access (HSUPA), to find out more about data rate and capacity enhancements for this link.

9.2 HSDPA Concept

Interference control and management in a code-division mutiple access (CDMA)-based system can be performed at the link level by some enhanced receiver structures called MUD (multiuser detection), which are used to minimize the level of interference at the receiver. At the network level, good management of the interference can be provided by an enhanced power control and associated call admission control (CAC) algorithms.

Note that the MUD technique is essentially used on the uplink in actual CDMA-based systems (UMTS, Interim Standard 95 [IS-95], etc.) since node B has knowledge of the CDMA codes used by all users. In the downlink, informing each user about the spreading codes used by other users would increase system complexity. Detection complexity increases with the number of users in the system. This has a direct impact on the terminal energy consumption (battery lifetime) and on the response time of the detector (detection delay). Consequently, in UMTS, the MUD technique is used only in the uplink. In the downlink, interference is managed at the network level using fast power control and CAC.

This philosophy of simultaneously managing the interference at the network level for dedicated channels leads to limited system efficiency. Fast power control used to manage the interference increases the transmission power during the received signal fades. This causes peaks in the transmission power and subsequent power rises that reduce the total network capacity. Power control imposes provision of a certain headroom, or margin, in the total node B transmission power to accommodate variations [6]. Consequently, system capacity remains insufficient and unable to respond to the growing need in bit rates due to the emergence of Internet applications.

A number of performance-enhancing technologies must be included in the UMTS standard to achieve higher aggregate bit rates in the downlink and to increase the spectral efficiency of the entire system. These techniques include AMC (adaptive coded modulation), fast link adaptation, hybrid ARQ (automatic repeat request), and fast scheduling. MUD and MIMO (multiple-input multiple-output) antenna solutions can also be included—which is expected in later releases of UMTS—to further improve system performance and efficiency.

The use of higher-order modulation and coding increases the bit rate of each user but requires more energy to maintain decoding performance at the receiver. Hence, the introduction of fast link adaptation is essential to extract any benefit from introducing higher-order modulation and coding into the system. The standard link adaptation used in current wireless system is power control. However, to avoid power rise as well as cell transmission power headroom requirements, other link adaptation mechanisms to adapt the transmitted signal parameters to the continuously varying channel conditions must be included. One approach is to tightly couple AMC and scheduling. Link adaptation to radio channel conditions is the baseline philosophy in HSDPA, which serves users having favorable channel conditions. Users with bad channel conditions should wait for improved conditions to be served. HSDPA adapts in parallel the modulation and the coding rates according to the instantaneous channel quality experienced by each user.

AMC still results in errors due to channel variations during packet transmission and in feedback delays in receiving channel quality measurements. A hybrid ARQ (HARQ) scheme can be used to recover from link adaptation errors. With HARQ, erroneous transmissions of the same information block can be combined with subsequent retransmission before decoding. By combining the minimum number of packets needed to overcome the channel conditions, the receiver minimizes the delay required to decode a given packet. There are three main schemes for implementing HARQ: chase combining, in which retransmissions are a simple repeat of the entire coded packet; incremental redundancy (IR), in which additional redundant information is incrementally transmitted; and self-decodable IR, in which additional information is incrementally transmitted but each transmission or retransmission is self-decodable.

The link adaptation concept adopted in HSDPA implies the use of time-shared channels. Therefore, scheduling techniques are needed to optimize the channel allocation to the users. Scheduling is a key feature in the HSDPA concept and is tightly coupled to fast link adaptation. Note that the time-shared nature of the channel used in HSDPA provides significant trunking benefits over DCH (dedicated channel) for bursty high-data-rate traffic. The HSDPA shared channel does not support soft handover due to the complexity of synchronizing the transmission from various cells. Fast cell selection can be used in this case to replace the soft handover. It could be advantageous to be able to rapidly select the cell with the best signal-to-interference ratio (SIR) for the downlink transmission.

MIMO antenna techniques could also provide higher spectral efficiency. Such techniques exploit spatial or polarization decorrelations over multiple channels to achieve fading diversity gain. There are two types of MIMO: blast MIMO and space–time coding.

HSDPA can be seen as a mixture of enhancement techniques tightly coupled and applied on a combined CDMA–TDMA (time-division multiple-access) channel shared by users [6]. This channel, called HS-DSCH (high-speed downlink shared channel), is divided into slots called transmit time intervals (TTIs), each equal to 2 ms. The signal transmitted during each TTI uses the CDMA technique. Since link adaptation is used, "the variable spreading factor is deactivated because its long-term adjustment to the average propagation conditions is not required anymore" [6]. Therefore, the spreading factor is fixed and equal to 16. The use of relatively low spreading factor addresses the provision for increased application bit rates.

Finally, the transmission of multiple spreading codes is also used in the link adaptation process. However, a limited number of Wash codes is used due to the low spreading adopted in the system. Since all these codes are allocated in general to the same user, MUD can be used at the user equipment to reduce the interference between spreading codes and to increase the achieved data rate. This is in contrast to traditional CDMA systems, in which MUD techniques are used in the uplink only.

9.3 HSDPA Structure

As mentioned in Section 9.2, HSDPA relies on a new transport channel, called the HS-DSCH, which is shared between users. Fast link adaptation combined with time domain scheduling takes advantage of the short-term variations in the signal power received at the mobile so that each user is served on favorable fading conditions. The TTI value, fixed at 2 ms in the 3GPP standard, allows this fast adaptation to the short-term channel variations.

To avoid the delay and the complexity generated by the control of this adaptation at the radio network controller (RNC), the HS-DSCH transport channel is

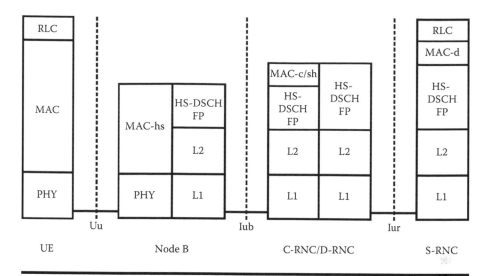

Figure 9.1   Radio interface protocol architecture of HSDPA system.

terminated at node B, unlike the transport channels in UMTS, which are terminated at the RNC. To handle the fast link adaptation combined with scheduling and HARQ at node B, a new media-access control (MAC) entity, called MAC-hs (high-speed MAC), has been introduced at node B.

The general architecture of the radio protocol is depicted in Figure 9.1 [1]. The MAC-hs is located below the MAC-c/sh entity in the controlling RNC. The MAC-c/sh provides functions to HSDPA that already exist in UMTS. MAC-d is still included in the serving RNC. The HS-DSCH frame protocol (HS-DSCH FP) handles the data transport from the serving RNC (SRNC) to the controlling RNC (CRNC) and between the CRNC and node B. Another alternative configuration, presented in Figure 9.2 [1], is also proposed in the 3GPP standards; the SRNC is directly connected to node B—that is, without any user plane of CRNC. Note that in both configurations, the HSDPA architecture does not affect protocol layers above the MAC layer.

9.4 Channel Structure

HSDPA consists of a time-shared channel between users and is consequently suitable for bursty data traffic. HSDPA is basically conceived for non-real-time data traffic. Research is actually ongoing to handle streaming traffic over HSDPA using improved scheduling techniques.

In addition to the shared data channel, two associated channels, HS-SCCH (high-speed shared control channel) and HS-DPCCH (high-speed dedicated

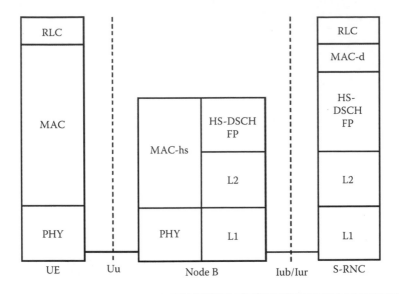

Figure 9.2 Radio interface protocol architecture of HSDPA system.

physical control channel) are used in the downlink and the uplink to transmit signaling information to and from the user. These three channels in HSDPA—HS-DSCH, HS-SCCH, and HS-DPCCH—are now described in more detail.

9.4.1 HS-DSCH Channel

The fast adaptation to the short-term channel variations requires handling of fast link adaptation at node B. Therefore, the data transport channel HS-DSCH is terminated at node B. This channel is mapped onto a pool of physical channels called HS-PDSCH (high-speed physical downlink shared channel) to be shared among all the HSDPA users in a time- and code-multiplexed manner [7, 8] (see Figure 9.3).

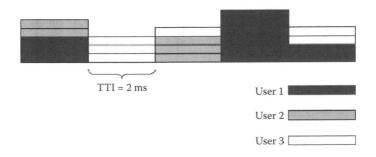

Figure 9.3 Code-multiplexing example over HS-DSCH.

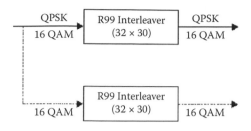

Figure 9.4   Interleaver structure for HSDPA.

Each physical channel uses one channelization code of fixed spreading factor equal to 16 from the set of 15 spreading codes reserved for HS-DSCH transmission.

Multicode transmission is allowed, which translates to a mobile user's being assigned multiple codes in the same TTI depending on the user's equipment capability. Moreover, the scheduler may apply code multiplexing by transmitting separate HS-PDSCHs to different users in the same TTI.

The transport channel coding structure is reproduced as follows: One transport block is allocated per TTI, so that no transport block concatenation—such as in UMTS DCH-based transmission—is used. The size of transport block changes according to the modulation and coding scheme (MCS) selected using the AMC technique. To each transport block, a cyclic redundancy check (CRC) sequence with 24 bits is added.

Since errors occur in bursts, one CRC sequence per transport block (i.e., per TTI) is sufficient. Once the CRC sequence is attached, the transport block bits are bit scrambled and are segmented into blocks to apply turbo block encoding. The code block size depends on the turbo coding rate and can reach a maximum value of 5114 bits [9]. The coding rate changes according to the MCS selected by the link adaptation technique. At the turbo encoder output, rate matching is applied by the physical-layer HARQ functionality.

After matching the number of bits to the number of bits in the allocated HS-DSCH physical channels, segmentation divides the resulting bits among the HS-DSCH physical channels. The bit sequence obtained for each physical channel is then interleaved using one step interleaver with a fixed size 32 × 30 (i.e., 32 rows and 30 columns, see Figure 9.4 [9]). Finally, the resulting bit sequence is modulated using 16-ary quadrature amplitude modulation (16QAM) or quadrature phase-shift keying (QPSK) according to the MCS selected [10,11].

9.4.2 HS-SCCH Channel

The downlink signaling related to the HS-DSCH is transmitted over the HS-SCCH.

The signaling information carried by the HS-SCCH contains essentially the transport format resource indicator (TFRI) and the HARQ information of the

HS-DSCH channel. The TFRI includes the channelization codes used by the HS-DSCH, the modulation scheme, and the transport block size.

The HARQ information consists of the HARQ new data indicator, the HARQ process identifier, and the redundancy and constellation version. Since the HS-DSCH channel is shared among users, the user equipment identity is sent over the HS-SCCH to indicate the identity of the user for which the HSDSCH is allocated during the TTI. Note that the user equipment identity is given by the 16-bit HS-DSCH radio network temporary identifier (H-RNTI) defined by the RRC (Radio Resource Control) [12,13].

The information carried by this channel is split into two parts [9]: Part I contains channelization codes (7 bits) and modulation scheme (1 bit). Part II includes transport block size (6 bits), HARQ process identifier (3 bits), HARQ new data indicator (1 bit), and redundancy and constellation version (3 bits). The HS-SCCH also contains a CRC attachment, which consists of 16 bits calculated over parts I and II and appended to part III, as in Figure 9.5 [9], where:

- $Xccs$: 7 bits of channelization code set information
- Xms : 1 bit of modulation scheme
- $Xtbs$: 6 bits of transport block size information
- Xhp: 3 bits of HARQ process information
- Xrv : 3 bits of redundancy version information
- Xnd: 1 bit of new data indicator
- $X1$: 8 bits of input to rate 1/3 convolutional encoder
- $Z1$: 48 bits of output
- $R1$: 40 bits after puncturing

Parts I and II are encoded separately using Release 99 convolutional code with coding rate equal to 1/3 and 8 tail bits. After convolutional coding, interleaving and rate matching to 120 HS-SCCH channel bits (three slots) is applied. The interleaving and rate matching are carried out separately for the two parts of the coded HS-SCCH to allow for early extraction of the time-critical information of part I of the HS-SCCH information [9].

Note that the postconvolution of part I is scrambled by a code generated from the user equipment identity, which is encoded using Release 99 convolutional code with coding rate equal to 1/2 and 8 tail bits. Of the resulting 48 bits, 8 are punctured using the same rate-matching rule as for part I of the HS-SCCH.

The resulting part I (40-bit sequence) is mapped into the first slot of the HS-SCCH TTI. The resulting part II is carried over the second and the third slots. Note that the HS-SCCH is spread with a fixed spreading factor of 128 so that the bit rate over this channel is fixed at 60 Kbps (i.e., 120 bits per TTI), as in Figure 9.6 [8].

Finally, this channel can be transmitted at a fixed power or can use power control. The decision to use power control is left entirely to the implementation.

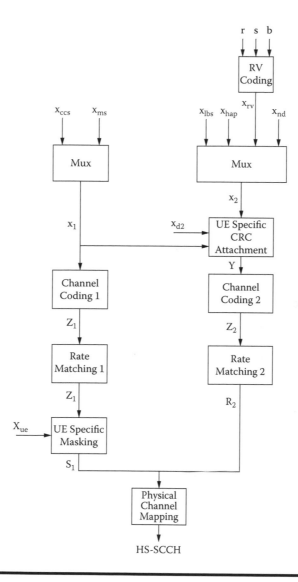

Figure 9.5   Overall coding chain for HS-SCCH.

9.4.3 *HS-DPCCH Channel*

In the uplink, signaling information has to be transmitted for the HARQ acknowledgment and the feedback measurement. The use of fast link adaptation on the HS-DSCH channel requires knowledge of the channel quality during the transmission. The user equipment measures the channel quality on the CPICH (common pilot channel) and sends the result to node B. This procedure is explained in Section 9.6.

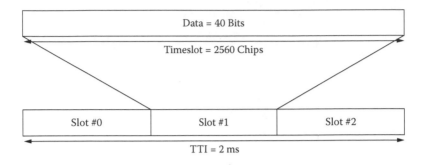

Figure 9.6 **Frame structure of HS-SCCH channel.**

The use of HARQ requires an acknowledgment message from the user to node B so that node B retransmits the erroneously received packet or a new packet.

This signaling information is carried by the HS-DSCH-associated uplink dedicated control channel (HS-DPCCH). This channel is spread, with a spreading factor of 256 (i.e., 30 bits per TTI), and is code multiplexed with the existing dedicated uplink physical channels (DPCHs). The HARQ acknowledgment is a 1-bit acknowledgment/nonacknowledgment (ACK/NACK) indication repeated 10 times and transmitted in one slot. The HARQ acknowledgment field is gated off when no ACK/NACK information is sent. The measurement feedback information contains a channel quality indicator (CQI) that may be used to select transport format and resource by the HS-DSCH serving node B, according to CQI tables specified in [11]. It is essential information needed for fast link adaptation and scheduling. The channel quality information, consisting of 5 bits, is coded using a (20,5) code transmitted over two slots (see Figure 9.7 [8]).

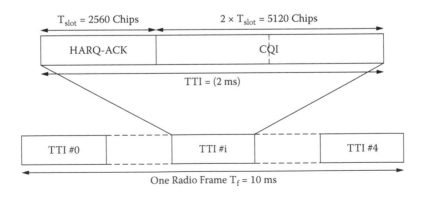

Figure 9.7 **Frame structure of HS-DPCCH channel.**

9.4.4 Timing of HSDPA Channels

The timing relation between the HSDPA channels is presented in Figure 9.8 [8]. In the downlink, the HS-SCCH is received two slots (1.334 ms) before the HS-DSCH channel. Therefore, the user gets the time to decode the first part of the HS-SCCH, which essentially contains the channelization codes of the HS-DSCH channel scrambled by the user equipment identity. Once this first part is decoded, the user is able to start decoding the HS-DSCH.

In the uplink, the HS-DPCCH starts $m*256$ chips after the start of the uplink DPCCH with m selected by the user equipment such that the ACK/NACK transmission (of duration 1 time slot) commences within the first 0–255 chips after 7.5 slots following the end of the received HS-DSCH. The 7.5 slots represent the user equipment processing time after the reception of the HS-DSCH channel. This time is required due to timing constraints imposed by detection with relatively high complexity, such as the MUD and HARQ mechanism, or soft combining. The 0- to 255-chip delay is needed because the HS-DPCCH is code multiplexed with other physical channels transmitted by the same user. HS-DPCCH and the other DPCCH channels use the same scrambling codes, and a physical synchronization between these channels is required to reduce the interference at the receiver.

The term *physical synchronization* means that the spreading sequences of these channels must start at the same instant. Since the spreading factors of HS-DPCCH and DPCCH are fixed at 256, the synchronization is achieved by introducing a delay of 0–255 chips. Note that the physical synchronization is different from the transport synchronization when the transport channels start their transmissions at the same time.

9.5 MAC-hs

The use of fast adaptation to the short-term channel variations requires handling of the HSDPA transport channels by node B. Therefore, a MAC-hs entity has been introduced in node B. The MAC-hs also stores the user data to be transmitted across the air interface. This imposes some constraints on the minimum buffering capabilities of node B. The transfer of the data queues to node B creates the need for a flow control mechanism (HS-DSCH FP) to handle data transmission from the SRNC to the CRNC if the Iur interface is involved and between the CRNC and node B. The overall MAC architecture at the UMTS, UTRAN (universal terrestrial radio access network) side and the user equipment side is presented, respectively, in Figures 9.9 and 9.10 [1,14].

9.5.1 MAC Architecture at the UTRAN Side

At the UTRAN side, the data to be transmitted on the HS-DSCH channel is transferred from MAC-c/sh to the MAC-hs via the Iub interface in case of configuration

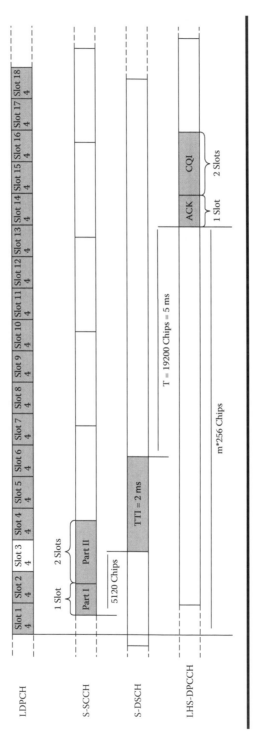

Figure 9.8   Channels transmission timing in HSDPA.

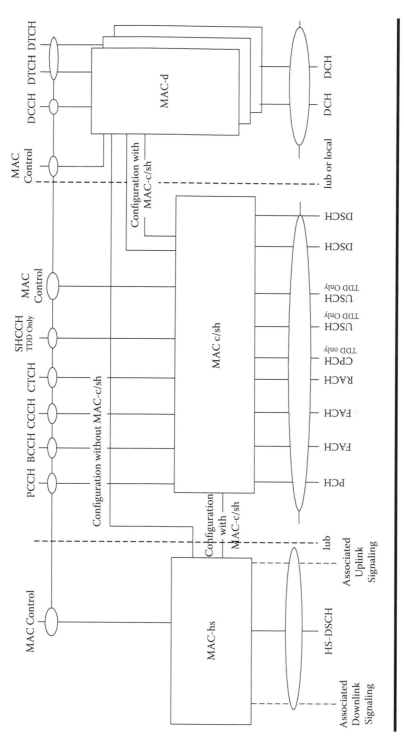

Figure 9.9   UTRAN side MAC architecture with HSDPA.

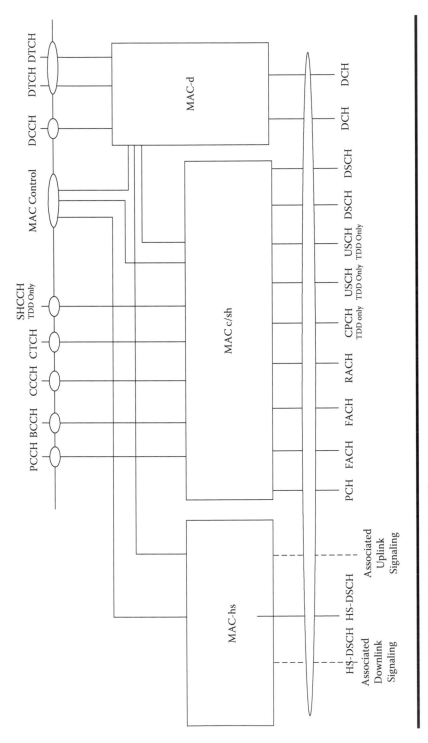

Figure 9.10 UE side MAC architecture with HSDPA.

with MAC-c/sh or from the MAC-d via Iur/Iub in case of configuration without MAC-c/sh. As specified by the 3GPP [1, 14], the MAC-hs entity is in charge of four logical functionalities:

1. Flow control: The presence of data queues in node B creates the need for a flow control mechanism (HS-DSCH FP) to handle data transport between MAC-c/sh and MAC-hs (configuration with MAC-c/sh) or MAC-d and MAC-hs (configuration without MAC-c/sh), taking the transmission capabilities of the air interface into account in a dynamic manner. The design of such flow control is a nontrivial task since such functionality, in cooperation with the packet scheduler, is intended ultimately to regulate the user's perceived service to fulfill the quality of service (QoS) allocated to the user depending on service class.

2. Scheduling/priority handling: This function manages HS-DSCH channel allocation between users as well as between HARQ entities and data flows of the same user according to service priority and wireless channel conditions. Based on status reports, from associated uplink signaling, the new transmission or retransmission of a given user is determined. To maintain proper transmission priority, a new transmission can be initiated on an HARQ process at any time. However, it is not permitted to schedule new transmissions, including retransmissions originating in the RLC layer, within the same TTI, along with retransmissions originating from the HARQ layer.

3. HARQ: This entity handles the HARQ functionality for users. In other words, it supports multiple HARQ processes of stop-and-wait HARQ protocols. One HARQ entity per user is needed, and one HARQ process is active during each TTI. The HARQ technique is explained in detail in Section 9.8.

4. TFRI selection: This entity selects the appropriate transport format and resource combination (i.e., the MCSs, described in Section 9.6) for the data to be transmitted on HS-DSCH. The selected TFRI changes from TTI to TTI according to wireless channel conditions.

9.5.2 MAC Architecture at the User Equipment Side

At the user equipment-side, the MAC-d entity is modified with the addition of a link to the MAC-hs entity. The links to MAC-hs and MAC-c/sh cannot be configured simultaneously in one user equipment. The mapping between MAC-d and the reordering buffer in MAC-hs is configured by higher layers. The two essential entities of the MAC-hs are [1, 14]:

1. The HARQ entity, which is responsible for handling all the tasks of the HARQ protocol, such as generating ACK and NACK messages. HSDPA uses the N-channel stop-and-wait HARQ protocol. This results in N HARQ

processes to be handled by the HARQ peer entities in the user equipment and the UTRAN. Only one HARQ process is active per HS-DSCH per TTI. The configuration of the HARQ protocol is provided by the RRC over the MAC-control service access point (SAP) [12].

2. The reordering entity, which organizes received data blocks and delivers them in sequence to higher layers on reception. One reordering entity per priority class can be used as specified by the 3GPP. The use of this reordering entity may increase the overall delay of the data reception at the upper layers. However, this entity has an important impact on the reduction of the Transmission Control Protocol (TCP) performance degradation over wireless links. Out-of-order received data at the transport layer are misinterpreted by TCP as triple duplicate congestion.

This erroneous interpretation results in retransmissions and reduction of congestion window size, in other words, in throughput reduction. In this MAC-hs reordering entity, out-of-sequence data blocks can be delivered to the upper layers when the delay of receiving the appropriate block (i.e., having the adequate sequence number) exceeds a certain limit. This limit is determined by a timer configured by the RRC. When the timer corresponding to a given block (i.e., MAC-hs packet data unit [PDU]) expires, the HARQ entity in the user equipment delivers the data blocks in the buffer, including the subsequent received blocks, to higher layers since the missing blocks will not be retransmitted by the HARQ entity in the UTRAN.

9.6 Fast Link Adaptation

The wireless channel in cellular systems is a composite multipath/shadowing channel. The radio waves follow several paths from the transmitter to reach the destination. Fading occurs when many paths interfere, adding up to a composite signal exhibiting short time signal power variations at the receiver. This power could be weaker or stronger than the required power needed to achieve a given user QoS. The link quality between the transmitter and the receiver is also affected by slow variations of the received signal amplitude mean value due to shadowing from terrains, buildings, and trees.

To deal with the problems caused by multipath fast fading, the existing wireless systems use diversity techniques such as long interleaving, robust channel coding, frequency hopping, and direct-sequence spread spectrum. These techniques are based on one concept: averaging the temporary fading effect over all the transmission time and the bandwidth so that bad conditions are compensated for by good conditions.

The spread-spectrum technique—frequency hopping and direct sequence—spreads the signal bandwidth over a wider frequency spectrum so that only a part of the spectrum is affected by the fading. Interleaving can be seen as spreading technique over time. By reordering the bits before transmission, the information message is spread out over time. Therefore, bursty errors caused by the fading channel are spread out in time so that the decoder receives distributed nonbursty random errors that are easier to detect and to correct. The channel coding includes redundancy in the transmitted signal to increase robustness against errors. Introducing more redundancy increases robustness but decreases the effective information bit rate. In current systems, the channel coding rate is fixed to deal with the worst case, and the transmission power is adapted to the channel conditions to achieve the application QoS [15].

Because fading results from the addition of different waves from multiple paths, it can potentially be predicted. Channel parameters (i.e., amplitude, phase, frequency) remain stationary for time windows on the order of half a wavelength. For example, at a carrier frequency of 2 GHz in case of UMTS and at a mobile speed of 36 km/h, the fading pattern could be predicted for a time window of approximately 7.5 ms. Estimation of the channel is feasible over a few milliseconds, and power consequently can be adjusted over such timescales.

In UMTS Release 99, channel estimation is used to adapt the transmission power of each user and every slot (corresponding to a rate of 1500 Hz) to the short-term channel variations. This is achieved at the cost of some power rise and higher interference, as explained in Section 9.3.

In HSDPA, the idea is to avoid power adaptation, and hence power control, by approaching the radio resource allocation and sharing from a different angle. Why continue using averaging techniques such as long interleaving and a fixed channel-coding rate to counteract the fast fading if these techniques require high-performance power control? Instead, one can use AMC tightly coupled with fast scheduling so that modulation orders and coding rates are adapted according to estimated channel fading. In addition, the HS-DSCH channel can be allocated to the user with favorable channel conditions. To this avail, a CQI has been introduced in HSDPA (for details, see the next section) to enable such intelligent allocation of resources to users.

The idea is to measure the channel quality over the CPICH and to transmit the measurement report over the HS-DPCCH channel to node B so that scheduling and AMC can act according to the CQI and hence can optimize channel resource allocation. The time window between the channel condition measurement and the resource allocation should not exceed half a wavelength, as indicated.

In the 3GPP specifications [11], 30 CQIs have been standardized (described in the next section). The timing relation between the channel measurement over the CPICH and the resource allocation over the HS-DSCH are depicted in Figure 9.11 [16]. The time window between the channel measurement and the start of the transmission over the HS-DSCH channel is seven time slots. By considering the transmission time of the HS-DSCH channel (one TTI), the overall delay between the radio channel measurement and the end of the packet transmission over the HS-DSCH channel is 10 timeslots. Therefore, HSDPA is

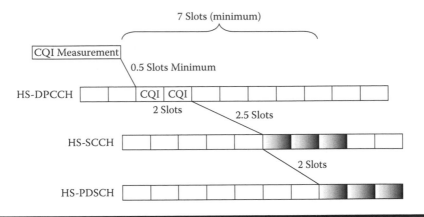

Figure 9.11 Timing relation between CQI measurement and HS-DSCH transmission.

suitable for an urban area where mobile users move at low speed (less than 40 km/h). For mobiles at higher speeds, the environment conditions change rapidly. Fortunately, in this case the mobiles have a low degree of randomness since they move along known paths (traveling by train, driving down a freeway, etc.). Therefore, special solutions can be performed to predict the channel-fading pattern in these cases [15].

In 3GPP Release 6, enhancements were added to the CQI reporting method by introducing tunable reporting rates through additional CQI reports during periods of downlink activity and fewer reports at other times [4]. These additional CQIs can be initiated on demand of fast layer 1 signaling. In addition, a certain number of successive CQI values may be averaged with respect to channel quality at the user equipment. The averaged value is reported and used with the instantaneous CQI measured to select the MCS. The motivation for this technology is to improve the selection of MCS so that the delay due to HARQ retransmissions can be reduced. In addition, the uplink signaling overhead may be reduced, and this decreases the uplink interference. Finally, the use of the feature requesting extra CQI transmissions by fast layer 1 signaling improves the performance of the first packets of a packet call.

9.7 Adaptive Modulation and Coding

As explained, the link adaptation in HSDPA is performed using adaptive modulation and coding. According to the channel quality, the modulation and coding are assigned to the user change so that higher peak data rate and average throughput can be provided. A CQI has been introduced to inform the system about the channel conditions. To guarantee a block error rate (BLER) lower than 10% on the HS-PDSCH, each CQI is mapped onto a specific modulation and coding scheme corresponding to a given transport format. The selection of the transport format is performed by the MAC-hs located in node B. Each transport format or MCS is defined by a [11]

1. Modulation format, which can be either QPSK or 16QAM.
2. Turbo encoder rate, which varies between 0.17 and 0.89. The encoding rate depends on the user equipment capabilities (i.e., maximum number of HS-DSCH codes it can handle) and the desired transport block size (TBS). The different code rates are obtained through puncturing of bits in the turbo encoder of rate 1/3.
3. Number of HS-PDSCH codes allocated to the user, which ranges from 1 to the maximum number of codes supported by the user equipment, dependent on its category. Note that for any user equipment category, this number cannot exceed 15 codes. In fact, the spreading factor used in HSDPA is fixed at 16. Therefore, 16 branch codes are available, from which at least 1 code branch is reserved for signaling and control channels, thus leaving a maximum of 15 codes at best to allocate to a given user.

In the 3GPP specifications [11], 12 user equipment categories are defined according to the maximum number of HS-DSCH the user equipment can handle simultaneously. The maximum transport format that can be allocated to the user according to the user equipment capabilities is depicted in Table 9.1. The transport formats and the corresponding CQIs for user equipment category 10 (i.e., maximum number of 15 simultaneous HS-PDSCH codes it can handle) are shown in Table 9.2 [11].

9.8 HARQ

Fast link adaptation provides the flexibility to match the MCS to the short-term channel variations for each user. However, link adaptation is sensitive to measurement errors, delays in the CQI procedure, and unexpected channel variations. Therefore, the use of ARQ, which is insensitive to CQI measurement errors, is indispensable for tolerating higher error rates to save cell resources, to use higher-order MCS, and to increase the user and the average cell throughput.

The ARQ technique used in UMTS is selective repeat. The retransmissions are performed by the RLC layer in the RNC. Introducing ARQ induces delays in receiving error-free information and unfortunately interacts with higher-layer protocols such as the TCP used to handle end-to-end internet protocol (IP) packet communications between end hosts. If these interactions are not addressed and handled properly, drastic degradation in application flow rates is experienced.

Table 9.1 Maximal AMC Transport Format for Different User Equipment Categories

UE Category	Transport Block Size	Number of HS-PDSCH	Modulation
1	7268	5	16-QAM
2	7168	5	16-QAM
3	7168	5	16-QAM
4	7168	5	16-QAM
5	7168	5	16-QAM
6	7168	5	16-QAM
7	14411	10	16-QAM
8	14411	10	16-QAM
9	17237	12	16-QAM
10	25558	15	16-QAM
11	3319	5	QPSK
12	3319	5	QPSK

In HSDPA, the ARQ protocol is performed by the MAC-hs entity in node B. HSDPA uses the ARQ protocol combined with the forward error correction (FEC) code so that an erroneous packet is not discarded but instead is softly combined with its retransmissions to reduce the average delay in receiving error-free information. In addition, the tight coupling of HARQ and fast link adaptation limits the excessive use of ARQ—that is, the delay—since retransmissions occur if link adaptation fails to cope with the instantaneous channel conditions.

9.8.1 HARQ Types

In HSDPA, three types of HARQ have been studied and standardized in the 3GPP specifications [1, 14]: Chase combining, incremental redundancy (IR), and self-decodable IR.

In [17], Chase showed that the sequence resulting from combining two copies of the same sequence presents a lower error rate than the original sequences. Therefore, instead of discarding erroneous packets, the user equipment proceeds to soft combining, called Chase combining, multiple retransmissions of the same packets before decoding. This concept, developed and standardized in the 3GPP

Table 9.2 CQI Mapping for User Equipment Category 10

CQI Value	Transport Block Size	Number of HS-PDSCH	Modulation
1	137	1	QPSK
2	173	1	QPSK
3	233	1	QPSK
4	317	1	QPSK
5	377	1	QPSK
6	461	1	QPSK
7	650	2	QPSK
8	792	2	QPSK
9	931	2	QPSK
10	1262	3	QPSK
11	1483	3	QPSK
12	1742	3	QPSK
13	2279	4	QPSK
14	2583	4	QPSK
15	3319	5	QPSK
16	3565	5	16-QAM
17	4189	5	16-QAM
18	4664	5	16-QAM
19	5287	5	16-QAM
20	5887	5	16-QAM
21	6554	5	16-QAM
22	7168	5	16-QAM
23	9719	7	16-QAM
24	11418	8	16-QAM
25	14411	10	16-QAM
26	17237	12	16-QAM
27	21754	15	16-QAM
28	23370	15	16-QAM
29	24222	15	16-QAM
30	25558	15	16-QAM

specifications of Release 5 [1], reduces the delays compared to the ARQ used in the UMTS Release 99. This HARQ high-speed algorithm does not interact optimally with the AMC and the fast link adaptation since the multiple retransmissions are the same copies of the first one; that is, the same MCS is used even if the channel conditions change. Consequently, enhanced HARQ algorithms have been introduced in the 3GPP specifications [1,14]. These new schemes rely on IR.

In HARQ IR, instead of retransmitting the same copy of the erroneous packet, redundant information is incrementally added to retransmission copies. This represents better protection of the information packets and copes more with channel conditions and AMC. In this type of HARQ, only a fraction of the information sequence is sent in the retransmission packet—according to the degree of redundancy. The retransmitted packet is not self-decodable and should be combined with the first transmission before decoding. To counteract this problem, a self-decodable IR scheme has been studied and developed by the 3GPP [1,14]. To obtain a self-decodable scheme, incremental redundant information is added to the first sequence, and incremental puncturing is also used so that the receiver can reconstruct the information sequence and can decode each retransmission before soft combining the retransmissions in case of unsuccessful decoding.

Note that in cases of retransmissions, node B can select a combination for which no mapping exists between the original TBS and the selected combination of MCS. When such cases occur, the TBS index (TBSi) value signaled to the user equipment shall be set as TBSi = 63. TBS is the transport block size value, according to Table 9.2, corresponding to the MCS signaled on the HS-SCCH. Let S be the sum of the two values, $S = $ TBSi + TBS. The transport block size L(S) can be obtained by accessing the position S in Table 9.3 [14].

9.8.2 HARQ Protocol

In UMTS Release 99, selective repeat (SR) ARQ is used so that only the erroneous blocks are retransmitted. To achieve selective retransmission, a sequence number is required to identify each block. This requirement increases the system and receiver complexity. By combining the SR protocol and the HARQ techniques in HSDPA, more requirements and complexity are introduced compared to the SR-ARQ in UMTS Release 99.

These requirements can be summarized by the following points [18,19]: First, HARQ requires that the receiver must know the sequence number of each block before combining separate retransmissions. Therefore, the sequence number must be encoded separately from the data. In addition, it should be highly protected to cope with the worst channel conditions. Consequently, more signaling bandwidth is consumed. Second, the MAC-hs entity, in charge of handling the HARQ, should deliver data to the upper layers in sequence. Therefore, this entity should store the multiple retransmissions of an erroneous block and the correctly received data that

Table 9.3 Transport Block Size for HSDPA

Index	TBS*	Index	TBS	Index	TBS	Index	TBS	Index	TBS	Index	TBS	Index	TBS	Index	TBS
1	137	33	521	65	947	97	1681	129	2981	161	5287	193	9377	225	16630
2	149	34	533	66	964	98	1711	130	3035	162	5382	194	9546	226	16931
3	161	35	545	67	982	99	1742	131	3090	163	5480	195	9719	227	17237
4	173	36	557	68	1000	100	1773	132	3145	164	5579	196	9894	228	17548
5	185	37	569	69	1018	101	1805	133	3202	165	5680	197	10073	229	17865
6	197	38	581	70	1036	102	1838	134	3260	166	5782	198	10255	230	18188
7	209	39	593	71	1055	103	1871	135	3319	167	5887	199	10440	231	18517
8	221	40	605	72	1074	104	1905	136	3379	168	5993	200	10629	232	18851
9	233	41	616	73	1093	105	1939	137	3440	169	6101	201	10821	233	19192
10	245	42	627	74	1113	106	1974	138	3502	170	6211	202	11017	234	19538
11	257	43	639	75	1133	107	2010	139	3565	171	6324	203	11216	235	19891
12	269	44	650	76	1154	108	2046	140	3630	172	6438	204	11418	236	20251
13	281	45	662	77	1175	109	2083	141	3695	173	6554	205	11625	237	20617
14	293	46	674	78	1196	110	2121	142	3762	174	6673	206	11835	238	20989
15	305	47	686	79	1217	111	2159	143	3830	175	6793	207	12048	239	21368

(continued)

Table 9.3 Transport Block Size for HSDPA (Continued)

Index	TBS*	Index	TBS	Index	TBS	Index	TBS	Index	TBS	Index	TBS	Index	TBS	Index	TBS
16	317	48	699	80	1239	112	2198	144	3899	176	6916	208	12266	240	21574
17	329	49	711	81	1262	113	2238	145	3970	177	7041	209	12488	241	22147
18	341	50	724	82	1285	114	2279	146	4042	178	7168	210	12713	242	22548
19	353	51	737	83	1308	115	2320	147	4115	179	7298	211	12943	243	22955
20	365	52	751	84	1331	116	2362	148	4189	180	7430	212	13177	244	23370
21	377	53	764	85	1356	117	2404	149	4265	181	7564	213	13415	245	23792
22	389	54	778	86	1380	118	2448	150	4342	182	7700	214	13657	246	22422
23	401	55	792	87	1405	119	2492	151	4420	183	7840	215	13904	247	24659
24	413	56	806	88	1430	120	2537	152	4500	184	7981	216	14155	248	25105
25	425	57	821	89	1456	121	2583	153	4581	185	8125	217	14411	249	25558
26	437	58	836	90	1483	122	2630	154	4664	186	8272	218	14671	250	26020
27	449	59	851	91	1509	123	2677	155	4748	187	8422	219	14936	251	26490
28	461	60	866	92	1537	124	2726	156	4834	188	8574	220	15206	252	26969
29	473	61	882	93	1564	125	2775	157	4921	189	8729	221	15481	253	27456
30	485	62	898	94	1593	126	2825	158	5010	190	8886	222	15761	254	27952
31	497	63	914	95	1621	127	2876	159	5101	191	9047	223	16045		
32	509	64	931	96	1651	128	2928	160	5193	192	9210	224	16335		

cannot be delivered to upper layers due to prior erroneous blocks. Consequently, more memory is required in the user equipment.

The increase in complexity and in requirements of the SR-HARQ leads to the adoption of simpler HARQ strategies. The stop-and-wait SWARQ protocol is quite simple to implement, but since it is inefficient, a trade-off between the simple SW and the SR, called N-channel SW, has been developed and standardized for HSDPA [1,14].

The N-channel SW consists of activating N-HARQ processes in parallel, each using the SW protocol. This way, one HARQ process can transmit data on the HS-DSCH while other instances wait for the acknowledgment on the uplink. Using this strategy, the retransmission process behaves as if the SR-HARQ were employed. The advantage of the N-channel SW strategy with respect to the SR protocol is that a persistent failure in a packet transmission affects only one channel, allowing data to be transmitted on the other channels. In addition, compared to the simple SW, the N-channel SW provides the MAC-hs entity with the flexibility to allocate the HS-DSCH channel to the same user if radio conditions are favorable. However, this HARQ strategy imposes timing constraints on the maximum acceptable retransmission delay. The transmitter must be able to retransmit the erroneous packet $(N-1)$TTIs after the previous transmission. An example of N-channel SW-HARQ protocol is presented in Figure 9.12, in which the HS-DSCH channel is shared by two users having, respectively, four-channel and one-channel SW-HARQ entities [20].

Increasing the number of processes to N relaxes the timing constraints but simultaneously increases the memory required to buffer the soft samples of each partially received block [13]. Therefore, the number of HARQ instances must be limited. This limit is configured by the upper layers in HSDPA to eight parallel and simultaneous HARQ channels according to the 3GPP specifications [1,14].

9.8.3 HARQ Management

The parallel HARQ instances are handled by the MAC-hs entity in the node B. The MAC-hs manages the data flow control and the reassembly of the RLC PDUs with fixed size into a MAC-hs PDU with variable size. In addition, it performs HARQ and scheduling functionalities.

Once the RLC PDUs are reassembled in an MAC-hs PDU, the resulting PDU is assigned an eight-bit transmission sequence number (TSN) and an HARQ Id, or one of the three available bits. The TSN is carried in the header of the MAC-hs PDU and is transmitted over the HS-DSCH channel; however, the HARQ Id is carried over the HS-SCCH channel. The use of the HARQ could generate situations in which a packet is decoded error free before one or more previous received packets. Therefore, the TSN is used in the reordering entity, in the MAC-hs layer, to deliver data in sequence to the upper layer. Note that the number of MAC-hs PDUs transmitted is limited by the transmitter window size, up to 32 [21].

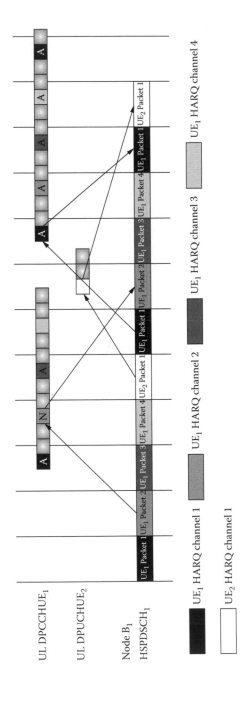

Figure 9.12 Example of *N*-channels stop- and -wait HARQ for two active users over HS-DSCH channel.

The reordering entity could result in a stall avoidance when a persistent failure occurs in the transmission of one block on a given channel of the *N*-channel SW. This can interact with higher-layer protocols and can drastically affect the performance of the data traffic carried over HSDPA (e.g., streaming). To control the stall avoidance in the user equipment reordering buffer, a timer T is introduced in the MAC-hs reordering entity and is configured by upper layers (see [14] for details).

The timer T is started when an MAC-hs PDU with a TSN higher than the expected TSN is correctly received. One timer is active at a given time. Once the MAC-hs PDU with the expected TSN is correctly received before the timer expiration, the timer is stopped. If the timer expires before correctly receiving the expected MAC-hs PDU, all correctly received MAC-hs up to the expected MAC-hs PDU and all correctly received packets with TSN higher than the expected TSN are delivered to the disassembly entity in charge of data delivery to the upper layer [14].

9.9 Packet Scheduling

The shared time structure of the HS-DSCH channel supports the use of time scheduling. Fast link adaptation based on AMC tightly coupled with scheduling provides higher transmission rates and average throughput. By allocating the HS-DSCH channel to the user with favorable channel conditions, higher-order MCS are selected, and higher achievable data rate and average throughput are provided. Introducing the MAC-hs entity in node B for scheduling and using a low TTI value of 2 ms allow better tracking of the short-term variations of the radio channel. Users with temporary good channel conditions are more easily selected.

With the growing demand on data application services—especially non-real-time services such as interactive and background—HSDPA, as any wireless system, should provide the capability of supporting a mixture of services with different QoS requirements. Even if the interactive and background services are seen as best-effort services with no service guarantees, these users still expect to receive data within a certain period of time. The starvation of these users can have a drastic effect on the performance of higher layers, such as the TCP layer. Therefore, a minimum service guarantee should be introduced for these services, and HSDPA should achieve some fairness in sharing resources among users and services. Fairness can be defined as meeting the data rate and the delay constraints of the different applications [5,6].

Consequently, the scheduler has two tasks to accomplish: to increase the average throughput by allocating the HS-DSCH channel to the user with favorable channel conditions and to achieve fairness between services. These two objectives are in contradiction, and there is a risk in achieving one at the expense of the other. A trade-off between fairness and efficiency (e.g., increasing the cell throughput) should be performed by the scheduler.

Scheduling over a time-shared channel has been addressed widely in the literature, and many proposals have been elaborated. The most famous algorithms are

Max C/I [2] and proportional fair (PF) proposed in [22,23]. In Max C/I, node B tracks the channel quality of each user by measuring the SIR on the CPICH and allocates the HS-DSCH to the user with the best SIR. This algorithm maximizes the cell capacity but presents a problem of fairness between users, especially for users at the cell border. In this context, work in [24] presented a performance comparison of several scheduling algorithms, concluding that algorithms providing the highest average cell throughput tend to present the largest variations in throughput per user.

References

1. 3GPP TS 25.308 V6.3.0, *HSDPA Overall Description. Stage 2*. Release 6, December 2004.
2. 3GPP TR 25.858 V5.0.0, *High-Speed Downlink Packet Access, Physical Layer Aspects*. Release 5, March 2002.
3. 3GPP TR 25.877 V5.1.0, *High-Speed Downlink Packet Access (HSDPA) Iub/Iur Protocol Aspects*. Release 5, June 2002.
4. 3GPP TR 25.899 V6.1.0, *High-Speed Download Packet Access (HSDPA) Enhancements*. Release 6, September 2004.
5. Holma, H., and Toskala, A., *WCDMA for UMTS: Radio Access for Third Generation Mobile Communications*, 3rd ed. London: Wiley, 2004.
6. Ameigeiras Gutierrez, P.J., Packet scheduling and QoS in HSDPA. Doctoral dissertation, Aalborg University, Aalborg, Denmark, October 2003, 59.
7. 3GPP TS 25.201 V6.2.0, *Physical Layer—General Description*. Release 6, June 2005.
8. 3GPP TS 25.211 V6.6.0, *Physical Channels and Mapping of Transport Channels onto Physical Channels (FDD)*. Release 6, September 2005.
9. 3GPP TS 25.212 V6.6.0, *Multiplexing and Channel Coding (FDD)*. Release 6, September 2005.
10. 3GPP TS 25.213 V6.4.0, *Spreading and Modulation (FDD)*. Release 6, September 2005.
11. 3GPP TS 25.214 V6.5.0, *Physical Layer Procedures (FDD)*. Release 6, March 2005.
12. 3GPP TS 25.331 V6.7.0, *Radio Resource Control (RRC) Protocol Specification*. Release 6, September 2005.
13. Tanner, R., and Woodard, J., *WCDMA Requirements and Practical Design*. London: Wiley, 2004.
14. 3GPP TS 25.321 V6.5.0, *Medium Access Control (MAC) Protocol Specification*. Release 6, June 2005.
15. Ericsson, N.C., On scheduling an adaptive modulation in wireless communications. Technical licentiate thesis, Uppsala University, Uppsala, Sweden, June 2001.
16. 3GPP TSG RAN WG1#28, *Duration of CQI Measurement*. Seattle, WA: Philips, August 2002. Tdoc R1-02-1084.
17. Chase, D., Code combining—A maximum-likelihood decoding approach for combining an arbitrary number of noisy packets. *IEEE Trans. Commun.*, 33(5), 385–393, May 1985.
18. 3GPP TSG RAN WG1#18, *Clarifications on Dual-Channel Stop-and-Wait HARQ*. Motorola, 2001. Tdoc R1-01-0048.

19. 3GPP TSG RAN WG1#18, Considerations on HSDPA HARQ Concepts. Nokia, 2001. Tdoc R1-01-0007.
20. 3GPP TSG RAN WG1#20, *Further Buffer Complexity and Processing Time Considerations on HARQ*. Busan, Korea: Nokia, May 2001. Tdoc R1-01-0553.
21. Bestak, R., Reliability mechanisms (protocols ARQ) and their adaptation in 3G mobile networks. Doctoral dissertation, ENST, Paris, December 2003. Available at http://pastel.paristech.org/archive/00000514/01/RBestakThese.pdf.
22. Holtzman, J.M., CDMA forward link waterfilling power control. *Proc. IEEE Veh. Technol. Conf. (VTC)*, 3, 1663–1667, May 2000.
23. Jalali, A., Padovani, R., and Pankaj, R., Data throughput of CDMAHDR: A high efficiency–high data wireless personal communication system. *Proc. 50th Annual IEEE VTC*, 3, 1854–1858, May 2000.
24. Elliott, R.C., and Krzymien, W.A., Scheduling algorithms for the CDMA2000 packet data evolution. *Proc. 52nd Annual IEEE VTC*, 1, 304–310, Fall 2002.

Chapter 10

IEEE 802.16-Based Wireless MAN

Sahar Ghazal and Jalel Ben-Othman

Contents

10.1 Introduction

The Institute of Electrical and Electronics Engineers (IEEE) 802.16 standard is a real revolution in wireless metropolitan-area networks (WirelessMANs) that enables high-speed access to data, video, and voice services. Worldwide Interoperability for Microwave Access (WiMAX) is the industry name given to the 802.16–2004 amendment by the vendor interoperability organization. The standard supports point-to-multipoint (PMP) as well as mesh mode. In the PMP mode, multiple subscriber stations (SSs) are connected to one base station (BS). The access channel from the BS to the SS is called the downlink (DL) channel, and the one from the SS to the BS is called the uplink (UL) channel. Two duplexing techniques are used: frequency division duplex (FDD), by which ULUL and DL operate on separate frequency channels and sometimes simultaneously, and time division duplex (TDD), by which UL and DLUL share the same frequency band. IEEE 802.16 defines connection-oriented media-access control (MAC).

The physical (PHY) layer specification works in both license bands and license-exempt bands and thus covers a range of 2 to 66 GHz. IEEE 802.16 is designed to support quality of service (QoS) mainly through the differentiation and classification of four types of service flows: unsolicited grant service (UGS), real-time polling service (rtPS), non-real-time polling service (nrtPS), and best effort (BE). Uplink scheduling is supported by the standard for only UGS flows. In this chapter, the QoS specified by the IEEE 802.16 standard is detailed, and some previous works on scheduling UL service flows are discussed.

To support mobility, the IEEE has defined the IEEE 802.16e amendment, which is also known as mobile WiMAX. Battery life and handover are essential issues in managing mobility between subnets in the same network domain (micromobility) and between two different network domains (macromobility).

10.2 Introduction

IEEE 802.16 is mainly aimed at providing broadband wireless access (BWA), and thus it may be considered as an attractive alternative solution to wired broadband technologies like digital subscriber line (xDSL) and cable modem access. Its main advantage is fast deployment, which results in cost savings. Such installation can be beneficial in very crowded geographical areas like cities and in rural areas where there is no wired infrastructure [1, 2]. The IEEE 802.16 standard provides network access to buildings through external antennas connected to radio BSs. The frequency band supported by the standard covers 2 to 66 GHz. In theory, the IEEE 802.16 standard, known also as WiMAX, is capable of covering a range of 50 km with a bit rate of 75 Mbps. However, in the real world, the rate obtained from WiMAX is about 12 Mbps with a range of 20 km. The Intel WiMAX solution for fixed access operates in the licensed 2.5-GHz and 3.5-GHz bands and the license-exempt 5.8-GHz band [3]. Research for a new standard of WirelessMAN dates from early 1998. In 2001, the 802.16 standard was finally approved by the IEEE. Here is a brief history of the standard:

1. IEEE 802.16-2001: IEEE 802.16 was formally approved by the IEEE in 2001 [1]. It is worth mentioning that many basic ideas of 802.16 were based on the Data Over Cable Service Interface Specification (DOCSIS). This is mainly due to the similarities between the hybrid fiber-coaxial (HFC) cable environment and the BWA environment. Utilizing the 10- to 66-GHz frequency spectrum, 802.16 is thus suitable for line-of-sight (LOS) applications. With its short waves, the standard is not useful for residential settings because of the non-line-of-sight (NLOS) characteristics caused by rooftops and trees.

2. IEEE 802.16a-2003: This extension of the 802.16 standard covers fixed BWA in the licensed and unlicensed spectrum from 2 to 11 GHz. The entire standard along with the latest amendments was published on April 1, 2003 [3]. This amendment was mainly developed for NLOS applications; thus, it is a practical solution for the last-mile problem of transmission where obstacles like trees and buildings are present. IEEE 802.16a supports PMP network topology and optional mesh topology BWA. It specifies three air interfaces: single-carrier modulation, 256-point transform orthogonal frequency-division multiplexing (OFDM), and 2048-point transform orthogonal frequency-division multiple access (OFDMA). The IEEE 802.16a standard specifies channel sizes ranging from 1.75 to 20 MHz. This protocol supports low-latency applications such as voice and video.

3. IEEE 802.16-2004: This standard revises and consolidates 802.16–2001, 802.16a2003, and 802.16c–2002 [1]. WiMAX technology based on the 802.16-2004 standard is rapidly proving itself as a technology that will play a key role in fixed broadband WirelessMANs. The MAC layer supports mainly PMP architecture and mesh topology as an option. The standard is specified to support fixed wireless networks. The mobile version of 802.16 is known as mobile WiMAX or 802.16e.

4. IEEE 802.16e (mobile WiMAX): This is the mobile version of the 802.16 standard. This new amendment aims at maintaining mobile clients connected to a MAN while moving around. It supports portable devices from mobile smart phones and personal digital assistants (PDAs) to notebook and laptop computers. IEEE 802.16e works in the 2.3-GHz and 2.5-GHz frequency bands [2].

In broadband wireless communications, QoS is still an important criterion. The WiMAX standard is designed to provide QoS through classification of different types of connections as well as scheduling. The standard supports scheduling only for fixed-size, real-time service flows. The admission control and scheduling of both variable-size real-time and non-real-time connections are not considered in the standard. Thus, WiMAX QoS is still an open field of research and development for both constructors and academic researchers. The standard should also maintain connections for mobile users and guarantee a certain level of QoS. IEEE 802.16e is designed specifically to support mobility.

The rest of this chapter is organized as follows: Section 10.3 presents the IEEE 802.16 MAC and Section 10.4 presents PHY layers. In Section 10.5, QoS and mobility issues are discussed, concentrating on QoS scheduling methods and mechanisms proposed by some authors. We conclude in Section 10.6.

10.3 IEEE 802.16 MAC Layer

The MAC layer is a common interface that interprets data between the lower PHY layer and the upper data link layer. In IEEE 802.16, the MAC layer is designed mainly to support the PMP architecture with a central BS controlling the SSs connected to it. The 802.16 MAC protocol is connection oriented. On entering the network, each SS creates one or more connections over which data are transmitted to and from the BS. The application must establish a connection with the BS as well as associated service flows.

The BS assigns the transport connection with a unique 16-bit connection identification (CID) [4]. DL connections are either unicast or multicast, while UL connections are always unicast. Every service flow is mapped to a connection, and the connection is associated with a QoS level. The MAC layer schedules the usage of air link resources and provides QoS differentiation. At SS initialization, three pairs of management connections are established between the SS and the BS in both the UL and DL direction:

- Basic connection is used to exchange short time-urgent MAC management messages (e.g., DL burst profile change request/DL burst profile change response).
- Primary management connection is used to exchange longer, more delay-tolerant management messages (e.g., registration request and registration response messages).
- Secondary management connection is used for higher-layer management messages and SS configuration messages (e.g., Dynamic Host Configuration Protocol [DHCP], Trivial File Transfer Protocol [TFTP]).

10.3.1 MAC Sublayers

The 802.16 MAC layer is divided into three sublayers: the service-specific convergent sublayer (CS), which interfaces to higher layers; the common part sublayer (CPS), which carries out the key MAC functions; and the privacy sublayer (PS), which provides authentication, secure key exchange, and encryption [1] (see Figure 10.1).

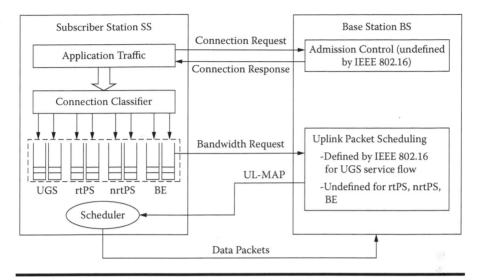

Figure 10.7   QoS architecture.

is admitted, both the SS and BS can allocate resources for this connection. QoS parameter changes are requested through exchanging dynamic service flow messages (DSA, DSC, DCD) between the BS and the SS and approved by the authorization module [6].

Figure 10.7 shows the existing QoS architecture defined by the IEEE 802.16 standard. As seen in Figure 10.7, admission control and the UL scheduler are both situated in the BS.

Admission control is responsible for accepting or rejecting the connection according to the available bandwidth that satisfies the connection and guarantees the required QoS without degrading the QoS for other existing connections. Admission control is not defined in the standard, although many propositions are made by different authors to establish admission control in the BS [4,7,8]. Since the IEEE 802.16 MAC protocol is connection oriented, the application first establishes the connection with the BS as well as the associated service flow (UGS, rtPS, nrtPS, or BE). Connection request/response messages are exchanged between the SS and the BS to establish the connection. The SS sends a connection request message asking for its bandwidth needs. If the required resources (bandwidth) are available, then the BS replies with a connection response message and the connection is established for further communication processes. Once the connection is established between the SS and the BS, the SS sends a bandwidth request message to the BS. The UL-MAP message is then generated by the UL packet scheduler in the BS. This message carries IEs about transmission opportunities or the time slot in which the SS can transmit during the UL subframe.

On the SS side, the scheduler retrieves packets from the queue and sends them on the UL channel according to the IE received in the UL-MAP message sent by the BS. UL packet scheduling, which is found on the BS side, controls all UL packet transmissions. The IEEE 802.16 standard defines UL packet scheduling in the BS for UGS service flows; rtPS, nrtPS, and BE service flows are undefined.

To provide priority to certain flows, they first must be classified. The classifier is a set of matching criteria applied to each packet entering the IEEE 802.16 network [1]. All data packets from the application layer in the SS are associated with a unique CID. The classifier then classifies these connections depending on their CIDs and forwards them to the appropriate queue. At the SS, the scheduler retrieves the packets from the queues and transmits them to the network in the appropriate time slots as defined by the UL-MAP message. DL classifiers are applied by the BS to packets it is transmitting, while UL classifiers are applied at the SS.

10.5.1.1 Service Flow Concepts

The concept of service flows on a transport connection is central to the operation of the MAC protocol, and thus service flows provide a mechanism for both UL and DL QoS management [2]. In the 802.16 standard, all service flows have a 32-bit service flow identifier (SFID). Active service flows also have a 16-bit CID. Since multiple service flows may need to share a common set of QoS parameters, the 802.16 standard defines the concept of service classes or service class names. A service class is normally defined in the BS to have a particular QoS parameter set. A service flow with a certain QoS parameter set being referenced to a certain service class may increase or even override the QoS parameter setting of that service class.

The IEEE 802.16 MAC layer enables classification of traffic flow and maps them to connections with specific scheduling services. Each connection is associated with a single scheduling data service, and each data service is associated with a set of QoS parameters that quantify aspects of its behavior. Four types of scheduling services are defined by the 802.16 standard [1]:

1. UGS: Supports real-time UL service flows that transport fixed-size data packets on a periodic basis, such as voice-over Internet protocol (VoIP). The service offers fixed-size grants on a real-time periodic basis.
2. rtPS: Supports real-time UL service flows that transport variable-sized data packets on a periodic basis, such as Moving Picture Experts Group (MPEG) video. The service offers real-time, periodic, unicast request opportunities that meet the flows' real-time needs and allow the SS to specify the size of the desired grant. This service requires more request overhead than UGS but supports variable grant sizes for optimum data transport efficiency.
3. nrtPS: Supports non-real-time flows such as File Transfer Protocol (FTP). The BS polls the nrtPS CIDs periodically so that the UL service flow requests opportunities even during network congestion.

4. BE: Provides efficient service for BE traffic in the UL, such as Hypertext Transfer Protocol (HTTP). For this service to work correctly, the SS is allowed to use contention request opportunities.

10.5.1.2 Uplink Scheduling

Scheduling services represent the data-handling mechanisms supported by the MAC scheduler for data transport on a connection. Each connection is associated with a single scheduling service. A scheduling service is determined by a set of QoS parameters that quantify aspects of its behavior. These parameters are managed using the DSA and DSC message dialogs [2].

Uplink request/grant scheduling is performed by the BS with the intent of providing each subordinate SS with bandwidth for UL transmissions or opportunities to request bandwidth. By specifying a scheduling type and its associated QoS parameters, the BS scheduler can anticipate the throughput and latency needs of the UL traffic and provide polls or grants at appropriate times.

The access control, which is based on the concepts of service flows, specifies a QoS signaling mechanism for both bandwidth request and bandwidth allocation in both the UL and DL channels. IEEE 802.16 defines only UL scheduling for UGS service flows and leaves the QoS scheduling algorithm for other service flows to be defined by the constructor.

IEEE 802.16 provides scheduling services for both UL and DL traffic. In the DL direction, the flows are simply multiplexed, and therefore the standard scheduling algorithms can be used. UL scheduling is complex as it needs to be in accordance with UL QoS requirements provided by the standard. In fact, the standard defines the required QoS signaling mechanism, such as bandwidth request (BW-REQ) and UL-MAP, but it does not define the UL scheduler. The UL scheduling mechanism, which is situated in the BS, is responsible for fair and efficient allocation of assigned time slots in the UL direction. To guarantee a certain level of QoS, the scheduling mechanism is applied to different types of traffic in the network, including real-time and non-real-time applications. In recent years, many scheduling mechanisms have been proposed for real-time and non-real-time traffic. It should be mentioned here that the scheduling algorithm can guarantee the QoS only if the number of connections is limited by the admission control. The sum of the minimum bit rates of all connected flows must be lower than the capacity allocated to the applications within the same QoS level [9].

10.5.1.3 Proposed Scheduling Methods

In computer networks, admission control and scheduling are very necessary to allocate sufficient resources for users while satisfying the QoS. In the IEEE 802.16 standard, admission control is not defined, while scheduling is defined only for UGS flows. MAN research has focused on scheduling service flows and defining mechanisms of admission control to enhance the QoS and to provide wireless end users with a similar level of QoS as wired users.

WiMAX supports all types of service flows, including real-time and non-real-time data traffic. Real-time traffic issued at periodic intervals consists of either variable-size data packets, such as MPEG video, or fixed-size data packets, such as VoIP. Non-real-time data traffic requires a decreased degree of QoS, such as FTP or HTTP data packets. In recent years, WiMAX network QoS research has focused on defining the admission control mechanism and UL scheduling algorithms. The rest of this section presents UL scheduling mechanisms proposed by several authors.

Hawa and Petr [10] propose a new mechanism for scheduling service flows. They use three types of queues for queuing service flows, with type 1 having the highest priority and type 3 having the lowest. Type 1 is used to store not only UGS flows, but also unicast rtPS and nrtPS flows. Semiprimitive priority is assigned to type 1 queues, which means that the service grant is not interrupted unless the type 1 grant arrives with a dangerously early deadline. In the work of Hawa and Petr, the proposed QoS architecture addresses only the UL channel. Scheduling of data traffic on the DL channel is not considered. On the other hand, the presented solution is a hardware implementation that is difficult to apply and costly in terms of money, although it is much faster than a software solution.

Wongthavarawat and Ganz [4] define a hierarchical QoS architecture. This architecture consists of two layers. A strict priority is used in the first layer, where the entire bandwidth is distributed between the different service flows. UGS has the highest priority, then rtPS, nrtPS, and finally BE. In the second level of this hierarchal architecture, different mechanisms are used to control the QoS for each class of service flow. The UL packet scheduler allocates fixed bandwidth to the UGS connection based on their fixed bandwidth requirements. Earliest deadline first (EDF) is used to schedule rtPS service flows, in which packets with the earliest deadline are scheduled first. The nrtPS service flows are scheduled using the weight fair queue (WFQ) based on the weight of the connection. The remaining bandwidth is equally allocated to each BE connection.

The disadvantage of this hierarchical structure is that high-level connections (like real-time data traffic) may starve low-priority ones. Chen, Jiao, and Wang [7] propose a new hierarchical scheduling algorithm that also consists of two layers. To overcome the stated problem, the deficit fair priority queue (DFPQ) replaces the strict priority mechanism in the first layer. The DFPQ schedules bandwidth application services in an active list. If the queue is not empty, it stays in the active list; otherwise, it is removed. This new algorithm serves different types of service flows in both the UL and DL and provides more fairness to the system. Priority in the second layer is distributed between different service flows using a similar method to the one presented by Wongthavarawat and Ganz [4]. Since UGS service flow is assigned a fixed bandwidth, it does not appear in the scheduling architecture. The round robin (RR) method is used to schedule BE service flows.

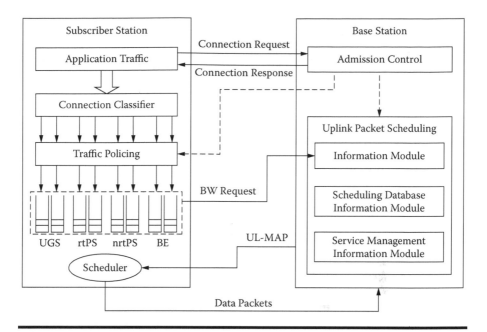

Figure 10.8   Proposed QoS architecture.

We noticed that most authors present a new QoS architecture by modifying the MAC layer in both the SS and BS. They add new modules that are not defined in the IEEE 802.16 standard. Because of the similarity of the proposed QoS architectures and the modules added, only the work done by Wongthavarawat and Ganz [4] is explained here (see Figure 10.8).

First, the authors [4] define the admission control that decides whether a new connection can be established based on the available resources while maintaining the same QoS of current connections. On the BS side, they define a new scheduling mechanism for rtPS, nrtPS, and BE by adding different modules responsible for collecting the queue size information from the bandwidth requests received during the previous time frame and then to generate the UL-MAP message. On the SS side, a traffic-policing module has been added. The role of this module is to enforce traffic based on the connection traffic contract and to control the delay violation for rtPS.

In the presented work, we noticed that some authors [4,7] define an admission control module in the BS that has to control the number of admitted service flows. While the scheduling mechanism presented by Hawa and Petr [10] does not take into consideration the DL burst, most of proposed solutions that came after pay more attention to this point. Another important issue is that the implemented UL scheduler is always found on the BS side, which means that the SS will only send its bandwidth requirements for each connection, and the BS has to schedule the different service flows following the criteria of priority predefined by the mechanism.

10.5.2 Mobility Management Support

Ongoing development and industry support for the new standard led to the development of a mobile version of IEEE 802.16 (known as mobile WiMAX). This version is based on amendment 802.16e and provides support for handoff and roaming [11]. This new amendment aims to keep mobile clients connected to the MAN and supports portable devices from smart mobile phones and PDAs to notebook and laptop computers. IEEE 802.16e can also use the network to provide fixed network services. Mobile WiMAX profiles cover 5-, 7-, 8.75-, and 10-MHz channel bandwidths for licensed worldwide spectrum allocations in the 2.3-, 2.5-, 3.3-, and 3.5-GHz frequency bands [12].

Mobility management solutions can be classified into two categories [13]: macromobility and micromobility management solutions. The first category refers to the movement of mobile stations (MSs) between two network domains, while in the second case the MS moves between two subnets within the same network domain.

In a mobile WiMAX system, in which MSs are moving within the sector of the BS, battery life and handoff are essential criteria for mobile applications. To extend battery life for mobile devices, the system is designed to support power saving. Sleep mode operation for power saving is one of the most important features to extend battery life for MSs.

In addition to minimizing MS power usage, sleep mode decreases usage of BS air interface resources. In the sleep period, the MS is considered unavailable to the serving BS [2]. Implementation of sleep mode is optional for the MS and mandatory for the BS.

The 802.16e amendment defines two operational modes for the MS when registering with the serving BS: sleep mode and awake mode [14]. In the awake mode, the MS is available to the serving BS and can send or receive data. The MS is considered unavailable to the serving BS during the sleep mode. Under sleep mode operation, the MS initially sleeps for a fixed interval called the *sleep window*. The initial maximum window size is negotiated between the MS and the BS. After waking, if the MS finds that there is no buffered DL traffic destined to it from the BS, then it doubles the sleep window size up to the maximum sleep window size. On the other hand, if the MS has packets to send on the UL channel, it can wake up prematurely to prepare for the UL transmission (e.g., bandwidth request), and then it can transmit its pending packets in the allocated time slots assigned by the BS. When the MS enters the sleep mode, the BS does not transmit data to the MS, which may power down or perform other operations that do not require connection with the BS.

To manage power usage in a more efficient way, the IEEE 802.16e standard also defines the idle mode [2]. In this mode, the MS becomes periodically available to receive DL broadcast traffic messaging without the need to register with a specific BS as the MS traverses an air link environment populated by multiple BSs. This mode allows the MS to conserve power and resources by restricting its activity to scanning at discrete intervals and thus eliminates the active requirement for handover operation and other normal operations. On the BS and network side, idle mode provides a simple and timely method for alerting the MS to pending DL traffic directed to the MS and thus eliminates air interface and network handover traffic from essentially inactive MS.

10.5.2.1 Handover Mechanism

Handover is the process by which an MS moves from the air interface provided by one BS to the air interface provided by another BS [2]. The MS needs to perform this process if it moves out of the transmission range of the serving BS or if the serving cell is overloaded. A serving BS periodically broadcasts an advertisement message (MOB-NBR-ADV), which is decoded by the MS to obtain information about the characteristics of the neighboring BS. A BS may allocate a time interval to an MS called a *scanning interval.* An MS then sends a scan request message (MOB-SCN-REQ). In response to this message, the serving BS indicates a group of neighbor BSs through a scanning response message (MOB-SCN-RSP).

The MS then selects one suitable target BS for handover (recommended by the serving BS) from this group. Through ranging, the MS can acquire the timing, power, and frequency adjustment information of the neighboring BS. The target BS–MS association information is reported to the serving BS.

The MAC layer (L2) handover is divided into two phases: the handover preregistration phase and the real handover phase [15]. During handover preregistration, the target BS is selected and preregistered with the MS. However, the connection to the currently serving BS is maintained, and packets may be exchanged during the preregistration phase. In the real handover, the MS releases the serving BS and reassociates with the target BS.

Either the MS or BS can start the handover process, which consists of several stages, starting with the cell reselection, followed by a decision to make the handover. After that, the MS synchronizes to the target BS DL and performs handover ranging followed by termination of MS contact as a final step in the handover process. Handover cancellation may be done by the MS at any time [2] (see Figure 10.9).

In the first step, cell reselection, the MS acquires information about the neighboring BSs from the advertisement message. The MS then selects a BS as a handover target. The serving BS may schedule scanning intervals or sleep intervals to conduct cell reselection activity. A handover begins when an MS decides to migrate from the

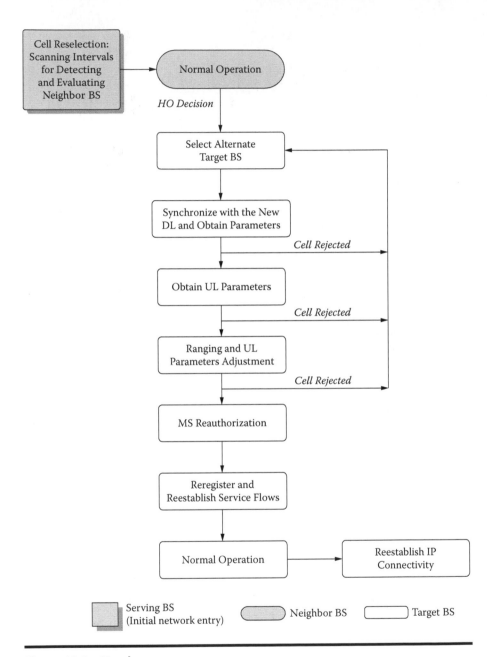

Figure 10.9 Handover process.

serving BS to the target BS. This decision may originate at the MS, the serving BS, or the network backbone. The serving BS may notify potential target BSs or send them MS information over the backbone network to expedite handover. Once the target BS is selected, the MS must synchronize with the DL transmission of the target BS. During this phase, the MS receives DL and UL transmission parameters. If the MS previously received information about this BS (through network topology acquisition), the length of this process may be shortened. After the synchronization step, the MS needs to perform initial ranging or handover ranging. Ranging is a procedure by which the MS receives the correct transmission parameters, such as time offset and power level. The target BS may acquire MS information from either the serving BSs or the backbone network. Depending on the knowledge of the target BS about the MS, some parts of the network reentry process may be omitted. The final step in handover is the termination of MS contact. The serving BS terminates all connections belonging to the MS; in other words, information in the queue, automatic repeat request (ARQ) state machine, counters, timers, and so on is all discarded.

The MS can cancel the handover process at any time and resume normal operation with the serving BS. The only condition is that the cancellation process must take place before expiration of the resource retain time interval after transmission of the handover indication message (MOB-HO-IND) [2].

10.6 Conclusion

This chapter presented the IEEE 802.16 standard and the different versions developed since its appearance in 2001. The standard modifies the MAC and PHY layers. The MAC layer PDU format defines two headers: the generic MAC header, which may be followed by payload or CRC data, and the signaling MAC header with no payload or CRC data. The MAC is connection oriented and provides the concept of service flows, with each associated with a QoS level. Admission control and UL scheduling mechanisms are not specified by the standard. Research has been aimed at enhancing the QoS and defining the admission control mechanism and the UL and DL scheduling architectures for the different types of connections. The scheduling mechanisms presented in this chapter are mainly situated in the BS, which is responsible for fair and efficient allocation of resources in the UL direction. While some scheduling mechanisms propose scheduling of connections in the UL direction only, other more recent proposals work in both the UL and DL direction. Implementation of the scheduler on the SS side is still possible and may provide the system with more advantages in terms of QoS. To provide mobility and handover in the WirelessMAN, the IEEE 802.16e amendment has been defined. This new version of the standard is designed to support mobility and handover in broadband wide-area networks. Management of mobility and battery energy consumption of the MS are important issues that need more research and improvement.

References

1. IEEE, *IEEE Standard for Local and Metropolitan Area Networks—Part 16: Air Interface for Fixed Broadband Wireless Access Systems, Standard 802.16–2004*. Washington, DC: IEEE, 2004.
2. IEEE, *IEEE Standard for Local and Metropolitan Area Networks—Part 16: Air Interface for Fixed and Mobile Broadband Wireless Access Systems Amendment 2: Physical and Medium Access Control Layers for Combined Fixed and Mobile Operation in Licensed Bands, Standard 802.16e–2005*. Washington, DC: IEEE, 2006.
3. Intel, *Understanding Wi-Fi and WiMAX as Metro-Access Solutions, White Paper*. Intel Corporation, 2004.
4. Wongthavarawat, K., and Ganz, A., Packet scheduling for QoS support in IEEE 802.16 broadband wireless access systems. *Int. J. Commun. Sys.*, 16, 81, 2003.
5. Eklund, C., Marks, R.B., and Stanwood, K.L., IEEE standard 802.16: A technical overview of the WirelessMAN air interface for broadband wireless access. *IEEE Commun. Mag.*, 40(6), 98, 2002.
6. Wood, M.C., *Analysis of the Design and Implementation of QoS over IEEE 802.16*. 2006.
7. Chen, J., Jiao, W., and Wang, H., *A Service Flow Management Strategy for the IEEE 802.16 Broadband Wireless Access Systems in TDD Mode, White Paper*. Washington, DC: IEEE, 2005.
8. Wang, H., Li, W., and Agrawal, D.P., *Dynamic Admission Control and QoS for 802.16 Wireless MAN, Wireless Telecommunications Symposium*. Washington, DC: IEEE, 2005, p. 60.
9. Bostic, J., and Kandus, G., *MAC Scheduling for Fixed Broadband Wireless Access Systems, COST 263 TCM*. 2002.
10. Hawa, M., and Petr, D.W., Quality of service scheduling in cable and broadband wireless access systems. In *Tenth International Workshop on Quality of Service*. Washington, DC: IEEE, 2002, p. 247.
11. WiMAX Forum, *Fixed, Nomadic, Portable and Mobile Applications for 802.16–2004 and 802.16e WiMAX Networks*. Beaverton, OR: WiMAX Forum, 2005.
12. WiMAX Forum, *Mobile WiMAX—Part 1: A Technical Overview and Performance Evaluation*. Beaverton, OR: WiMAX Forum, 2006.
13. Hu, J.-Y., and Yang, C.-C., On the design of mobility management scheme for 802.16 based network environment. *IEEE 62nd Veh. Technol. Conf. (VTC-2005Fall)*, 2720, 2005.
14. Han, K., and Choi, S., Performance analysis of sleep mode operation in IEEE 802.16e mobile broadband wireless access systems. *IEEE 63rd Veh. Technol. Conf. (VTC-2006-Spring)*, 3, 1141, 2006.
15. Kim, K., Kim, C., and Kim, T., A seamless handover mechanism for IEEE 802.16e broadband wireless access. *Int. Conf. Comput. Sci.*, 2, 5, 2005.

Chapter 11

3GPP Long-Term Evolution

Gábor Fodor, András Rácz, Norbert Reider,
András Temesváry, Gerardo Gómez,
David Morales-Jiménez, F. Javier López-Martínez,
Juan J. Sánchez, and José Tomás Entrambasaguas

Contents

11.1 Introduction

The long-term evolution (LTE) physical layer is targeted to provide improved radio interface capabilities between the base station and user equipment (UE) compared to previous cellular technologies such as Universal Mobile Telecommunications System (UMTS) or high-speed downlink (DL) packet access (HSDPA).

In this chapter, we discuss the radio resource management (RRM) functions in LTE. The term *radio resource management* is generally used in wireless systems in a broad sense to cover all functions that are related to the assignment and the sharing of radio resources among the users (e.g., mobile terminals, radio bearers, user sessions) of the wireless network. The types of required resource control, required resource sharing, and assignment methods are primarily determined by the basics of the multiple-access technology such as frequency-division multiple access (FDMA), time-division multiple access (TDMA), or code-division multiple access (CDMA) and their feasible combinations. Likewise, the smallest unit in which radio resources are assigned and distributed among the entities (e.g., power, time slots, frequency bands/carriers, or codes) also varies depending on the fundamentals of the multiple-access technology employed on the radio interface. This chapter presents a detailed description of the LTE radio interface physical layer. For that purpose, Section 11.2 provides an introduction to the physical layer, focusing on the physical resources structure and the set of procedures defined within this layer.

The placement and the distribution of the RRM functions to different network entities of the radio access network (RAN), including the functional distribution between the terminal and the network as well as the protocols and interfaces between the different entities, constitute the *RAN architecture*. Although the required RRM functions determine, to a large extent, the most suitable RAN architecture, it is often an engineering design decision how a particular RRM function should be realized. For example, whether intercell interference coordination (ICIC) or handover control is implemented in a distributed approach (in each base station) or in a centralized fashion can both be viable solutions. We discuss such design issues throughout this chapter.

In LTE, the radio interface is based on the orthogonal frequency-division multiplexing (OFDM) technique. In fact, OFDM serves both as a modulation technique and as a multiple-access scheme. Consequently, much of the RRM functions can be derived from the specifics of the OFDM modulation. In the rest of this section, we give an overview of the LTE RAN architecture, including an overview of the OFDM-based radio interface. Subsequently, we define and introduce the notion of radio resource in LTE and present the requirements that the Third Generation Partnership Project (3GPP) has set on the spectral efficient use of radio resources, which entail the presence of certain RRM functions in the system. For a comprehensive overview and detailed description of the overall LTE system, refer to [1].

11.1.1 The LTE Architecture

Before discussing the details of the LTE architecture, it is worth looking at the general trends in radio link technology development, which drive many of the architectural changes in cellular systems today. The most important challenge in any radio system is to combat the randomly changing radio link conditions by adapting the transmission and reception parameters to the actual link conditions (often referred to as the *channel state*). The better the transmitter can follow the fluctuations of the radio link quality and adapt its transmission accordingly (modulation and coding, power allocation, scheduling), the better it will utilize the radio channel capacity. The radio link quality can change rapidly and with large variations, which are primarily due to the fast fading fluctuations on the radio link, but other factors such as mobility and interference fluctuations also contribute to these. Because of this, the various RRM functions have to operate on a timescale matching that of the radio link fluctuations. As we discuss, the LTE requirements on high (peak and average) data rates, low latency, and high spectrum efficiency are achieved partly due to the Radio Resource Control (RRC) functions being located close to the radio interface where such instantaneous radio link quality information is readily available. Besides the fast-changing radio link quality, the bursty nature of typical packet data traffic also imposes a challenge on the radio resource assignment and requires dynamic and fast resource allocation that takes into account not only the instantaneous radio link quality but also the instantaneous packet arrivals. As a consequence, a general trend in the advances of cellular systems is that the radio-specific functions and protocols get terminated in the base stations, and the rest of the RAN entities are radio access technology agnostic. Thereby, the RAN exhibits a distributed architecture without a central RRC functionality.

The LTE architecture is often referred to as a *two-node architecture* as logically there are only two nodes involved—in both the user and control plane paths—between the UE and the core network. These two nodes are (1) the base station, called eNode B, and (2) the serving gateway (S-GW) in the user plane and the mobility management entity (MME) in the control plane, respectively. The MME and the GW belong to the core network, called the evolved packet core (EPC) in 3GPP terminology. The GW executes generic packet-processing functions similar to router functions, including packet filtering and classification. The MME terminates the so-called nonaccess stratum signaling protocols with the UE and maintains the UE context, including the established bearers and the security context, as well as the location of the UE. To provide the required services to the UE, the MME talks to the eNode Bs to request resources for the UE. It is important to note, however, that the radio resources are owned and controlled solely by the eNode B, and the MME has no control over the eNode B radio resources. Although the MME and the GW are LTE-specific nodes, they are radio agnostic.

The LTE architecture is depicted in Figure 11.1, which shows both the control plane and the user plane protocol stacks between the UE and the network. As can be seen, the radio link-specific protocols, including radio link control (RLC) [2] and medium-access control (MAC) [3] protocols, are terminated in the eNode B. The Packet Data Convergence Protocol (PDCP) layer [4], which is responsible for header compression and ciphering, is also located in the eNode B. In the control plane, the eNode B uses the RRC protocol [8] to execute the longer timescale RRC toward the UE. For example, the establishment of radio bearers with certain quality-of-service (QoS) characteristics, the control of UE measurements, or the

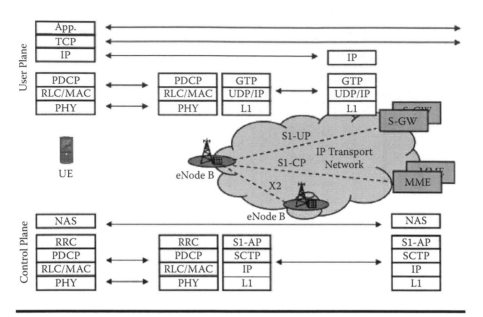

Figure 11.1 The 3GPP long-term evolution (LTE) RAN architecture. SCTP is stream control transmission protocol. L1 is Layer-1.

control of handovers are supported by RRC. Other short timescale RRC toward the UE is implemented via the MAC layer or the physical layer control signaling (e.g., the signaling of assigned resources and transport formats via physical layer control channels).

The services are provided to the UE in terms of evolved packet system (EPS) *bearers*. The packets belonging to the same EPS bearer get the same end-to-end treatment in the network. A finite set of possible *QoS profiles*, in other words packet treatment characteristics, are defined, which are identified by so-called labels. A label identifies a certain set of packet treatment characteristics (i.e., scheduling weights, radio protocol configurations such as RLC acknowledge or unacknowledge mode, hybrid automatic repeat request [HARQ] parameters.). Each EPS bearer is associated with a particular QoS class, that is, with a particular QoS label. There are primarily two main bearer types, guaranteed bit rate (GBR) and non-GBR bearers. For GBR bearers, the network guarantees a certain bit rate to be available for the bearer at any time. The bearers, both GBR and non-GBR, are further characterized by a maximum bit rate (MBR), which limits the maximum rate that the network will provide for the given bearer.

The end-to-end EPS bearer can be further broken down into a radio bearer and an access bearer. The radio bearer is between the UE and the eNode B, while the access bearer is between the eNode B and the GW. The access bearer determines the QoS that the packets get on the transport network, while the radio bearer determines the QoS treatment on the radio interface. From an RRM point of view, the radio bearer QoS is in our focus since the RRM functions should ensure that the treatment that the packets get on the corresponding radio bearer is sufficient and can meet the end-to-end EPS bearer-level QoS guarantees.

In summary, we can formulate the primary goal of RRM to control the use of radio resources in the system such that the QoS requirements of the individual radio bearers are met, and the overall used radio resources on the system level are minimized. That is, the ultimate goal of RRM is to satisfy the service requirements at the smallest possible cost for the system.

11.1.2 The Notion of Radio Resource in LTE

The radio interface of LTE is based on the OFDM technology, in which the radio resource appears as one common shared channel, shared by all users in the cell. The scheduler, which is located in the eNode B, controls the assignment of time–frequency blocks to UEs within the cell in an orthogonal manner such that no two UEs can be assigned the same resource, thereby avoiding intracell interference. One exception is multiuser spatial multiplexing, also called multiuser MIMO (multiple-input multiple-output), when multiple UEs having spatially separated channels can be scheduled on the same time–frequency resource, which is supported in the uplink (UL) of LTE. There needs to be such a scheduler function both for the UL and for the DL shared channels. We note that the LTE physical layer has been designed such that it is compatible with both frequency-domain and time-domain duplexing (FDD/TDD) modes.

Figure 11.2 shows the resource grid of the UL and DL shared channels. The smallest unit in the resource grid is the resource element (RE), which corresponds to one subcarrier during one symbol duration. These REs are organized into larger blocks in both time and frequency, where seven of such symbol durations

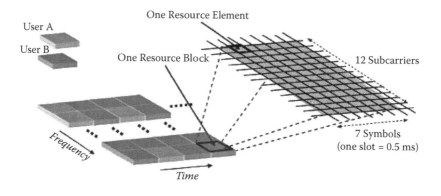

Figure 11.2 **Uplink/downlink resource grid.**

constitute one *slot* of length 0.5 ms, and 12 subcarriers during one slot forms the so-called resource block (RB). Two consecutive time slots are called a *subframe,* and 10 such subframes create a *frame,* which is 10 ms long. The scheduler can assign RBs only in pairs of two consecutive RBs (in time); that is, the smallest unit of resource that can be assigned is two RBs. There is, however, one important difference between the feasible assignments on the UL and DL shared channels. Since in the UL the modulation uses the single-carrier FDMA (SC-FDMA) concept, the allocation of RBs per UE has to be on consecutive RBs in frequency. We note that the SC-FDMA modulation basically corresponds to a discrete Fourier transform (DFT) precoded OFDM signal for which the modulation symbols are mapped to consecutive OFDM carriers. The primary motivation for using the SC-FDMA scheme in the UL is to achieve better peak-to-average power ratios (PAPRs). For more details on the layer 1 (L1) radio interface parameters, modulation, and coding schemes, see [5].

Since the LTE physical layer is defined such that it supports various multiantenna MIMO schemes [6], such as transmit diversity and spatial multiplexing, the virtual space of radio resources is extended, with a third dimension corresponding to the antenna port, beside the classical time and frequency domains. This essentially means that a time–frequency resource grid is available per antenna ports. In the DL, the system supports multistream transmission on up to four transmit antennas. In the UL, no multistream transmission is supported from the same UE, but multiuser MIMO transmission is possible.

Based on this discussion, we can define the abstract RE in LTE as the three-tuple of *[time, frequency, antenna port].* Thus, the generic radio resource assignment problem in LTE can be formulated to find an optimal allocation of the *[time, frequency, antenna port]* resource units to UEs such that the QoS requirements of the radio bearers are satisfied while minimizing the use of the radio resources. A closely related function to resource assignment is link adaptation (LA), which selects transport format (i.e., modulation and coding scheme, MCS) and allocates power to the assigned *[time, frequency, antenna port]* resource. It is primarily the scheduler in the eNode B that executes this resource assignment function, although the antenna configuration can be seen as a somewhat separated function from the generic scheduler operation. The scheduler selects the time–frequency resource to assign to a particular UE based on the channel conditions and the QoS needs of that UE. Then, the LA function selects the MCS and allocates power to the selected time–frequency resources. The antenna configuration, such as the MIMO mode and its corresponding parameters (e.g., the precoding matrix) can be controlled basically separately from the time–frequency assignments of the scheduler, although the two operations are not totally independent.

In an ideal case, the assignment of *[time, frequency, antenna port]* resources and the allocation of MCS and power setting would need to be done in a network-wide manner on a global knowledge basis to obtain the network-wide optimum assignment. This is, however, infeasible in practical conditions due to obvious reasons as such a solution would require a global "superscheduler" function operating based on global information. Therefore, in practice, the resource assignment is performed by distributed entities operating on a cell level in the individual eNode Bs. However, this does not preclude having some coordination between the distributed entities in neighboring eNode Bs, which is an important aspect of the RRM architecture that needs to be considered in LTE. Such neighbor eNode B coordination can be useful in the case of various RRM functions such as ICIC. These aspects are discussed in the sections focusing on the particular RRM function.

We can differentiate the following main RRM functions in LTE:

- radio bearer control (RBC) and radio admission control (RAC)
- dynamic packet assignment—scheduling
- LA and power allocation
- handover control
- ICIC
- load balancing
- MIMO configuration control
- MBMS (multimedia broadcast multicast services) resource control

11.1.3 Radio Resource-Related Requirements

Prior to the development of the LTE concept, the 3GPP defined a number of requirements that this new system should fulfill. These requirements vary depending on whether they are related to the user-perceived performance or to the overall system efficiency and cost. Accordingly, there are requirements on the peak user data rates, user plane and control plane latency, and spectrum efficiency. The requirements on the spectral efficiency or on the user throughput, including average and cell edge throughputs, are formulated as relative measures to baseline HSPA (high-speed packet access, i.e., the 3GPP Release 6 standards suite, performance). For example, it is required to achieve a spectral efficiency and user throughput of at least two to three times that of the HSPA baseline system. The DL and UL peak data rates should reach at least 100 Mbps and 50 Mbps (in a 20-MHz band), respectively. For the full set of requirements, see [7].

It is clear that fulfilling such requirements can only be possible with highly efficient RRM techniques that are able to squeeze out the most from the instantaneous radio link conditions by adapting to the fast fluctuations of the radio link and by exploiting various diversity techniques. With respect to adapting to radio link fluctuations, fast LA and link quality-dependent scheduling have high importance, while in terms of diversity, the various MIMO schemes, such as transmit diversity, spatial multiplexing, and multiuser MIMO play a key role.

11.2 Physical Layer Overview

According to the initial requirements defined by the 3GPP (3GPP 25.913), the LTE physical layer should support peak data rates of more than 100 Mbps over the DL and 50 Mbps over the UL. A flexible transmission bandwidth ranging from 1.25 MHz to 20 MHz will provide support for users with different capabilities. These requirements will be fulfilled by employing new technologies for cellular environments, such as OFDM or multiantenna schemes (3GPP 36.201). In addition, channel variations in the time–frequency domain are exploited through LA and frequency-domain scheduling, giving a substantial increase in spectral efficiency. To support transmission in paired and unpaired spectrum, the LTE air interface supports both FDD and TDD modes.

11.2.1 Physical Resources Structure

Physical resources in the radio interface are organized into radio frames. Two radio frame structures are supported: type 1, applicable to FDD, and type 2, applicable to TDD. Radio frame structure type 1 is 10 ms long and consists of 10 subframes of length $T_{subframe} = 1$ ms. A subframe consists of two consecutive time slots, each of 0.5 ms duration, as depicted in Figure 11.3. One slot can be seen as a time–frequency resource grid, composed of a set of OFDM subcarriers along several OFDM symbol intervals. The number of OFDM subcarriers (ranging from 128 to 2,048) is determined by the transmission bandwidth, whereas the number of OFDM symbols per slot (seven or six) depends on the cyclic prefix (CP) length (normal or extended). Hereafter, we assume the radio frame type 1 with normal CP (i.e., seven OFDM symbols per slot).

In case of multiple-antenna schemes, a different resource grid is used for each transmit antenna. The minimum resource unit allocated by the scheduler to a user is delimited by a physical resource block (PRB). Each PRB corresponds to 12 consecutive subcarriers for one slot; that is, a PRB contains a total of 7×12 REs in the time–frequency domain (as shown in Figure 11.4).

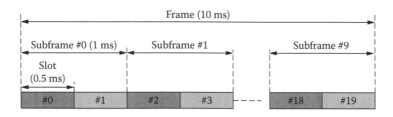

Figure 11.3   Radio frame structure type 1.

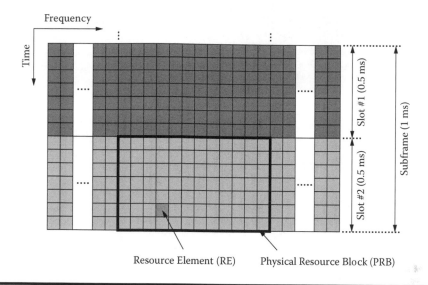

Frequency

Time

Slot #1 (0.5 ms)

Slot #2 (0.5 ms)

Subframe (1 ms)

Resource Element (RE)   Physical Resource Block (PRB)

Figure 11.4   Subframe structure.

11.2.2 Reference Signals

In LTE DL, special reference signals are used to facilitate the DL channel estimation procedure. In the time domain, reference signals are transmitted during the first and third last OFDM symbols of each slot. In the frequency domain, reference signals are spread every six subcarriers. Therefore, an efficient channel estimation procedure may apply a two-dimensional (2D) time–frequency interpolation to provide an accurate estimation of the channel frequency response within the slot time interval.

When a DL multiantenna scheme is applied, there is one reference signal transmitted from each antenna in such a way that the mobile terminal is able to estimate the channel quality corresponding to each path. In that case, reference signals corresponding to each antenna are transmitted on different subcarriers so that they do not interfere with each other. Besides, REs used for transmitting reference signals on a specific antenna are not reused on other antennas for data transmission. An example of reference signal allocation for two-antenna DL transmission is illustrated in Figure 11.5.

The complex values of the reference signals in the time–frequency resource grid are generated as the symbol-by-symbol product of a 2D orthogonal sequence and a 2D pseudorandom sequence. This 2D reference signal sequence also determines the cell identity to which the terminal is connected. There are 504 reference signal sequences defined in the LTE specification, corresponding to 504 different cell identities. In addition, frequency shifting is applied to the reference signals to provide frequency diversity.

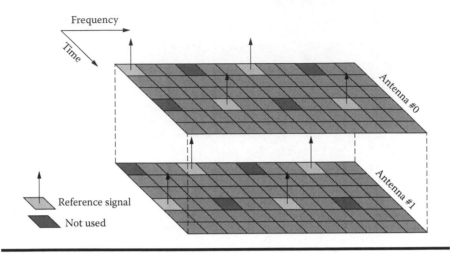

Figure 11.5 Example of reference symbol allocation for two-antenna transmission in downlink.

Reference signals are also used in the UL to facilitate coherent demodulation (then called *demodulation signals*) as well as to provide channel quality information for frequency-dependent scheduling (referred to as *sounding signals*). Demodulation signals are transmitted on the fourth OFDM symbol of each UL slot along the transmission bandwidth allocated to a particular user, whereas sounding signals utilize a larger bandwidth to provide channel quality information on other frequency subcarriers. The UL reference signals are based on constant-amplitude zero autocorrelation (CAZAC) sequences.

11.2.3 Synchronization Signals

The base station periodically sends synchronization signals in the DL so that the mobile terminals may always be synchronized. These signals also help the terminal during the cell search and handover procedures. Synchronization signals consist of two portions:

- *Primary synchronization signal*: Used for timing and frequency acquisition during cell search. The sequence used for the primary synchronization signal is generated from a frequency-domain Zadoff-Chu sequence and transported over the primary synchronization channel (P-SCH).
- *Secondary synchronization signal*: Used to acquire the full cell identity. It is generated from the concatenation of two 31-length binary sequences. Secondary synchronization signals are allocated on the secondary synchronization channel (S-SCH).

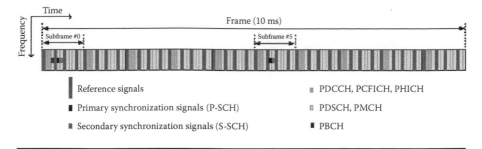

Figure 11.6   Downlink radio frame structure.

Synchronization signals are transmitted on the 72 center subcarriers (around DC subcarrier) within the same predefined slots, in the last two OFDM symbols in the first slot of subframes 0 and 5 (twice per 10 ms). Figure 11.6 shows the time allocation of both the synchronization and reference signals within the DL radio frame.

11.2.4 Physical Channels

The LTE supports a wide set of physical channels that are responsible for carrying information from higher layers (both user data and control information). The complete set of DL/UL physical channels, together with a brief explanation of their purpose, is listed in Table 11.1. Simplified diagrams showing the location of the physical channels and signals in the radio frame are provided in Figure 11.7 (UL) and Figure 11.8 (DL). For each physical channel, specific procedures are defined for channel coding, scrambling, modulation mapping, antenna mapping, and RE mapping. A general structure of the whole DL processing sequence is illustrated in Figure 11.9. This sequence is determined by the following processes:

- *Cyclic redundancy check (CRC)*: The first step in the processing sequence is the CRC attachment. A CRC code is calculated and appended to each transport block (TB), thus allowing for receiver-side detection of residual errors in the decoded TB. The corresponding error indication, reported via UL, can be used by the DL HARQ protocol to perform a retransmission.
- *Channel Coding*: The goal of this process is to increase reliability in the transmission by adding redundancy to the information vector, resulting in a longer vector of coded symbols. This functionality includes code block segmentation, turbo or convolutional coding (depending on the channel type), rate matching, and code block concatenation.
- *Scrambling*: Scrambling of the coded data helps to ensure that the receiver-side decoding can fully utilize the processing gain provided by the channel coding. Scrambling in LTE DL consists of multiplying (exclusive-or operation) the sequence of coded bits (taken as input) by a bit-level scrambling sequence.

Table 11.1   LTE Physical Channels

Direction	Channel Type		Description
Downlink	PDSCH	Physical downlink shared channel	Carries downlink user data from upper layers as well as paging signaling.
	PMCH	Physical multicast channel	Used to support point-to-multipoint multimedia broadcast/multicast service (MBMS) traffic.
	PBCH	Physical broadcast channel	Used to broadcast a certain set of cell or system-specific information.
	PCFICH	Physical control format indicator channel	Determines the number of OFDM symbols used for the allocation of control channels (PDCCH) in a subframe.
	PDCCH	Physical downlink control channel	Carries scheduling assignments, uplink grants, and other control information. The PDCCH is mapped onto resource elements in up to the first three OFDM symbols in the first slot of a subframe.
	PHICH	Physical HARQ indicator channel	Carries the HARQ ACK/NAK (acknowledgment/nonacknowledgment).
Uplink	PUSCH	Physical uplink shared channel	Carries uplink user data from upper layers. Resources for the PUSCH are allocated on a subframe basis by the scheduler.
	PUCCH	Physical uplink control channel	PUCCH carries uplink control information, including channel quality indication (CQI), HARQ ACK/NACK, and uplink scheduling requests.
	PRACH	Physical random-access channel	Used to request a connection setup in the uplink.

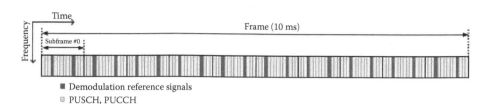

■ Demodulation reference signals
▨ PUSCH, PUCCH

Figure 11.7   Uplink radio frame structure.

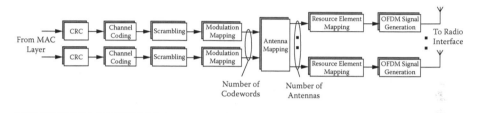

Figure 11.8   Physical layer processing sequence in the base station (downlink).

■ *Modulation mapping*: The block of scrambled bits is modulated into a block of complex-valued modulation symbols. Since LTE uses adaptive modulation and coding (AMC) to improve data throughput, the selected modulation scheme is based on the instantaneous channel conditions for each user. The allowed modulation schemes for DL and UL are shown in Table 11.2.

■ *Antenna Mapping*: Signal processing related to multiantenna transmission is performed at this stage. This procedure is responsible for mapping and pre-coding the modulation symbols to be transmitted onto the different antennas. The antenna mapping can be configured in different ways to provide different multi-antenna schemes including transmit diversity, beam forming, and spatial multiplexing.

■ *RE mapping*: Modulation symbols for each antenna shall be mapped to specific REs in the time–frequency resource grid.

■ *OFDM signal generation*: The last step in the physical processing chain is the generation of time-domain signal for each antenna.

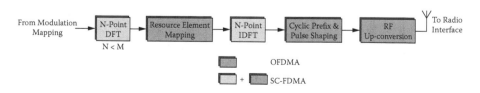

Figure 11.9   Processing sequence in the transmission modem (uplink and downlink).

Table 11.2 Modulation Schemes

Direction	Channel	Modulation Schemes
Downlink	PDSCH	QPSK, 16QAM, 64QAM
	PMCH	QPSK, 16QAM, 64QAM
	PBCH	QPSK
	PCFICH	QPSK
	PDCCH	QPSK
	PHICH	BPSK
Uplink	PUSCH	QPSK, 16QAM, 64QAM
	PUCCH	BPSK, QPSK
	PRACH	u^{th} root Zadoff-Chu

BPSK, binary phase-shift keying; QAM, quadrature amplitude modulation; QPSK, quadrature phase-shift keying.

In addition to the functionalities previously described, HARQ with soft combining is jointly used with CRC and channel coding to allow the terminal to request retransmissions of erroneously received TBs. When a (re)transmission fails, incremental redundancy is used to enable the combination of successively received radio blocks until the full block is correctly decoded.

Previous functionalities may be dynamically configured depending on the type of physical channel being processed. As an example, the DL control channels (physical downlink control channel, PDCCH) are convolutionally encoded and use a transmit diversity configuration in the antenna mapping functionality. However, for DL data channels (i.e., physical downlink shared channel, PDSCH or physical multicast channel, PMCH) turbo coding is applied, and other multiantenna schemes such as spatial multiplexing may be employed.

Besides, some of the physical-layer functionalities are adapted to the higher-layer (e.g., scheduling) decisions to perform LA or channel aware scheduling. The MCS, resource allocation, and multiantenna-related information are employed to configure the corresponding functionalities. Therefore, a cross-layer design is required to support efficient and adaptive physical-layer functionalities.

11.2.5 OFDM/SC-FDMA Signal Generation

The LTE DL transmission scheme is based on orthogonal frequency-division multiple access (OFDMA), which is a multiuser version of the OFDM modulation scheme. In the UL, SC-FDMA is used, which can also be viewed as a linearly precoded OFDM

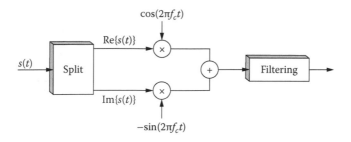

Figure 11.10 Modulation and upconversion.

scheme known as DFT-spread OFDM. However, SC-FDMA has been selected for the UL due to the lower peak-to-PAPR of the transmitted signal compared to OFDM. Low PAPR values benefit the terminal in terms of transmit power efficiency, which also translates into increased coverage. The processing sequence in the signal generation process is quite similar in DL and UL, as shown in Figure 11.10. The main difference comes from the elimination of the antenna mapping process and the addition of a DFT-spread block, which is the key process for the PAPR reduction.

The following processing blocks are involved in the signal generation sequence:

- *N-point DFT*: This block performs the DFT-spread OFDM operation, which provides a high degree of similarity with the DL OFDM scheme and the possibility to use the same system parameters.
- *RE mapping*: Modulation symbols for each antenna shall be mapped to specific REs in the time–frequency resource grid.
- *M-point IDFT*: OFDM symbols are converted into the time domain to be transmitted through the air interface.
- *CP*: For each OFDM symbol to be transmitted, the last N_{cp} samples of the symbol are prepended to the symbol as a CP with the aim of improving the robustness to multipath effect.
- *RF upconversion*: Complex-valued OFDM baseband signal for each antenna is mapped to the carrier frequency f_c, as depicted in Figure 11.11.

In general, the use of OFDM in both DL and UL (in a DFT-spread OFDM form) is an adequate transmission scheme for several reasons:

- OFDM improves the spectral efficiency of the system.
- OFDM provides a high degree of robustness against channel frequency selectivity.
- Frequency-domain LA and scheduling are allowed.
- LTE can easily support different spectrum allocations by varying the number of OFDM subcarriers.

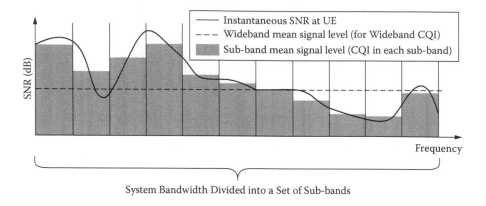

Figure 11.11 Wideband channel quality indicator (CQI) versus subband CQI. SNR, signal-to-noise ratio.

Table 11.3 summarizes the main OFDM modulation parameters for different spectrum allocations. As mentioned, LTE supports a scalable transmission bandwidth that leads to a different number of occupied subcarriers (assuming the same subcarrier spacing). Two different CP lengths are allowed. A longer CP is more suitable for environments with very extensive delay spread (e.g., very large cells). Equivalently, in case of multimedia/multicast single-frequency network (MBSFN)-based transmission, all eNodeBs are transmitting the same signal to the terminals, which increases the delay spread of the equivalent channel. In this particular case, the extended CP is therefore typically needed. However, a longer CP is less efficient from an overhead point of view, so we have to consider these two factors simultaneously.

11.3 Summary

In this chapter, we discussed that a number of advanced RRM functions are needed in today's wireless systems; these functions are being developed and standardized in particular for the 3GPP LTE networks to fulfill the ever-increasing capacity demands by utilizing the radio interface more efficiently. Considering the facts that the available radio spectrum is a limited resource and the capacity of a single radio channel between the UE and the network is also limited by the well-known theoretical bounds of Shannon, the remaining possibilities to increase the capacity is to increase the number of such "independent" radio channels plus trying to approach the theoretical channel capacity limits on each of these individual channels. Advanced RRM techniques play a key role in achieving these goals.

A straightforward way of increasing the number of such independent channels is to increase the system bandwidth or the number of cells in a given deployment.

Table 11.3 OFDM Modulation Parameters

Transmission BW	1.25 MHz	2.5 MHz	5 MHz	10 MHz	15 MHz	20 MHz
Slot durativos	0.5 ms (2 slots compound 1-ms subframe)					
Subcarrier spacing	15 kHz					
Sampling frequency	1.92 MHz (1/2 × 3.84 MHz)	3.84 MHz	7.68 MHz (2 × 3.84 MHz)	15.36 MHz (4 × 3.84 MHz)	23.04 MHz (6 × 3.84 MHz)	30.72 MHz (8 × 3.84 MHz)
DFT size	128	256	512	1,024	1,536	2,048
Number of occupied subcarriers[a]	76	151	301	601	901	1,201
Number of OFDM symbols per subframe (short/long CP)	7 symbols (short CP) or 6 symbols (long CP) per slot					
CP length (μs/samples) Short	(4.69/9) × 6, (5.21/10) ×1	(4.69/18) × 6, (5.21/20) ×1	(4.69/36) × 6, (5.21/40) ×1	(4.69/72) × 6, (5.21/80) ×1	(4.69/108) × 6, (5.21/120) × 1	(4.69/144) × 6, (5.21/160) ×1
Long	(16.67/32)	(16.67/64)	(16.67/128)	(16.67/256)	(16.67/384)	(16.67/512)

[a] Including DC subcarrier.

In LTE, the maximum supported system bandwidth size has been increased to 20 MHz, and a variety of flexible system bandwidth configurations are possible, while the number of cells in a network is more of a deployment issue than a system design principle. The other and less-straightforward possibility for increasing the number of independent radio channels is to employ various spatial multiplexing techniques and advanced receiver structures that can better separate out the radio channels in the spatial domain. As we have seen in this chapter, LTE employs all of these methods, including MIMO, beam-forming, and advanced receiver methods to increase system capacity and coverage.

The other component of increasing the capacity comes from the better utilization of such single radio channels, trying to approach the theoretical limits of the channel capacity. This is primarily achieved by fast LA and dynamic scheduling methods, all part of RRM functions of LTE, which try to follow the fast fluctuations of the radio link and try to exploit the time, frequency, or multiuser diversity of the radio channel. This chapter has shown methods and examples of how these advanced RRM functions are realized in LTE and how these functions altogether make LTE a high-performance, competitive system for many years to come.

References

1. Dahlman, E., Parkvall, S., Skold, J., and Beming, P., *3G Evolution, HSPA and LTE for Mobile Broadband*. New York: Academic Press, 2007.
2. 3GPP TS 36.322, *E-UTRA Radio Link Control (RLC) Protocol Specification*. ftp:// ftp.3gpp.org/Specs/archive/36_series/36.322/.
3. 3GPP TS 36.321, *E-UTRA Medium Access Control (MAC) Protocol Specification*. ftp:// ftp.3gpp.org/Specs/archive/36_series/36.321/.
4. 3GPP TS 36.323, *E-UTRA Packet Data Convergence Protocol (PDCP) Specification*. ftp://ftp.3gpp.org/Specs/archive/36_series/36.323/.
5. 3GPP TS 36.211, *E-UTRA Physical Channels and Modulation*. ftp://ftp.3gpp.org/ Specs/archive/36_series/36.211/.
6. Oestges, C., and Clerckx, B., *MIMO Wireless Communications: From Real-World Propagation to Space-Time Code Design*. New York: Academic Press, 2007.
7. 3GPP TS 25.913, *Requirements for Evolved UTRA (E-UTRA) and Evolved UTRAN (E-UTRAN)*. ftp://ftp.3gpp.org/Specs/archive/25_series/25.913/.
8. 3GPP TS 36.331, E-UTRA Radio Resource Control (RRC) Protocol Specification. ftp://ftp.3gpp.org/Specs/archive/36_series/36.331/.

Chapter 12

Mobile Internet Protocol

Andrianus Yofy and Watit Benjapolakul

Contents

12.1 Introduction

In the past few years, we have seen an explosion and increasing variety of wireless and mobile devices offering Internet Protocol connectivity, such as notebook computers, personal digital assistants, handhelds, and digital cellular phones [1,2]. However, the application of Internet in these mobile devices, such as video conferencing, Internet telephony, and other real-time applications, introduces some technical obstacles. The most basic technical obstacle is the way the Internet Protocol sends the data packets from the source to the destinations according to Internet Protocol addresses [2]. In the original design, an Internet Protocol address is assumed to be static. If a source wants to send packets to a destination, these packets are routed based on the static Internet Protocol address of the destination. When the destination is a mobile device, it means that the device can move from one network to another. If during the movement the mobile is receiving data packets from another device in the Internet, this ongoing communication can be terminated since the Internet

Protocol address of the mobile device in the previous network is different from the Internet Protocol address of the new network. The mobile device has to change its Internet Protocol address in the new network to continue the communication. This problem makes the seamless roaming of the mobile device impossible [2]. One of the solutions to this problem is the Mobile Internet Protocol (Mobile IP).

12.2 Mobile Internet Protocol

Mobile IP (RFC [Request for Comments] 3344) [3] is a proposed Internet standard that improves the previous Internet Protocol to support mobility [4]. It allows a mobile host to move from one network to another without breaking the ongoing communication by assigning two Internet Protocol addresses to the mobile host. The first is the home address, which is static, and the second is the care-of address, which changes at each new point of attachment. Mobile IP resides in the network layer [1] in providing Internet mobility. Thus, this makes Mobile IP not depend on the media in which the communication takes place. A mobile host can roam from one type of medium to another without breaking its connectivity.

Mobile IP consists of three main entities: mobile node, home agent, and foreign agent [1]. Figure 12.1 shows the entities and their relationship in Mobile IP. The mobile node is a device such as a cellular phone or a laptop that can roam from one network to another without losing connectivity and simply uses the Internet Protocol home address to maintain the ongoing communication. The home agent is a router on the home network that keeps the current location of the mobile node by recording the mobile node's care-of address. The home agent intercepts each packet sent by a device on the Internet, called a correspondent node, to the mobile node's home address and forwards the packets to the mobile node's care-of address. The foreign agent is a router in the foreign network that helps the mobile node to inform its home agent of its current location, provides a care-of address for the mobile node, forwards packets from the home agent to the mobile node, and

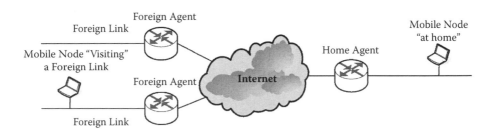

Figure 12.1   Mobile IP entities and relationship.

serves as a default router if the mobile node wants to send packets to another node in the Internet.

There are three main phases in Mobile IP: agent discovery, registration, and tunneling [1, 3]. The mobile node finds out its home agent or foreign agent, identifies its current point of attachment, and gets a care-of address if connected to the foreign agent during the agent discovery phase. The home agents and foreign agents broadcast their existence periodically through agent advertisement messages. A mobile node may optionally force any agents to broadcast agent advertisement messages by sending an agent solicitation message. This is useful when the mobile node is moving very fast from one agent to another while the transmission rate of agent advertisement sent by the agents is too low [1]. The mobile node accepts these agent advertisement messages and decides whether they are on its home network or a foreign network. If the mobile node notices that they are on its home network, it operates like any other fixed host on the home network. On the other hand, when the mobile node detects that it has moved to a foreign agent network, it obtains a care-of address on that network.

There are two types of care-of addresses [3]: foreign agent care-of address and colocated care-of address. The foreign agent provides a foreign agent care-of address by means of broadcasting agent advertisement messages. The mobile node gets the colocated care-of address through some methods, such as conducting the Dynamic Host Configuration Protocol (DHCP) or using a long-term address owned by the mobile node that is utilized only when the mobile node is on the foreign network. When a mobile node hears another foreign agent's agent advertisement and decides to move to this new foreign agent service area, the mobile node starts the registration process.

The mobile node registers its recent location with the home agent during registration [3]. The mobile node operating away from the home network will register its new care-of address every time it changes its point of attachment. Registration is also utilized by the mobile node to request forwarding services from the foreign agent, to renew a registration that is due to expire, and to deregister when the mobile node goes back to its home agent network. The mobile node, foreign agent, and home agent exchange information through registration messages, which are valid for a specified lifetime. The mobile node conducts registration by exchanging registration request and registration reply messages with the home agent through the current foreign agent where the mobile node is now located. Using these messages, the home agent can create and modify the mobility-binding information of that mobile node, such as new lifetime and new care-of address. Other scenarios of registration, such as registration without foreign agent if the mobile node uses colocated care-of address and direct registration when the mobile node returns to its home agent, are also possible.

A registration request message is first sent by the mobile node to start the registration process. The foreign agent processes this message and then forwards it to the

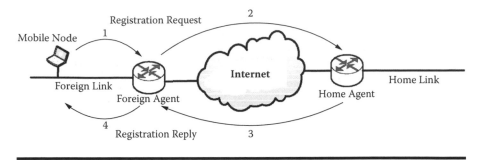

Figure 12.2   A mobile node registers through foreign agent.

home agent. The home agent later sends a registration reply message to the foreign agent to inform whether the registration is successful. Finally, the foreign agent processes the registration reply and delivers it to the mobile node. One example of processes at the foreign agent and home agent is the authentication of the registration messages for security purposes. Figure 12.2 illustrates the registration process.

When a node on the Internet, called a *correspondent node*, sends data packets to the mobile node, the data packets are first delivered to the home network of the mobile node. The home agent then intercepts the data packets. A tunnel [3] is then set up by the home agent to the care-of address of the mobile node to route the data packets. Before sending the data packets through the tunnel, the home agent encapsulates the data packets to reach the endpoint of the tunnel. At the tunnel endpoint, the foreign agent decapsulates the data packets and delivers the data packets to the mobile node. The default tunnel mode is Internet Protocol encapsulation within Internet Protocol encapsulation. In this type of encapsulation, an Internet Protocol packet is put inside the payload portion of another Internet Protocol packet. The header of the encapsulated packet contains the information that is used to route the packet to the endpoint of the tunnel. Optionally, generic routing encapsulation and minimal encapsulation within the Internet Protocol may be used.

In the reverse direction, the mobile node delivers the data packets directly to their destination using standard Internet Protocol routing mechanisms, not necessarily sending them through the home agent. The mobile node sends data packets using its home Internet Protocol address as the source address within the Internet Protocol packets. This keeps the information in the correspondent node's record that the mobile node is located in its home network although in the real condition the mobile node is located in another network. Although the mobile node moves from one foreign agent to another, the corresponding node sends the packet with the home address Internet Protocol address as the destination address. Figure 12.3 shows the routing of data to and from the mobile node when it is in a foreign network and has conducted the registration process through the foreign agent.

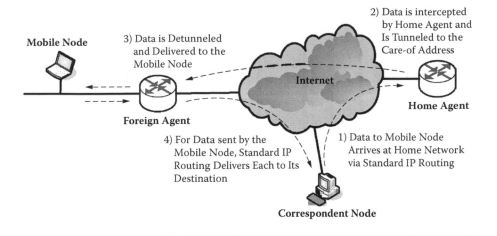

Figure 12.3   Routing operation in Mobile IPv4 [3].

The security aspect is important in the Mobile IP since the mobile computing environment is vulnerable to many security problems, such as passive eavesdropping and active replay attacks [3]. The mobile node and home agent conduct authentication of registration messages using the Hash-Based Message Authentication Code Message Digest 5 (HMAC-MD5) as the default algorithm. The receiver of the message compares the authenticator value it computes over the registration message over the value in the extension to verify the authenticity. The foreign agent must also support authentication using the HMAC-MD5 algorithm. Replay protection using the identification field in the registration message is used by the home agent to prevent the attacker from replaying the registration message. The mobile node and home agent implement timestamp-based or nonce-based for replay protection. Using a timestamp-based method, the node that is generating the message includes the recent time of day, and the node that is receiving the message verifies that this timestamp is close enough to its own time of day. If a nonce-based method is applied, node A inserts a new random number in every message to node B and checks whether node B returns that same number in its message to node A.

12.3 Mobile IP Route Optimization Scheme

The Mobile IP Route Optimization scheme [5, 6] is designed to improve the performance of the triangle routing scheme in Mobile IP. As we can see in Figure 12.4, in the triangle routing scheme all packets that the correspondent node sends to the mobile node must be sent first to the home agent, and then the home agent sends packets to the mobile node, although the mobile node has moved close to

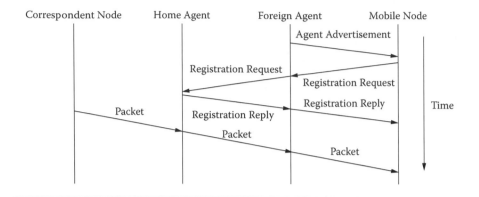

Figure 12.4 Time diagram of packet transmission using triangle routing scheme.

the correspondent node [7]. Here, packets to the mobile node are often routed along paths that are considerably longer than optimal. This indirect routing slows the delivery of the packets to the mobile nodes and puts an unnecessary load on the networks and routers along their paths through the Internet [7]. Moreover, the mobile node needs to update its location to the home agent every time it moves to a new foreign agent, although the mobile node has moved far away from its home agent. This gives rise to high signaling cost and a packet delay problem because before the mobile node receives packets it has to receive a location update message (registration reply) from its home agent. Figure 12.4 shows the time diagram of packet transmission using a triangle routing scheme.

A route optimization scheme improves the efficiency of routing in Mobile IP by using the basic idea that the correspondent node knows the latest information of the mobile node's location. Here, the correspondent node updates the binding information of the mobile node in its routing table every time the mobile node roams from one network to another [5]. The route optimization scheme creates a binding message that is sent from the home agent to the correspondent node to let the correspondent node know where the current mobile host (current foreign agent) of the mobile node is. When the correspondent node knows where the current mobile host of the mobile node is, the correspondent node can send packets to the mobile node's care-of address directly instead of relaying the packets to probably a far-away home agent. Figure 12.5 illustrates the direction of packet transmission in the Mobile IP route optimization scheme.

The advantages of the Mobile IP route optimization scheme are lower communication cost and shorter delivery delay time than the Mobile IP triangle routing scheme. Figure 12.6 shows the time diagram of the Mobile IP route optimization scheme. All processes of route optimization that alter the routing of Internet Protocol packets to the mobile node are authenticated using the same type of procedures described in the basic Mobile IP.

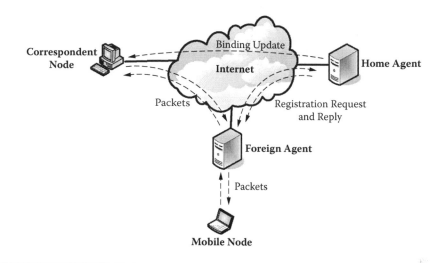

Figure 12.5 **The direction of packet transmission in the Mobile IP route optimization scheme.**

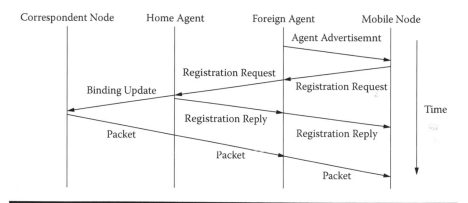

Figure 12.6 **Time diagram of packet transmission in the Mobile IP route optimization scheme.**

12.4 Conclusion

The Mobile IP is a network-layer solution to the problem of Internet application in mobile and wireless devices. This protocol assigns two addresses to the mobile node: the home address and the care-of address. These addresses guarantee the mobile node will roam seamlessly from one network to another. There are three main entities in Mobile IP: mobile node, foreign agent, and home agent. These entities conduct three main phases in Mobile IP: agent discovery, registration, and tunneling.

The mobile node finds its current location in the network during agent discovery. The mobile node registers its latest location with the foreign agent and the home agent during the registration phase. In tunneling, the corresponding node delivers packets to the mobile node through the home agent and the foreign agent. Within these phases, security plays an important role because the mobile computing environment is vulnerable to many security problems, such as eavesdropping and replay attack. The Mobile IP route optimization scheme proposes a solution to the triangle routing problems in Mobile IP.

Acknowledgment

We wish to thank the Thesis Supporting Fund of Chulalongkorn University and the AUN/SEED-Net's Collaborative Research Support Project (JFY2002–2005) for research support.

References

1. Solomon, J.D., *Mobile IP. The Internet Unplugged*. Englewood Cliffs, NJ: Prentice Hall, 1998.
2. Perkins, C.E., Mobile networking through Mobile IP. *IEEE Internet Comput.*, 2, 58–60. January–February 1998.
3. Perkins, C.E., IP mobility support for IPv4. *Request for Comments (RFC) 3344*, August 2002.
4. Dixit, S., and Prasad, R., *Wireless IP and Building the Mobile Internet*. Boston: Artech House, 2003.
5. Perkins, C.E., Mobile IP. *IEEE Commun. Mag.*, 66–82, May 2002.
6. Johnson, D.B., and Perkins, C.E. Route optimization in Mobile IP. *Internet Draft*, draft-ietf-mobileip-optim-09.txt, February 2000.
7. Bejerano, Y., and Cidon, I., An anchor chain scheme for IP mobility management. *Wireless Network*, 9(5), 409–420, September 2003.

Chapter 13

SIP: Advanced Media Integration

Marisol Hurtado, Andre Rios, Antoni Oller,
Jesus Alcober, and Sebastia Sallent

Contents

13.1 Introduction

The explosive growth of Internet Protocol (IP) networks capable of providing sophisticated convergent services based on multimedia applications has increased the development of new services as an alternative to traditional communication systems. In this way, the service convergence is growing under an IP-based provisioning framework and controlled mainly by lightweight application-level protocols like Session Initiation Protocol (SIP) [1].

SIP, a key technology enabling this evolution, is creating new possibilities to develop multimodal communications. SIP increases the possibilities of developing integrated multimedia services and offers advantages such as independence from the underlying transport network, session establishment with the involvement of multiple devices, and the support for mobility.

End users need request, find, and consume advanced media services, regardless of terminal capabilities and location. SIP is an effective way to do this, not by providing services, but just by primitives that can be used to implement them.

Fundamentally, SIP is a control protocol that offers additional capacities such as user location, availability, and capabilities. SIP depends on relatively intelligent endpoints that require little or no interaction with servers. Each endpoint manages its own signaling, both to the user and to other endpoints. There is a separation between signaling and media—any kind of media supported in next-generation networks can be signaled, such as uncompressed high-definition (HD) or three-dimensional (3D) video applications.

But, the really powerful advantage of SIP is that, with the help of companion protocols, it can provide an Internet-style plethora of continuous media (CM) value-added services, such as conferencing, instant messaging, or even streaming and gaming. This concept is named SIP-CMI (SIP-based CM integration) [2].

On the other hand, the non-CM, value-added services can be provided by Service-Oriented Architecture (SOA) [3]. Furthermore, a combination of SIP and Web services can fulfill all the needs of the current and future Internet user. Web services and SIP services are the paradigm of converged communication services [4].

Users demand personalized services that can be accessed from any terminal within any network. Therefore, service providers must deploy more advanced services with shorter time-to-market and development cycles in heterogeneous networks. SIP and SOA have characteristics that satisfy these criteria, and this is the main reason SIP has been selected by the Third Generation Partnership Project (3GPP) as a major component of the IP Multimedia Subsystem (IMS).

This chapter explains the SIP capabilities of building advanced media services and the possibilities of performing advanced services in an SIP environment.

In Section 13.2, the SIP application server and the control of media servers are explained as the basis of value-added service implementation and delivery. In Section 13.3, the SIP SOA paradigm is described. In Section 13.4, important features of SIP, flexible control and management mechanisms based in SIP, such as cross layering, are drawn. Section 13.6 covers the foreseeable role of SIP in the Internet of the future.

13.2 Building Advanced Media Services

From the architectural point of view, an SIP platform can be composed of elements grouped in two categories: clients and servers.

SIP clients can act not only as user agent client (UAC) but also as user agent server (UAS). For instance, they can be an SIP terminal (e.g., a voice-over IP [VoIP] telephone in the case of IP telephony) and SIP gateways. These are devices that provide services such as format translation and communication procedures between SIP terminals and non-SIP terminals or transcoding between audio and video codecs.

Commonly, SIP servers work as proxy, redirect, and registrar. Besides these types, two other servers have special relevance to developing advanced media services: media servers and application servers.

Media servers allow the possibility of having immediate access to any audiovisual (audio and video) content stored in their repository. When users request this type of service, they should know, only and exclusively, what service to request or with whom they want to communicate. With current paradigms, when users ask for access to streaming content, they should know which server will provide them this content. They should know the characteristics and present conditions of their network to request the exact quality that it should be able to access. Users see the network as an information transport system that should be adapted if they want to use it.

In case of CM services, and in the special case of streaming services, SIP needs to be extended with media server control mechanisms. In this way, there are works in progress in the Internet Engineering Task Force (IETF) and World Wide Web Consortium (W3C) that propose an architectural framework of media server control. On the other hand, there are other proposals focused on exploiting the cooperation between SIP and RTSP (Real Time Streaming Protocol).

The application servers are components in charge of external application execution, which adds intelligence to the services. Examples of advanced services in a VoIP environment are call filtering, destination blockage, and call forwarding. The number of services deployed in this environment has no limit, and service providers have been forced to offer a high level of personalized services according to the necessities of a user or group of users. Furthermore, service providers look for a service delivery platform environment that minimizes the time to market before the incorporation of new services. Advanced mechanisms to implement such services can be provided by the Java and open source software communities.

In the following, approaches to media server control and application servers are explained in more detail.

13.2.1 Media Server Control Mechanisms

A *media server* is a media-processing network element that performs all the media processing for advanced services. It is controlled by the application server, which provides a service execution environment, application-specific logic, and all the signaling for one or more services.

Application servers and media servers work together in a client–server or master–slave relationship. The application servers provide the service logic for each specific application, and the media server acts as a media-processing resource for the applications.

There is a clear tendency to standardize this control. Currently, there are two main approaches: SIP based and Java based.

In the SIP domain, several SIP-based control markup languages extend SIP capabilities, such as VoiceXML (extensible markup language) and Call Control eXtensible Markup Language (CCXML) [5], Media Server Control Markup Language (MSCML) [6], Media Session Markup Language (MSML) [7], and Media Server Control (MEDIACTRL) [8]. In the Java-based approach, the main programming interface is the Media Server Control API (JSR 309) [9].

13.2.1.1 CCXML and VoiceXML

CCXML is a W3C Working Group XML-based standard designed to provide telephony support to VoiceXML. CCXML is designed to inform the voice browser how to handle the telephony control of the voice channel, whereas VoiceXML is designed to provide a voice user interface to a voice browser.

CCXML with SIP integrates the CCXML browser, dialog servers (interactive voice response platforms), and conference bridges.

CCXML defines an XML document that describes a VoIP service in terms of its state diagram and its transitions. When an event occurs in the signaling layer, a CCXML interpreter maps the diagram transitions in call control actions. CCXML offers a higher abstraction level for programming complex services. Figure 13.1 shows an example of service implementation that automatically answers incoming calls and ends its execution when the caller hangs up.

As depicted in Figure 13.2, several entities can appear in a CCXML session:

■ Connection. This represents the endpoints of a communication.
■ Conference. This manages the mixing of audiovisual streams when there are several participants in a multiconference.

```
<?xml version="1.0" encoding ="UTF-8"?>
<ccxml>
    <var name="theState" expr=" start" />
    <eventprocessor state ="theState">
        <transition state ="start" event="conn.alerting" name="evt">
            <assign name="theState" expr="connected"' />
            <accept connectionid="evt.conn.connectionid"/>
        </transition>
        <transition state="connected" event="conn.disconnected"
                    name="evt">
            <exit/>
        </transition>
    </eventprocessor>
</ccxml>
```

Figure 13.1   CCXML for an incoming call.

- ■ Voice Dialogs. These are associated with an existing connection or conference, and allow executing media actions in a media server, for example, music on hold, dual tone multifrequency (DTMF), IVR, Early Media, and so on.

CCXML brings Web and XML advantages to call control and provides advanced call control features that VoiceXML lacks. It allows flexible development of applications such as multiparty conferencing, call center integration, and follow me and find me services. Its main features are

- ■ Basic call control features such as answer, create call, disconnect, reject, and redirect
- ■ Java call control/Java for advanced intelligent networks (JCC/JAIN)-based call model
- ■ Flexible asynchronous event model
- ■ Multiparty conferencing
- ■ IVR/Dialog integration (VoiceXML, SALT, among others)

Figure 13.2   CCXML entities.

13.2.1.2 MSCML

The MSCML is an IETF Request for Comments (RFC) proposal aimed at enabling enhanced conference control functions such as muting individual callers or legs in a multiparty conference call, the ability to increase or decrease the volume, and the capability of creating subconferences. MSCML also addresses other feature requirements for large-scale conferencing applications, such as sizing and resizing of a conference. MSCML implements a client server model to perform the media control server.

13.2.1.3 MSML

The MSML is an IETF draft proposed as an extension to SIP that aims to control media servers with dialogs through a master–slave scheme. MSML makes use of XML content in the message bodies of SIP INVITE and INFO requests without modifying the SIP in any way.

13.2.1.4 MEDIACTRL

MEDIACTRL is a new IETF working group formed to create a standard protocol for controlling media servers. Control channels are negotiated using SIP standard mechanisms that would be used in a similar manner to create an SIP multimedia session. It highlights a separation of the SIP signaling traffic and the associated control channel that is established as a result of the SIP interactions.

The use of SIP for the specified mechanism provides many inherent capabilities, including

- Service location: Uses SIP proxies or back-to-back user agents (B2BUAs) for discovering control servers
- Security mechanisms: Leverage established security mechanisms such as transport layer security (TLS) and client authentication
- Connection maintenance: Provides ability to renegotiate a connection, ensure it is active, audit parameters, and so on
- Agnostic: Generic protocol allows for easy extension

A CONTROL message is used by the control client to invoke control commands on a control server.

13.2.1.5 Media Server Control API (JSR 2309)

The goal of JSR 309 is to specify an API that standardizes access to external media server resources and to provide their multimedia capabilities from services built on application servers. JSR 309 defines a mechanism to provide

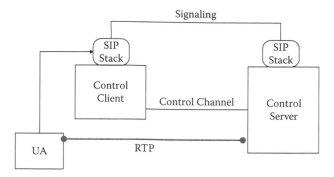

Figure 13.3 Simplified view of the proposed mechanism.

- Media network connectivity to establish media streams
- IVR functions to play/record/control multimedia contents from a file or streaming server
- A way to join/assemble IVR function to a network connection, following any topology, to compose conferences, bridges, and so on

The JSR 309 API may be structured following three main sets of functionalities, shown in Figure 13.4.

13.2.1.6 SIP and RTSP Synergy

SIP and RTSP [10] are related to each other, but they are independent protocols that allow initiating and controlling of stored, live, and interactive multimedia

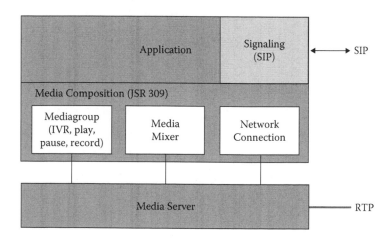

Figure 13.4 Media control architecture.

sessions in the Internet. The former is used mainly for inviting participants into a multimedia session to initiate, for instance, a multiparty conference. The latter is used to control playback and recording for stored CM to offer a service like video on demand. However, in both scenarios, these protocols can work separately. The improvement of the IP networks in terms of bandwidth capacity and quality of service (QoS) and their expansion in terms of reachability and value addition are increasing the number and variety of deployment scenarios for streaming media applications. These scenarios impose new requirements on signaling protocols to offer more flexible and scalable applications. In this sense, cooperation between SIP and RTSP can be key in creating new collaborative applications. Due to that, their synergy expands the number of media control possibilities not available with only one of them.

Currently, there is no integrated solution, but there are some approaches for SIP and RTSP convergence. For instance, in [11] a proxy architecture based on collaborative session transfer is proposed, and in [12] a merged scheme is described, by which the reuse of RTSP as a stream control is proposed. This control is negotiated by SIP/SDP (Session Description Protocol) to keep the compatibility with existing streaming solutions. Another option is to use a gateway to translate SIP messages in RTSP and vice versa. This last scheme is described in Figure 13.5.

13.2.2 SIP Application Servers

Also, SIP can be used to provide more advanced and rich services. Application servers are those that provide added functionalities. SIP application servers manage incoming

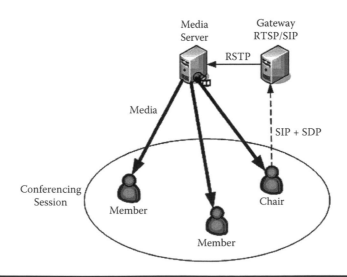

Figure 13.5 Internet conferencing example combining SIP and RTSP.

requests and run associated specific applications. Services can be customized by the parameters on specific fields of the request, origin information, or other variables.

Currently, several initiatives aim at providing advanced mechanisms to implement SIP services. Initiatives within the community of Java developers and in the open source community are highlighted.

13.2.2.1 Java-Based Approach

Java for Advanced Intelligent Networks [13] aims at standardizing different APIs that enable the fast development of next-generation Java-based communications. SIP has been covered by the JAIN initiative, producing JAIN-SIP API. This API encapsulates an SIP stack behind some well-defined interfaces.

Given the amazing evolution of dynamic Web services, the SIP community has also adopted one of the most widespread environments to create and deploy SIP services. It is called the servlet programming API, specified within the Java community process. The next section describes the JAIN initiative and the SIP servlet programming model.

The JAIN initiative has defined a set of Java technology APIs that enable the rapid development of next-generation Java-based communication products and services for the Java platform.

Within the JAIN community, JAIN SIP is the part in charge of implementing interfaces for the use of the SIP. The JAIN SIP specification provides functionalities from RFC 3261 (SIP) and the RFCs with SIP extensions. Some of these features are methods to format the messages, application ability to receive and send messages, analysis ability about incoming messages by accessing their fields, transport of control information generated at a meeting (RFC 2976), and instant messaging (RFC 3428). Other RFCs that complement the SIP specification are RFCs 3262, 3265, 3311, 3326, and 3515.

Jain SLEE [13] is the Java standard for service logic execution environment (SLEE). SLEE is defined as runtime services with low latency and high performance that are asynchronous event oriented (i.e., basic telecommunication service characteristics).

JAIN/JCC [13] is a specification (JSR 21) that abstracts the creation of VoIP services from the underlying protocol. It supports different signaling protocols, such as SIP, SS7, H.323, or MGCP. Its operation philosophy is based on subscription to the receipt of different events associated with the network roles discussed next.

The SIP servlet API is a specification of the Java community that defines a high-level API for SIP servers, enabling SIP applications to be deployed and managed based on the well-known HTTP (Hypertext Transfer Protocol) servlet model.

SIP servlets achieve their highest potential when they can be deployed along with HTTP servlets in the same application server (also known as the *servlet container*). This environment, which combines SIP and HTTP, truly realizes the power of converged voice–data networks. HTTP represents one of the most powerful data transmission protocols used in modern networks (think of the SOAP Web

Figure 13.6 SIP servlet API.

services protocol). At the same time, SIP is the protocol of choice in most modern and future VoIP networks for the signaling part. Therefore, an application server capable of combining and leveraging the power of these two APIs will be the most successful, as shown in Figure 13.6.

Convergence of SIP and HTTP protocols into the same application server offers, among others, a key advantage. It is not necessary to have two different servers (HTTP and SIP), so it reduces maintenance problems and eases user and configuration provisioning.

Several SIP application servers have been developed, such as WeSIP AS [14], implemented by Voztelecom company and Broadband Research Group from Universitat Politecnica de Catalunya (UPC), IBM Websphere, BEA WebLogic SIP Server, or Glassfish, among others.

13.2.2.2 Developing Advanced Media Services Using Open Source Tools

Open SIP express router (OpenSer) [15], Asterisk [16], and RTP Proxy are the three key components to build an open source VoIP platform.

OpenSer is one of the most widely used SIP proxies and is usually the entry point for most SIP networks, providing authentication, registration, and other basic functionalities. It is a VoIP server based on SIP used to build a large-scale IP telephony infrastructure. It can operate as registering, proxy, redirect, and so on. When operating with SIP standard, it makes interoperability with other SIP systems (from other vendors) easy. OpenSer is composed of a core system and several modules, with support of presence, authentication by authentication authorzation and accounting (e.g., RADIUS), remote calls (XML-RPC) and Call Processing Language (CPL). It also offers a Web-based interface application/server where it can monitor the state of the server and negotiate all features.

The OpenSer configuration file is more than just a typical configuration file. It combines both static settings and a dynamic programming environment. In fact, a configuration file is a program that is executed for each message received by the OpenSer.

Also, OpenSer uses advanced techniques for executing programmable logic. One of them is called Call Programming Logic (CPL). CPL scripts are XML-based documents. They describe operations that the server performs on a call setup event. CPL scripts can reside on an SIP proxy server, an application server, or intelligent agent. Figure 13.7 shows how open SER works with CPL.

Asterisk is an open source IP PBX (private branch exchange) composed of several modules that operate as a simple PBX, gateway, or media server. Asterisk has a

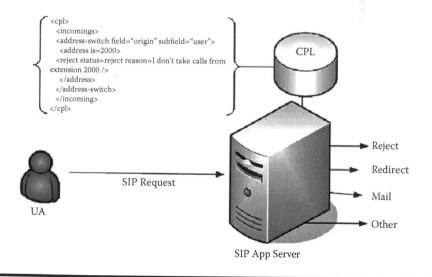

Figure 13.7 Call-processing logic.

general public license (GPL), and although it was originally developed for the Linux operating system, it currently works on other operating systems.

Asterisk has many features only available in expensive PBX systems, such as conferences, IVR, voice mailbox, and automatic distribution of calls. It is possible to add new features on its contexts collection, called *dialplan*, which is written in the characteristic script language of Asterisk, and to add modules written in C language or in another programming language supported by Linux. Asterisk can be installed on a traditional computational platform working as a gateway interconnecting the PSTN (public switched telephone network) and IP environments.

RTP Proxy is a server that operates together with the SER and any proxy server, solving the topic of network address translation (NAT) traversal with the appropriate handling of ports. One of the most used is Portaone RTP Proxy.

13.3 Service-Oriented Architecture

SOA is an architectural pattern for creating business processes organized as services. It provides an infrastructure that allows applications to exchange information and participate in business processes. While the SOA concept is general and is based on the idea of any type of message provision of service, it is most often used in Web environments and based on Web standards [17], such as HTTP, XML, SOAP, Business Process Execution Language (BPEL), Web Service Description Language (WSDL), or universal description discovery integration (UDDI).

SOA is characterized by the following properties:

- *A logic view.* It is an abstraction of services such as programs, databases, and business processes within the application. They are defined in terms of what they do. Thus, its basic components are the services that implement the business logic and the basic functionality of the system.
- *Message oriented.* As part of the description of the service, the exchanged messages between suppliers and applicants are defined. The internal structure of the service (programming language, internal processes, etc.) remains hidden at this level of abstraction.
- *Description oriented.* A service is described in processable metadata. The description supports the public nature of SOA. Only those public details are included in the description and are important for the use of the service. Thus, a simple Web service is characterized by four standards: XML, SOAP, WSDL, and UDDI, which work according to the basic model request/response.
- *Granularity.* Services tend to use a small number of operations with relatively complex messages.
- *Network oriented.* Services tend to be used over a network, although this is not a prerequisite.
- *Platform independent.* Messages are delivered in a standard format and are neutral to the platform (usually XML).

SOA is associated with a client–server relationship between software modules, in which services are subroutines serving customers. However, not all environments and software topologies fit this model, for example, advanced multimedia services. With SOA 2.0, an event-oriented architecture is deployed in which software modules are related to business components and provide alerts and notifications events.

13.3.1 SOA and SIP Convergence

The goal of the SOA and SIP convergence is to offer an infrastructure to implement services with high complexity using collaborative services or an orchestration of services (e.g., in the form of a BPEL process [18]). They can be implemented in different environments, programming languages, and platforms and can be located in different service providers.

In the case of advance media services, it takes advantage of the power of SOA and SIP convergence because of the SOA. This paradigm can be applied with Web service orchestration, transparent integration with other services, and integration with CM services such as session and stream control, multimedia service composition, and call signaling, as shown in Figure 13.8. SIP works as an application-layer

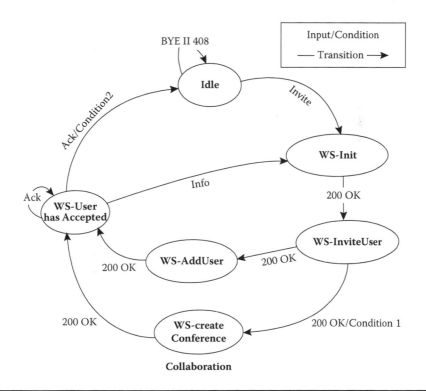

Figure 13.8   SIP/Web services collaboration.

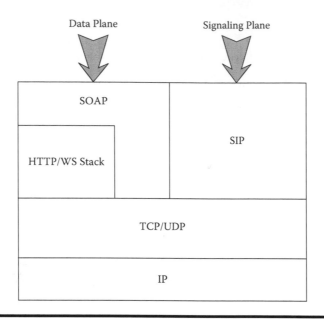

Figure 13.9 Data plane and control plane.

signaling protocol and media control in parallel with SOAP in the data plane, as depicted in Figure 13.9.

In the specific case of the Web services, the service provider develops its application and provides a WSDL interface that describes the capabilities of the service. The location of the service is stored in a UDDI repository that manages the publication, search, and discovery of services.

13.4 MPEG-21/SIP-Based Cross Layer

Network connections are becoming ubiquitous; therefore, people expect to access multimedia content anywhere and anytime. One approach to solve this requirement is to use a technique known as *session mobility*, which allows a user to maintain a media session even while changing terminals. For example, on entering the office, a person may want to continue a session initiated on a mobile device on his or her desktop personal computer (PC). However, to develop this solution, it is necessary to consider that the delivery and adaptation of multimedia contents in distributed and heterogeneous environments requires flexible control and management mechanisms in terminals and in control entities inside the network. On one hand, there is MPEG-21 digital item adaptation (DIA), which provides normative descriptions for supporting adaptation of multimedia content but does not define interactions with transport and control mechanisms. On the other hand, there is SIP, which is an application-layer protocol used for establishing and tearing down multimedia sessions. It uses SDP, which is a format for describing streaming media initialization parameters. An evolution of SDP, SDPng [19], was proposed to address the perceived shortcomings of SDP.

Both technologies come from different worlds, and to develop a solution for advanced signaling, it is important to reach interoperability between the IETF approaches on multimedia session establishment and control and the MPEG-21 efforts for multimedia streaming and adaptation to bring advanced multimedia service provisioning and adaptation services to the customer. Some proposals that join MPEG-21 and SDPng exist [20, 21], but they are at draft level, and their focus is general, without detailing implementation aspects in solutions such as session mobility. The main idea is to contribute to the XML model in which MPEG-21 and SDPng are based.

There are proposals for which MPEG-21 is used to provide a common support for implementing and managing end-to-end QoS [22]. As Figure 13.10 shows, MPEG-21 is proposed as a cross-layer adaptation.

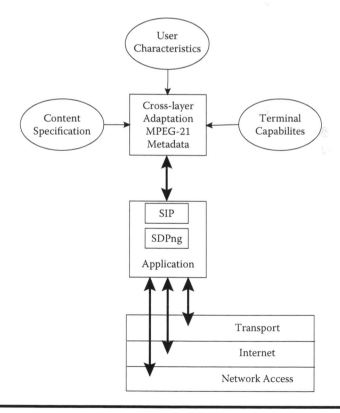

Figure 13.10   MPEG-21-enabled cross-layer interaction.

13.5 Service Convergence Platform of Telecom Operators

There are currently different initiatives for the integration of telecommunication networks. They share the common objective of defining platforms that allow the development of new and complex services that can be used by end users independent of the transport network used. In this way, software developer communities will be allowed to operate the characteristics of the present and future telecommunication networks for the benefit of telecom operators and service providers. Therefore, the goal is to define a new framework that makes it possible to offer new business models for application development and telecommunication services, creating new business models and sources of incomes for network operators, as shown in Figure 13.11. The current user demands services that cannot always be offered by the operator whose main product is connectivity. In this new model, application service providers (ASPs) can easily integrate their solutions on the network of the operator. All of them lead to a rise in traffic and to a higher number of user services. As a result, the business of telecommunications opens up to other actors.

The current trend for access to these value-added services is based on the concept of SOAs. The idea of these architectures is the definition of a framework that makes standardized access to the different available resources in the network possible. This allows for integration and interoperability among services built from different vendors, which can be built in different technologies and deployed in different networks. One of the most prominent technologies in this field is the so-called Web services.

In the case of implementing telephony services, video transfer, or real-time communications, called CM services, the help of specific protocols is necessary (e.g., SIP, SDP, RTP, and RTSP). These protocols provide the specific transport and signaling needs of the media content. In this sense, within the new-generation network (NGN) scenario, a new set of standards known as IMS is gaining popularity.

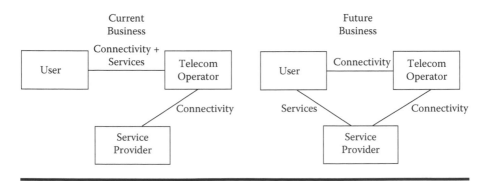

Figure 13.11   The telecommunication business model.

IMS is an evolving architecture for providing voice, video, and other advanced multimedia services to mobile and fixed environments. It promises to have an important impact on the whole of the telecommunications industry, including network providers, applications developers, and end users. Although IMS standards come from 3GPP, they have also been adopted by 3GPP2 and TISPAN (a European organization focused on next-generation fixed networks). IMS can be considered the basis for all variants of the NGN envisioned by the traditional telecommunication industry.

The key technology behind IMS is SIP, which was chosen as the underlying protocol in many of the important interfaces between elements in an IMS-based network. IMS specification is available from the 3GPP Web site [23]. The reason for this choice is that SIP presents several advantages to building new features or services. SIP is simple because it is based on a straightforward request–response interaction model, making life simple and comprehensible for developers. It is extensible because it can set up sessions for any media type. It is also flexible, making it possible to interact with individual protocol messages without breaking anything. Finally, it is well known because it is based on HTTP and other IETF standards, using many Web-based technologies to develop innovative SIP applications.

IMS basically has three layers: the connectivity layer, the control layer, and the service layer. Figure 13.12 shows a simplified view of the layered architecture of IMS. The connectivity layer comprises devices, such as routers or switches, for backbone and network access (multiple types of network access are allowed). The control layer comprises network control servers for managing call or session setup, modification, and release. Several roles of SIP servers, collectively called call session control function (CSCF) are used to process SIP signaling packets in the IMS. Among the security and user subscription functions are the authentication center (AuC) and the HSS (home subscriber server), which communicate with the CSCF to provide security management for subscribers. The interconnections with other network operators or other types of networks are handled by border gateways control functions (BGCFs). Finally, the service layer comprises application and content servers to execute value-added services for the user. Generic service enablers, as defined in the IMS standard, are implemented as services in an SIP application server, which is the native server in IMS.

13.6 SIP and the Future of the Internet

There are efforts toward an Internet redesign after 30 years of great success. GENI (Global Environment for Network Innovations) [24], FIND [25], and FIRE [26] are among the major initiatives that are set to address the urgent need of defining research programs for achieving this goal. A need for new architectures and protocols is envisioned. However, it is foreseen that SIP and its extensions will accomplish the major parts of the expected requirements for signaling and management planes.

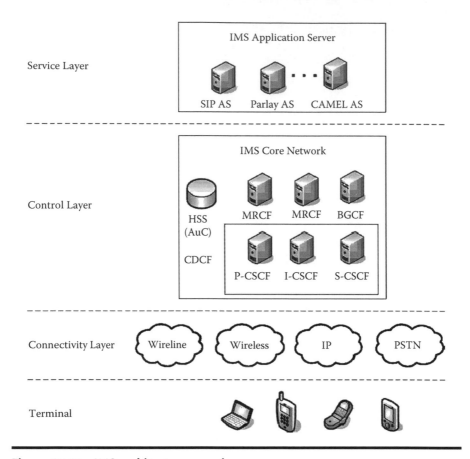

Figure 13.12   IMS architecture overview.

The content for future Internet will be able to be adapted to users with respect to their preferences, terminal capabilities, and access networks. The nature of this content will be three dimensional (3D), including haptic features. Now, of the five senses (hearing, sight, taste, touch, and smell), only the first two are used for the Internet; the rest will appear progressively. This multimodality will be interactive to achieve the maximum level of collaboration among users.

To accomplish this new featured content, the network should be content- and service-centric, regardless of the location of the end user, and must transport real-time 3D multimodal media to multifunctional devices.

SIP decouples the media from the signaling, and if the network is able to transmit this media, SIP can signal it. The underlying idea is because SIP uses companion protocols to perform its functions and if, until now, SIP used SDP to transport media information, in the future it will be SDPng with embedded XML that

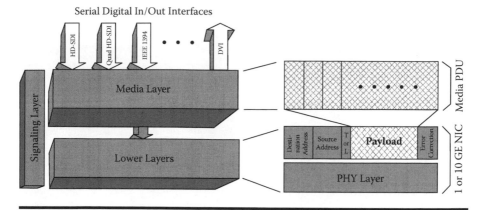

Figure 13.13 New media architecture and its signaling.

provides this function, with an evolution of MPEG-21 performing a five-senses content-exchange framework, for example.

On the other hand, the capacity of networks has grown during recent years. New generations of optical networks are enabling many different services that until now were not accessible to users. Video engineers need to transmit video signals, such as plugging wires locally, and users want to consume this studio-quality video at cultural events or in health scenarios. SIP can provide the signaling framework for the transmission of these standardized digital interfaces, such as HDSDI, SDTI, SDI, and Firewire (IEEE 1394), even if the transport of these signals is carried out by a new media network architecture (see Figure 13.13). In this case, SIP would be able to signal these current and future media contents due to its decoupled nature from media. In the case of SIP, the major involvement would emerge since IP addresses, routing, and DNS names, among others, are closely tied to an underlying structure evolving to a clean-state Internet [27].

13.7 Experimental Results

The advent of high-speed fiber-optic networks has increased the capacity in bandwidth, and the current supported high speed consolidates them as the ideal medium for high-bandwidth transmissions. Currently, the real-time transport of high-quality video over IP has been successfully tested in research networks, expanding the range of video formats to be used on IP networks and the improvement of new high-quality video services. Real-time transport of high-quality live video is one of the main current efforts in innovation and research of next-generation Internet organizations, such as the i2CAT Foundation. In i2CAT, a SIP-CMI test bed has been implemented as an add-on of its SIP platform (SIPCAT). The first one has provided the SIP-CMI subsystem and the second one the SIP infrastructure.

In the following section, several examples of real experiences are outlined, focusing on the concepts that have been explained in this chapter.

13.7.1 Example of Advanced Service: Online Transcoding

The goal of the platform is to deploy a system through which users can access high-quality contents from their terminal, regardless of its characteristics, by interacting with different modules in a transparent way from the users' point of view using Web services and SIP services. Within this context, a content transcoding module has been developed using the SIP-CMI approach. The main functionalities of the transcoding module are to provide

- Real-time transcoding to adapt to user specifications or QoS network limitations
- Full information for transcoders and media servers
- A list of available resources, such as high-quality videos or different coding schemes, among others
- Resource updating and inserting mechanisms

To develop this transcoding module, it has been necessary to implement different service elements and deploy the SIP-CMI platform to validate them. An adapter to be plugged into the media plane and onto the control plane has been implemented, as shown in Figure 13.14.

As a proof of concept, related to the media plane, VideoLAN (local area network) has been used as a media server, and a VideoLAN control interface has been implemented (VideoLAN service control).

Related to the control plane, a driver for the VideoLAN control service and a transcoding service have been implemented based on states machines. After that, they have been deployed on an SIP application server. As a result, the transcoding module has become a video-on-demand HDV (high definition video) system. Current video-on-demand HDV systems [28] solve this control of the server by using Web applications to initiate a video stream from the transmitting unit to the receiving unit with VideoLAN.

13.7.2 SIP Infrastructure

SIPcat is a value-added services integrated platform based on SIP. As Figure 13.3 shows, this platform offers the main SIP components needed to develop the SIP-CMI approach. The main goal of this implementation is to offer an SIP test platform where a developer community (ASPs) and user community test SIP value-added services.

The SIPcat platform offers advanced communication services of voice, video, and data using several communication protocols, such as H.323, PSTN, H.320, 3G domains, and SIP gateways. All core components have been implemented using

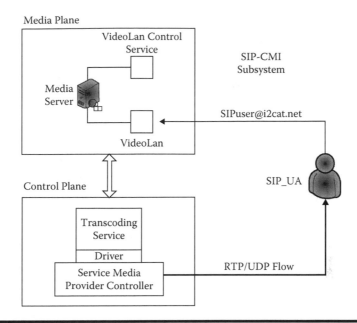

Figure 13.14   Transcoding service implemented on the SIP-CMI platform.

open source solutions, and its core architecture has been enriched with components from commercial vendors, such as Radvision, Polycom, and Tandberg.

The components that integrate this configuration are the classic elements of the SIP framework. Proxy server and registrar are implemented by OpenSer. In addition, the management services guarantee the proper operation of the platform. The management includes four fundamental aspects: location, accounting, security, and monitoring.

The location scheme is based on DNS, ENUM, LDAP, and H.350 protocols. The security includes the authorization and authentication performed by RADIUS. The monitoring is performed by automatic SNMP-based scripting and service control implemented using Nagios. The accounting is a proprietary design oriented to provide the information of resource utilization.

In SIPcat (see Figure 13.15), the dialing scheme was designed according to Internet2 test bed platforms: ViDeNet and SIP.edu. The H.323 dialing plan is compatible with the global dialing scheme (GDS), and the SIP-URI is compatible with the e-mail contact addresses. The GDS is a new numbering plan for the global video and VoIP network test bed, developed by ViDeNet [29].

This infrastructure is based on the above-mentioned Internet2 SIP.edu [30] experience. It promotes the convergence of voice and e-mail identities, increasing SIP reachability, and builds community schools that are developing and deploying campus SIP services in the Internet2 community. Another similar experience is the Global University Phone System (GUPS) [31], using mainly Asterisk.

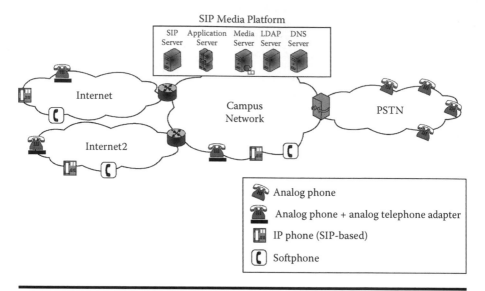

Figure 13.15 SIPcat scenario.

The service delivery is supported by providing different service platforms, such as Web services, SIP servlets [13], and CPL [32]. The main element of the application and media control is the application server implemented through an SIP servlets framework. A framework has been designed for SIP-based services development. The design of the framework is based on a three-tier architectonic model.

First, the core component is an SIP application server implemented by an SIP stack based in JAIN SIP [33]. Another main component is an SIP servlets container.

The next layer is composed of the CallControl, which is responsible for the main control of the calls and the redirection to the respective service. The implementation follows the front controller and the JAIN JCC specification [13]. The second part is a group of components designed to facilitate the design and implementation of value-added services. They provide a level of abstraction for the service implementation.

Over this middleware, the services are executed. A states machine-oriented deployment is proposed to diminish the development of the services.

13.7.3 Example of SIP Cross Layering: Light Path Establishment

As explained in the section on cross layering, the session information contained in SDP can be used by the application server to request, for example, a light path establishment between the endpoints. This idea was deployed in Provisioning and Monitoring of Optical Services (PROMISE) [34], a European cooperative project under the CELTIC-EUREKA cluster and focused on supporting a range of advanced high-capacity services over an adaptable optical network infrastructure, which generalized multiprotocol label switching (GMLS) protocols make possible. The main results of the project were a control plane–enabled test bed to showcase the provisioning and maintenance of optical services in a real-world context. One of these services was the provision of an optical path between an HD transmission unit and an HD receiving unit using the SIP-CMI approach. To establish this path, the SIP application server (SMPC) had to request, using Web services, the optical service before successfully establishing the SIP session (Figure 13.16) and initiating the HD transmission.

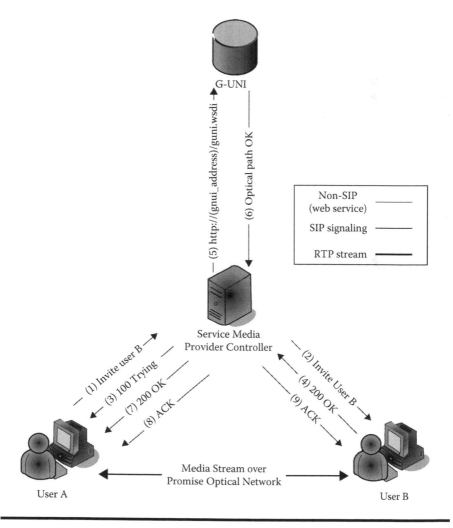

Figure 13.16   **SIP session setup using GMPLS PROMISE optical network.**

13.7.4 Examples of High-End Videoconferencing

Videoconferencing systems have been an important area of telecommunications research, resulting in a large number of products that can be used to transmit and receive video in real time over IP networks. These systems vary mainly in terms of quality and in the amount of bandwidth used to transmit video over networks. In this sense, the telecom vendors and manufacturers have focused on developing low- to medium-bandwidth conferencing systems, typically based on H.320 and H.323 standards. In addition to these commercial products and systems, academic researchers and the university community have also looked for a higher-quality and higher-bandwidth video system that operates over high-speed research and educational networks (e.g., Internet2-Abilene Network).

One of the main aspects of SIP signaling is the decoupling between signaling and media. As a consequence, any kind of media can be signaled by SIP if SDP can support it. Three examples are briefly explained concerning videoconferencing using new media:

- Tele-immersion
- Digital video (DV)
- Uncompressed HD video

Tele-Immersion for Applications Supporting New Interactive Services (TIFANIS) [34] was a European cooperative project under the CELTIC-EUREKA cluster, focused on designing and implementing a full tele-immersive system, with several trial scenarios. Basic SIP services to tele-immersive systems have been provided and tested by developing a transmission reception subsystem (TRS) module and integrating in one of the cubicles (see Figure 13.17).

With the new generation of camcorders, a user can manage high-quality video on its own. Editing tools are known and common among users. Users only need to plug the digital interface of the camcorder into the PC and capture its signal.

Digital Video Transport System (DVTS) [35] and UltraGrid [36] are examples of software that receive there signals and transmit them directly to the network. If users can get enough bandwidth, there are ways to transmit high-quality video; moreover, there is no compression delay. SIP can fill the gap to build a high-quality videoconferencing system (end to end).

DVTS is software that allows encapsulating DV format from a Firewire interface for transmission over IP networks, resulting in a high-quality DV stream that consumes roughly 30 Mbps of bandwidth.

UltraGrid is an HD video conferencing and distribution system. It is also considered one of the first systems capable of supporting uncompressed gigabit-rate HD video over IP. In fact, an UltraGrid node converts SMPTE 292M HD video signals into RTP/UDP/IP packets, which can then be distributed across an IP network, reaching transmission rates to 1.2 Gbps. There are few systems with these characteristics, such as iHDTV and i-Visto gateway [37].

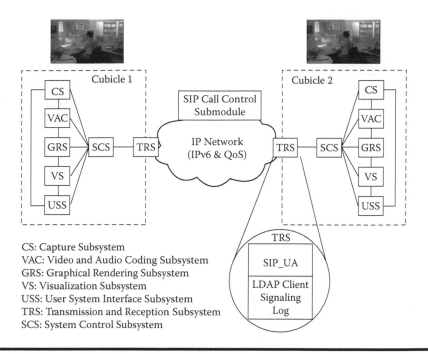

Figure 13.17 TRS subsystem and SIP-CMI subsystem in TIFANIS project.

The i2CAT Foundation research project called MACHINE has developed an open source graphical SIP UA to establish and control high-quality videoconferencing, becoming a DVTS and UltraGrid videoconferencing system. Figure 13.18 shows the SIP message exchange to establish and terminate a basic videoconferencing session using an OpenSer as an SIP server.

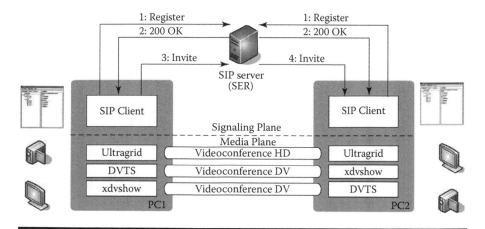

Figure 13.18 HD videoconferencing with SIP messages.

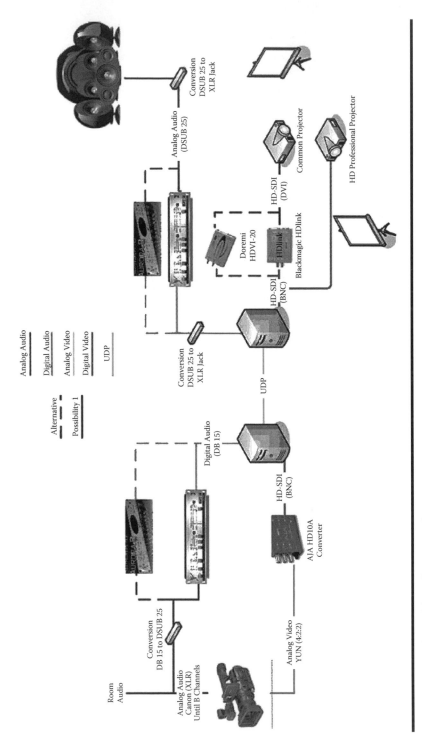

Figure 13.19 Ultragrid testing scenario.

Figure 13.19 shows the current testing scenario with UltraGrid, to demonstrate that regardless of the complexity of the media transport, SIP can signal it.

Acknowledgment

This research was supported by the i2CAT Foundation and the MCyT (Spanish Ministry of Science and Technology) under projects TSI2005-06092 and TSI2007-66637-C02-01, which are partially funded by FEDER.

References

1. Rosenberg, J., Schulzrinne, H., Camarillo, G., Johnston, A., Peterson, J., Sparks, R., Handley, M., and Schooler, E., *SIP: Session Initiation Protocol.* IETF RFC 3261. June 2002.
2. Hurtado, M., Oller, A., and Alcober, J., The SIP-CMI platform—an open testbed for advanced integrated continuous media services. TRIDENTCOM 2006. Barcelona, Spain, March 2006.
3. Hiroshi, W., and Junichi, S., A service-oriented design framework for secure network applications. In *AU: Proceedings of the 30th IEEE International Conference on Computer Software and Applications Conference (COMPSAC).* Chicago, September 17–21, 2006.
4. Liu, F., Chou, W., Li, L., and Li, J., WSIP—Web service sip endpoint for converged multimedia/multimodal communication over IP. In *Web Services, 2004. Proceedings. IEEE International Conference,* 690–697, June 6–9, 2004.
5. W3C, *Voice Browser Call Control: CCXML Version 1.0, W3C Working Draft.* http://www.w3.org/TR/ccxml/, June 29, 2005.
6. Dyke, J.V., Burger, E., and Spitzer, A., *Media Server Control Markup Language (MSCML) and Protocol.* IETF RFC 4722. November 2006.
7. Saleem, A., Xin, Y., and Sharratt, G., *Media Server Markup Language (MSML).* Internet draft (draft-saleem-msml-05). August 17, 2007.
8. Melanchuk, T., *An Architectural Framework for Media Server Control.* Internet draft (draft-ietf-mediactrl-architecture-02). February 5, 2008.
9. JSR 309, *Media Server Control API.* http://jcp.org/en/jsr/detail?id=309.
10. Schulzrinne, H., *Real Time Streaming Protocol (RTSP).* IETF RFC 2326. April 1998.
11. Kahmann, V., Brandt, J., and Wolf, L., Collaborative streaming in heterogeneous and dynamic scenarios. *Commun. ACM,* 49(11), 58–63, 2006.
12. Marjou, X., Whitehead, S., Ganesan, S., Montpetit, M., Ress, D., and Goodwill, D., *Session Description Protocol (SDP) Format for Real Time Streaming Protocol (RTSP) Streams.* Internet draft (draft-marjou-mmusic-sdp-rtsp-00). October 12, 2006.
13. Java Community Process. http://www.jcp.org/en/jsr/detail?id=21, January 22, 2002.
14. WeSIP. http://www.wesip.com, July 17, 2006.
15. OpenSer. http://www.openser.org/, March 2007.
16. Asterisk. http://www.asterisk.org/, January 2004.
17. Booth, D., Haas, H., McCabe, F., Newcommer, E., Champion, M., and Orchard, C.F.D., Web service architecture. In *W3C Recommendation* [Online]. http://www.w3.ord/TR/2004/NOTE-ws-arch-20040211/, November 2004.

18. Fu, X., Bultan, T., and Su, J., Analysis of interacting BPEL Web services. In *Proceedings of the 13th International World Wide Web Conference (WWW'04)*. USA: ACM Press.

19. Kutcher, Ott, Borman, *Session Description and Capability Negotiation*. Internet draft (draft-ietf-mmusic-sdpng-08). February 20, 2005.

20. Kassler, A., Guenkova-Luy, T., Schorr, A., Schmidt, H., Hauck, F., and Wolf, I., Network-based content adaptation of streaming media using MPEG-21 DIA and SDPng. *Seventh International Workshop on Image Analysis for Multimedia Interactive Services (WIAMIS06)*. Invited Paper to Special Session on Universal Multimedia Access (UMA), Seoul (Korea), April 2006.

21. Guenkova-Luy, T., Schorr, A., Hauck, F., Gomez, M., Timmerer, Ch., Wolf, I., and Kassler, A., Advanced multimedia management—Control model and content adaptation. *IASTED International Conference on Internet and Multimedia Systems and Applications (EuroIMSA 2006)*, Innsbruck, Austria, February 2006.

22. Toufik Ahmed, A., and Djama, I., Delivering audiovisual content with MPEG-21-enabled cross-layer QoS adaptation [Packet video] *J. Zhejiang Univ. Science A,* 7(5), 784–793, 2006.

23. 3GPP home page. http://www.3gpp.org/, October 2002.

24. GENI (Global Environment for Network Innovations) home page. http://www.geni.net/, May 2002.

25. Fisher, D. U.S. National Science Foundation and the future internet design. *SIGCOMM Comput. Commun. Rev.,* 37(3), 85–87, 2007.

26. Gavras, A., Karila, A., Fdida, S., May, M., and Potts, M., Future internet research and experimentation: The FIRE initiative. *SIGCOMM Comput. Commun. Rev.,* 37(3), 89–92, 2007.

27. Balakrishnan, H., Lakshminarayanan, K., Ratnasamy, S., Shenker, S., Stoica, I., and Walfish, M., A layered naming architecture for the internet. In *SIGCOMM '04: Proceedings of the 2004 Conference on Applications, Technologies, Architectures, and Protocols for Computer Communications*. New York: ACM Press, 2004, pp. 343–352.

28. *VOD Services for High Definition Video Streaming over IPv6*. Networked Media Lab, Gwangju Institute of Science and Technology (GIST). http://hdtv.nm.gist.ac.kr/v2/main.php, September 2004.

29. ViDe // Vide Development Initiative. http://www.vide.net/, June 2002.

30. SIP.edu. http://mit.edu/sip/sip.edu/index.shtml, March 2003.

31. GUPS. http://www.aboutreef.org/gups-press.html, August 2005.

32. Lennox, J., Wu, X., and Schulzrinne, H.M, *Call Processing Language (CPL): A Language for User Control of Internet Telephony Services*. IETF RFC 3880. October 2004.

33. *The Java Community Process (SM) Program—JSR 32: JAIN SIP API Specification*. http://www.jcp.org/en/jsr/detail?id=116.

34. Celtic Initiative. http://www.celtic-initiative.org/Projects/projectinfo.asp, May 2004.

35. *DVTS, DV Stream on IEEE1394 Encapsulated into IP*. http://www.sfc.wide.ad.jp/DVTS/, May 2004.

36. Perkins, C.S., and Gharai, L., *Ultragrid: A High Definition Collaboratory*. http://ultragrid.east.isi.edu/, November 2002.

37. *i-Visto Internet Video Studio System for HDTV Production*. http://www.i-visto.com/, April 2006.

Chapter 14

VoIP in a Wireless Mobile Network

Qi Bi, Yang Yang, and Qinqing Zhang

Contents

14.1 Introduction

14.1.1 Evolution of Circuit Voice Services in Mobile Networks

Voice-over Internet Protocol (VoIP) has rapidly become popular in wireline networks. The evolution of voice service from circuit-switched voice to VoIP has been due to the proliferation of IP networks that can deliver data bits cost-effectively. Similar to the wireline networks, voice in wireless mobile networks started with circuit-switched networks. Since the deployment of the first commercial wireless system, wireless mobile networks have evolved from first-generation analog networks to second-generation (2G) digital networks. Third-generation (3G) wireless mobile networks that are currently in use offer a highly efficient circuit-switched service and have double the spectral efficiency of 2G systems. After the deployment of the 3G wireless voice system, however, the number of voice subscribers, having reached a saturation point in many parts of the world, no longer increased. Consequently, the focus on wireless system design shifted to wireless data applications. This resulted in the emergence of high-speed packet-switched data (HSPD) networks, which have been gradually deployed alongside circuit-switched voice networks worldwide.

The growth of the HSPD networks has made it even more desirable to offer a voice service using VoIP together with data to support highly diversified multimedia services. When HSPD and VoIP are deployed together, a separate circuit-switched network need not be dedicated for voice services. Besides, as most of the HSPD networks were designed using the most up-to-date technology, the spectral efficiency of wireless mobile VoIP networks can be greatly improved over that of packet data services designed for 3G voice networks. This makes the wireless mobile VoIP service attractive to service providers who face constraints on the availability of the frequency spectrum; they also have the advantage of operating an integrated network offering both voice and data services.

14.1.2 Challenges of VoIP over Wireless

Although VoIP has become very popular and successful in wireline systems, it is still in its infancy in the wireless system, and many technical challenges remain. Here are some of the major challenges:

■ *Delay and jitter control.* In wireline systems, channels are typically clean, and end-to-end transmission can be almost error free, requiring no retransmissions. However, a wireless channel could be unfavorable, resulting in bit errors and corrupted packets. Packets may have to be retransmitted multiple times to ensure successful reception, and the number of retransmissions depends on the dynamic radio-frequency (RF) conditions. This could introduce significant delay and delay variations. Further, unlike the circuit channels, which have a dedicated fixed bandwidth for continuous transmission, packet transmissions are typically bursty and share a common channel that allows multiplexing for efficient channel utilization. This operation also results in loading-dependent delay and jitter.

■ *Spectral efficiency.* In a wireline VoIP system, bandwidth is abundant, and it is often used to trade off a shorter delay. In fact, more bandwidth-efficient circuit-switched transmissions have been abandoned in favor of the flexibility of packet-switched transmissions, even though packet transmissions incur extra overhead. In wireless systems, however, the spectrum resource is generally regarded as the most expensive resource in the network, and high-spectral efficiency is vitally important for service providers. Therefore, wireless VoIP systems must be designed such that they can control delay and jitter without sacrificing spectral efficiency. Packet transmission overheads must also be kept to a minimum over the air interface.

■ *Mobility management.* In many HSPD systems, mobility management has been designed mainly for data applications. When mobile users move among cell sites, the handoff procedure follows the break-before-make principle. This leads to a large transmission gap when the mobile is being handed off from one cell site to another. While a transmission gap is often acceptable in data applications, it is unacceptable for voice. To support VoIP applications, the handoff design must be optimized so that the transmission gap during the handoff is minimized and does not have an impact on voice continuity.

■ *Transmission power and coverage optimization.* With the packet overheads, a higher power is needed to transmit the same amount of voice information, which results in smaller coverage. In addition, bursty packet transmissions also cause a higher peak-to-average transmission power ratio, which in turn may lead to a higher power requirement in the short term and degraded performance in the outer reaches of the areas covered by the network. Therefore, more advanced techniques must be adopted to compensate for these shortcomings.

14.1.3 3G/4G HSPD System Overview

The two dominant bodies that define wireless mobile standards worldwide are 3GPP (Third Generation Partnership Project) and 3GPP2, which have produced their respective 3G standards: UMTS (Universal Mobile Telecommunications System) terrestrial radio access (UTRA) and cdma2000. Their respective fourth-generation (4G) evolution systems, namely, long-term evolution (LTE) and ultra mobile broadband (UMB), are currently being standardized. The technologies used by the two standard systems are similar; therefore, we use the 3GPP2 standards as the example in this chapter.

On the 3G side, cdma2000 standards include two components: 3G1x, focusing on circuit-switched voice service, and the cdma2000 evolution data optimized (EVDO) standard dedicated to high-speed packet data services. For our discussion of VoIP, we focus on the EVDO system.

The EVDO air interface design adopts a dynamic time-division multiplexing (TDM) structure on the forward link with a small time-slot structure (600 time slots per second) to enable fast and flexible forward packet scheduling. Each active user feeds the desired channel rate back in real time. To improve the transmission efficiency under dynamic RF fading conditions, a hybrid automatic repeat request (HARQ) technique is employed with incremental redundancy (IR). In EVDO Revision0 (Rev0) [1], 12 nominal data rates were defined, ranging from 38.4 kbps to 2.4 Mbps. A scheduler at the base transceiver station (BTS) decides which user's packet will be transmitted in each open time slot, and the packet must be transmitted with the matching data rate requested by the terminal.

On the reverse link, the design of EVDO Rev0 is similar to that of 3G1x. Code-division multiple access (CDMA) technology is used for data traffic as well as real-time feedback (overhead) information transmissions. Each user can transmit packets at any time with data rates from 9.6 kbps to 153 kbps. The packet transmission duration is 16 time slots. The reverse data rates are controlled jointly by each mobile and the BTS in a distributed fashion. The BTS monitors the aggregate interference level and broadcasts the overload information via a dedicated control channel on the forward link. The information is used by each mobile to adjust the reverse traffic channel rate upward or downward, subject to the available power headroom. When there is no packet to be sent, the reverse data channel is gated off, whereas the pilot and other overhead channels are continuously transmitted.

In EVDO RevisionA (RevA) [2], more forward data rates are defined up to 3.1 Mbps, which are further augmented, up to 4.9 Mbps, by EVDO RevisionB (RevB) [3]. The BTS has the flexibility of transmitting at either the data rate requested by the terminal or at one of several compatible data rates lower than the requested data rate; it can also multiplex packets from multiple users onto a single time slot.

The reverse link operation is significantly upgraded in EVDO RevA. Twelve encoder packet sizes are defined, with effective data rates ranging from 4.8 kbps to 1.8 Mbps. The HARQ technique is used with up to four transmissions; each transmission or subpacket lasts four time slots. Reverse data rate control (DRC) is done via a comprehensive resource management scheme that adjusts the allowable terminal transmission power, which in turn determines the transmission data rate that can be achieved.

Besides the channel operation-related improvements, EVDO RevA also specifies the quality-of-service (QoS) framework that enables the radio access network (RAN) to differentiate between applications with different QoS expectations. On the air interface, the forward link scheduler applies different scheduling policies and priorities to traffic flows with different QoS requirements either across users or to the same user. The reverse resource management scheme allows multiplexing data from different flows that share the same physical channel, and flow QoS requirements are taken into consideration to determine resource distribution.

UMB [4, 5] is an evolution technology of 1x EVDO. It is the 4G broadband high-speed data system developed by the 3GPP2 standard body. It operates on a much wider carrier bandwidth, ranging from 1.25 to 20 MHz, and uses orthogonal frequency-division multiple-access (OFDMA) technology for both forward and reverse link data transmissions. OFDMA enjoys advantages over CDMA in several aspects. The OFDM transmission symbol duration is much longer than that of the CDMA chip, which requires less-stringent time synchronization between transmitter and receiver to combat intersymbol interference in multipath fading environments. Compared with CDMA, orthogonal transmission between OFDMA subcarriers improves reverse link performance as it eliminates intracell interference. The RF transmission can be scheduled in time as well as frequency dimensions, hence improving scheduling efficiency.

In UMB, a carrier bandwidth is divided into a set of subcarriers with a spacing of 9.6 kbps. The basic transmission unit is defined as a tile that consists of 16 subcarriers in frequency and 8 OFDM symbols in time, which lasts about 1 ms depending on the cyclic prefix length configuration. Each tile can deliver an instantaneous data rate between 75 and 630 kbps under different coding and modulation combinations. Up to six HARQ transmissions are allowed for a packet, which helps it to exploit the time diversity gain. In a 5-MHz carrier, this leads to data rate support over a wide range: from 0.4 to 17 Mbps. Further, the UMB standard specifies multiple types of advanced antenna technologies that can improve spectral efficiency, including

■ Multiple-input multiple-output (MIMO) technology, by which multiple data streams are transmitted to and from a single user simultaneously via multiple transmitting and receiving antennae. Depending on the antenna configuration, MIMO can double or even quadruple spectral efficiency in a multipath-rich environment and under relatively benign RF conditions. With MIMO, the user peak data rate in UMB is increased to 35 Mbps for a 2 × 2 antenna configuration and to 65 Mbps for a 4 × 4 antenna configuration.

■ The SDMA scheme, by which the BTS simultaneously transmits data streams to different users using the same frequency resources, given these users are located at distinctively different locations and the signals of those streams cause minimum mutual interference via multiple antenna beam-forming techniques. With SDMA, the aggregate RF transmission efficiency can be significantly improved.

OFDMA is used in UMB for both forward and reverse link data transmissions. The orthogonal nature of OFDMA significantly improves the spectral efficiency on the reverse link compared with the CDMA-based EVDO system. However, to optimize mobility management, UMB employs a hybrid CDMA OFDMA structure on the reverse link operation. That is, a certain portion of the tile resources is configured to operate in CDMA fashion and is used by each user to feed back information, such as the change requests of the forward or reverse serving sector. CDMA and OFDMA transmissions maintain separate pilots. When there is no data transmission, the strong OFDMA pilot is not transmitted, whereas the weak CDMA pilot is, which the BTSs continue to monitor to track the reverse channel conditions and provide channel quality feedback information to the terminal. The hybrid operation and the split of pilots not only facilitate a fast handoff process but also improve the mobile battery power performance.

We now discuss the various techniques that have been designed in the EVDO and UMB systems to address the VoIP service challenges described.

14.2 Techniques to Improve VoIP Transmission Efficiency

14.2.1 Speech Codec and Silence Suppression

Due to the limited bandwidth over the air interface, the speech codec used in cellular systems is often in the category of a narrowband, low-bit-rate speech coding. In cdma2000 systems, the enhanced variable rate codec (EVRC) [6] for low-bit-rate speech is used to generate a voice frame every 20 ms. There are four different frame types defined in the EVRC (also called EVRC-A), full rate at 171 bits (equivalent to 8.55 kbps); 1/2 rate at 80 bits (equivalent to 4 kbps); 1/4 rate at 40 bits (equivalent to 2 kbps); and 1/8 rate at 16 bits (equivalent to 0.8 kbps). The 1/4 rate frame is actually not used in the EVRC voice coder (vocoder). The percentage of frames with each rate varies depending on the talk spurt structure and voice activity. The EVRC vocoder has been widely deployed in 3G1x voice networks.

A new speech codec, EVRC-B [7], has been standardized and implemented as the next-generation speech codec for cdma2000 networks in 3GPP2. The main design consideration of EVRC-B is to provide a smooth and graceful trade-off between network capacity and voice quality. The EVRC-B vocoder uses all four frame types. It defines eight modes (encoder operating points) with different source data rates and utilizes a reduction rate mechanism to select a mode, which trades average encoding rate for the network capacity.

Table 14.1 shows the estimated average source data rates for active voice encoding under different modes [7]. Mode0 of EVRC-B, like EVRC, does not use the 1/4 rate frame. It has a data rate similar to that of EVRC but is not compatible with EVRC and, in general, offers better voice quality than does EVRC. Other modes produce lower data rates than EVRC and show a gradual degradation of voice quality. Mode7 of EVRC-B uses only the 1/4 and 1/8 rate frames. The different modes provide a mechanism for service providers to dynamically adjust and prioritize voice capacity and quality in their networks.

Table 14.1   EVRC-B Target Rates for Different Modes

	Mode							
	0	*1*	*2*	*3*	*4*	*5*	*6*	*7*
Active source rate (kbps)	8.3	7.57	6.64	6.18	5.82	5.45	5.08	4.0

Among the different data rate frames, the 1/8 rate frames usually represent background noise during silent periods and during the speech gaps within talk spurts. They do not carry any actual speech signals and hence have minimum impact on the receiving voice quality. Therefore, most of the 1/8 rate frames can be dropped at the encoder side and need not be transmitted. Only a very small percentage of the 1/8 rate frames are transmitted to assist noise frame reconstruction at the receiver end to generate a more natural background noise environment. This technique is called *silence detection and suppression*, and it greatly improves voice transmission efficiency over the air interface by eliminating a significant portion of the packets containing 1/8 frames when packets are transmitted for a VoIP call.

14.2.2 Header Compression

Header compression, which works by exploiting redundancy in headers, is essential to providing a cost-effective VoIP service in wireless networks. A VoIP packet is generally carried over the RTP/UDP/IP (Real-Time Protocol/User Datagram Protocol/Internet Protocol) protocol stack. The uncompressed header is about 40 bytes per voice packet. In a full-rate EVRC voice frame of 22 bytes, the RTP/UDP/IP header adds an overhead of more than 180% over the voice information. Without any header compression, the overhead would become dominant and have a significant impact on the air interface capacity. RTP, UDP, and IP headers across the packets of a media stream have significant redundancy. The fields in a header can be classified into multiple categories:

- Static or known fields, which do not need to be sent with every packet, for instance, the source and destination address in the IP header, the source and destination port in the UDP header.
- Inferred fields, which can be inferred from other fields and thus do not need to be sent with every packet; for example, the time stamp in the RTP header, which increases by a fixed amount with every increase in the sequence number. The time stamp, hence, need not be sent during a talk spurt once the sequence number has been sent. Only at the beginning of a new voice spurt after a silence suppression will the time stamp have to be sent to the receiver again.
- Changing fields that can be sent in compressed form to save bandwidth.

Figure 14.1 shows the header field classification in the RTP/UDP/IP headers.

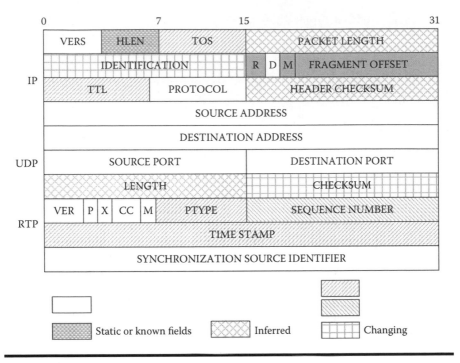

Figure 14.1   RTP/UDP/IP header field classification.

Header compression and suppression have been specified in both 3GPP and 3GPP2 standards. The most popular header compression scheme is the robust header compression (ROHC) [8], which is the standard specification of the Internet Engineering Task Force (IETF) for a highly robust and efficient header compression scheme. The ROHC protocol specifies several operating modes, each requiring a different level of feedback information. The selection of the mode depends on the type of the underlying communication channel between the compressor and the decompressor that may have different capability of providing such feedback information. With ROHC compression, RTP/UDP/IP headers are reduced from 40 to 1–4 bytes. Furthermore, ROHC design is also resilient to packet errors, making it suitable for use with the error-prone wireless link transmissions. As a result, ROHC greatly improves the transmission efficiency of VoIP packets over the air interface. The performance of ROHC in the EVDO RAN is analyzed and evaluated in [9].

It should be emphasized that ROHC is typically not used as an end-to-end header compression scheme as the full IP header information is still needed for general routing purposes. It is mostly used in the RAN. A typical implementation of ROHC is shown in Figure 14.2. The ROHC compressor and decompressor are located in the mobile handset and radio network controller (RNC) or packet data service node (PDSN).

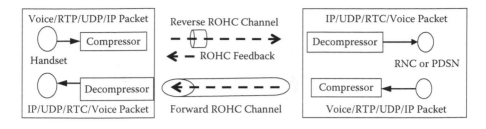

Figure 14.2 ROHC implementation architecture.

14.3 Techniques to Support VoIP in EVDO RevA and RevB

As mentioned, EVDO RevA and RevB standards have been enhanced to support QoS applications such as VoIP. In particular, the low volumes and directionally symmetric nature of VoIP traffic do not align well with the traffic assumptions in the original design of the EVDO Rev0, which was designed for Internet-oriented packet data services. As a result, a set of novel features, specially tailored to improve VoIP performance and capacity, have been introduced into the standard. This section describes some of the salient techniques used in EVDO RevA and RevB to support VoIP services.

14.3.1 QoS Enforcement in EVDO RevA/RevB

When a VoIP service is offered together with the general packet data service over the same air interface, the issue of QoS immediately arises. The original EVDO design focused on supporting general data services and lacked supporting mechanisms for QoS applications. All data from and to a user were mixed, with no means of differentiating between them. EVDO RevA fundamentally changed the picture by classifying various applications targeted to different users or even the same user. A QoS application traffic is mapped to a radio link protocol (RLP) flow, which bears a specific traffic characteristic and performance target. An abbreviated approach to negotiating QoS parameters is provided via the standardized FlowProfileID, which identifies the type of application, delay sensitivity, and the required data rates.

EVDO RevA and RevB allow a user to open multiple flows over the same air interface connection and multiplex the traffic from different flows on the same physical traffic channel. The radio link (RL) packet header identifies the flow the packet belongs to so that the mobile and the RAN can treat the packet appropriately, based on flow QoS requirements. For the VoIP service, QoS enforcement is focused on expedited packet delivery. Therefore, a VoIP packet is usually assigned a higher priority than a non-real-time data packet while scheduling traffic throughout the RAN. The specific approaches to the air interface QoS enforcement are different between the forward link and the reverse link.

On the forward link, the air interface scheduler is the central entity that evaluates the priority of the packet and determines the air interface transmissions for all users served by the sector. The specific QoS scheduling algorithm used in the BTS is typically proprietary to the infrastructure vendor, while the commonly used scheduling metrics, although not limited to, include

- the QoS requirements of the flow, such as packet latency and packet loss requirements, flow throughput expectation, and so on
- the current flow performance
- the data backlog of the flow
- the dynamic RF condition of the mobile user
- the fairness among flows to different users with similar QoS requirements

Since the objective of QoS enforcement is to satisfy the flow performance regardless of the user's RF condition, it often leads to conflict with the RF efficiency optimization, and it is up to the scheduler to strike a balance between the two objectives.

On the reverse link, the QoS enforcement is provided via the enhanced reverse traffic channel media-access control (RTCMAC) protocol in EVDO RevA. The RTCMAC protocol supports multiple media-access control (MAC) layer flows corresponding to the flow of different applications. Each MAC flow can be configured as a high-capacity (HiCap) or low-latency (LoLat) flow. A LoLat flow always has transmission priority over a HiCap flow, although data from both flows can be multiplexed on the same physical layer packet. A LoLat flow typically enjoys boosted traffic power to reduce the transmission latency via early termination of HARQ. On the aggregation side, the shared RF resource on the reverse link is the interference received by the BTS. It is measured using the *rise-over-thermal* (RoT) characteristic, which is defined as the total power received by the BTS normalized by the thermal noise level. RoT needs to be controlled within a target operating point to ensure the stability of CDMA power control. To that end, the BTS monitors the RoT level and compares it with the target operating point. Once the RoT exceeds the target, the BTS broadcasts an "overload" indication to the sector. The resource management scheme embedded in the terminal determines the distribution of traffic power for each active MAC flow based on multiple factors, such as

- the overload indications received from the BTS
- the flow type, HiCap or LoLat
- the traffic demand and the performance of the flow

In addition, the terminal considers other constraints, such as the available power headroom and the recent history of the packet transmission patterns, so that it does not produce an abrupt surge of interference to other users. Once the transmission power is determined, the attainable physical transmission data rate is determined.

More information about the QoS framework within the EVDO system can be found in [10].

14.3.2 Efficient Packet Format Design for VoIP

Compared with typical Internet application data packets, VoIP packets are much smaller in payload size. For example, the full-rate EVRC vocoder generates a 22-byte voice frame, each of a 20-ms interval, during a talk spurt. Using the header compression techniques described, the VoIP packet size presented to the RAN is usually brought down to within 26 bytes. To transmit the small VoIP packets efficiently over the air interface, EVDO RevA defines a new RLP packet format that enables an ROHC compressed full-rate EVRC VoIP packet to be placed perfectly into an RL physical packet of size 256 bits without any segmentation or padding so that the air interface transmission overhead is minimized.

On the forward link, a different technique is used to improve the VoIP transmission efficiency. In the original design, a physical packet carries information targeted to only a single user. To support VoIP traffic, significant padding would have to be appended to the physical packet, leading to poor RF efficiency. The EVDO forward link is partitioned into 600 time slots per second, and it would become a serious resource limit if each time slot carried only single-user packets. To overcome these issues, EVDO RevA defines a new MAC packet format on the forward link, namely, multiuser packet (MUP), for which up to eight RLP packets from different users can be multiplexed within a single physical packet. This not only improves the packet transmission efficiency greatly but also effectively removes the time slot resource as a capacity constraint.

14.3.3 Expediting VoIP Packet Delivery within EVDO RevA/RevB RAN

Besides the QoS enforcement that in general enables the EVDO RevA/RevB RAN to expedite VoIP packet delivery, there are certain techniques used on the packet delivery method to match the VoIP traffic characteristics.

VoIP packets arrive in 20-ms intervals; hence, it is desirable to deliver the VoIP packets within a 20-ms interval or within 12 time slots. The HARQ operation of the EVDO RevA/RevB reverse link allows up to four subpacket transmissions, which takes 16 time slots. To match VoIP requirements, a termination target is introduced, which can control the number of subpackets it takes to transmit a packet under a given packet error rate (PER). For VoIP flow, the termination target is set to three subpackets with a 1% PER. This eliminates reverse link queuing delay and expedites VoIP packet delivery. Figure 14.3 shows the worst-case (with 1% PER) timeline for the VoIP transmission on EVDO RevA reverse link.

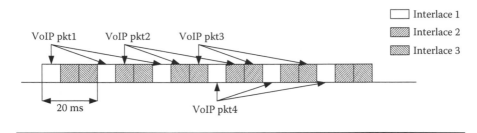

Figure 14.3   VoIP transmission timeline in EVDO RevA.

Beyond the air interface, EVDO RevA/RevB allows out-of-order RLP packet delivery to minimize VoIP transmission delay within the RAN. That is, the RLP receiver in the mobile or the network submits a VoIP packet to the upper layer as soon as it is received, even if there are RLP packets with smaller sequence numbers yet to be received. This is based on the consideration that a wireless VoIP client design typically includes a dejittering algorithm that is able to tolerate a certain degree of varying packet delay, and some can even make use of the information contained in an out-of-order arrived packet. The support of out-of-order packet delivery supports the client design and helps optimize the end-to-end VoIP application performance.

14.3.4 Smooth Mobility Support

One of the major challenges of supporting VoIP service in a mobility environment is to ensure service continuity when users are moving within the network. EVDO supports soft handoff in the reverse link as part of the CDMA operation, thereby providing a smooth make-before-break transition when the user moves across cell boundaries.

On the forward link, though, EVDO operates in the simplex mode, in which the terminal communicates with only a single sector (also known as the serving sector) at any moment of time. The terminal indicates the serving sector along with the desired channel rate through the reverse link DRC channel. If the terminal detects a stronger signal from another sector, it changes the serving sector indication on the DRC channel and switches immediately to the new serving sector to receive packets. In EVDO Rev0, the network detects the serving sector change based on DRC channel decoding and reroutes the packets if the new serving sector resides in a different BTS so that the user can be served again. This process is similar to the break-before-make transition incurred in a hard handoff, which creates a long traffic gap in the range of about 50–100 ms. For VoIP applications, such long gaps can cause a short-term voice break and have an impact on the service quality.

To minimize the service gap during serving sector change, especially during the serving BTS change, EVDO RevA introduces a new reverse control channel, namely, data source control (DSC) channel, dedicated to forward link handoff.

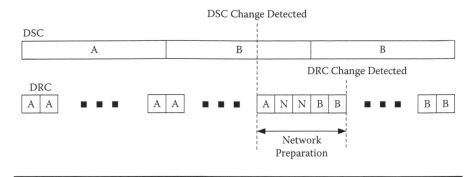

Figure 14.4   Serving sector switch timeline in EVDO RevA/RevB.

The terminal indicates the serving BTS on the DSC channel. Whenever the terminal determines that it has to change the serving BTS, it indicates the change on the DSC channel while continuing to point to the current serving sector on the DRC channel for a predefined duration. This staggered operation provides the crucial time required by the network to detect the imminent serving BTS change and arrange for the data to be rerouted accordingly. When the terminal finally switches to the new serving sector, the new serving BTS is ready to serve the user. In this way, the traffic gap is minimized, often to within 20 ms, a range undetectable by the VoIP application. From the service perspective, it effectively achieves a seamless handoff. Figure 14.4 illustrates the process.

14.3.5 Other VoIP Performance Improvement Techniques Used in EVDO

14.3.5.1 Mobile Receiver Diversity

Mobile receiver diversity (MRD) is a well-known technique to improve signal quality. Since the EVDO forward link operates in the simplex mode, MRD maintains the RF link quality at or above the acceptable level, leading to continuous voice packet transmissions within the network coverage. With MRD, the overall mobile received signal quality can improve significantly, up to 3 dB on average with balanced and uncorrelated antennae. It also greatly reduces the possibility of the simplex RF connection becoming extremely poor due to deep channel-fading conditions.

14.3.5.2 Interference Cancellation on EVDO Reverse Link

The EVDO reverse link air interface capacity is generally interference limited due to its CDMA operation, so naturally interference mitigation or avoidance techniques have been widely investigated to improve the capacity. One of the interference mitigation solutions, successive interference cancellation (SIC), has been considered in the latest EVDO BTS design. The concept works like this: The BTS first tries to decode all users' data. If some users' data can be successfully decoded, the BTS reconstructs the received signals and removes those signals from the total received signal. The BTS then tries again to decode the remaining users' data. With the removal of a part of the interference from those signals that have been successfully decoded, the effective signal quality of the remaining users' data improves and with it the possibility of being successfully decoded as well. Since the strongest interference source is typically from the transmissions within the same sector, SIC is able to eliminate a significant fraction of the interference from the decoding process. As a result, the same data can be transmitted with reduced power while maintaining the same decoding performance, which in turn lowers the interference received by the BTS. Under the same operational target of interference level, SIC provides an effective means to improve the overall RF capacity on the EVDO reverse link. Refer to [11] for more information.

14.3.5.3 Reverse Link Discontinuous Transmission

Each active EVDO terminal maintains an RF connection with the BTS. In EVDO Rev0 and RevA, the terminal continuously transmits through the pilot channel to facilitate channel estimation in the BTS receiver. The reverse-feedback channel signals are also transmitted continuously to assist forward link operations. For applications that involve large amounts of data transmissions, the transmission power overheads are minor. However, for VoIP applications, which generate a low volume of user data, the overheads become high in both terminal power consumption and the interference generated. Given the nature of the VoIP application and the enabling of silence suppression, almost no VoIP packets are sent over the air interface when a user is not talking, resulting in low reverse traffic channel activity. Under this condition, due to the continuous transmission of the pilot channel and all the overhead channels, the power overheads are quite heavy on the VoIP terminal.

To address the issue, EVDO RevB allows discontinuous transmission (DTX) on the reverse link. Figure 14.5 illustrates the DTX reverse link channel transmission. All feedback channels for a forward link operation are gated with a 50% duty cycle. If there is no data traffic during a subpacket interval, the terminal also gates off the pilot and the associated reverse rate indication (RRI) channel transmissions with a 50% duty cycle. To maintain the performance, the gated overhead channels are transmitted with boosted power. Meanwhile, to conserve the terminal power, the terminal also shuts off the forward signal reception during gated periods. From a VoIP capacity perspective, pilot gating using a DTX operation reduces interference to other users and hence improves the overall VoIP capacity on the reverse link.

With all the techniques described, the VoIP capacity in the EVDO RevA is estimated as 35 Erlang per sector-carrier with the EVRC vocoder [12]. The EVDO RevB system provides further improvement by about 35%, yielding a significantly higher capacity than the circuit-based voice capacity of the existing cdma2000 network, which operates at a target capacity of about 26 Erlang per sector-carrier.

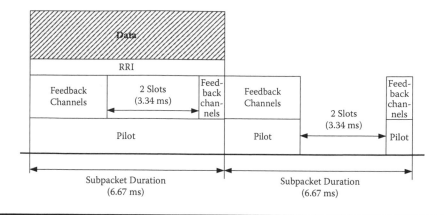

Figure 14.5   EVDO RevB reverse channels with DTX operation.

14.4 Techniques and Performance to Support VoIP in UMB

Unlike EVDO, the support for multimedia applications, including VoIP, is an inherent part of the UMB standard. Many techniques have been specified to optimize the performance of VoIP over the air interface.

14.4.1 Fine Resource Allocation Unit

In UMB, the resource allocation unit for a packet transmission is a tile, which consists of a group of 16 contingent subcarriers for one frame duration. Each subcarrier occupies a bandwidth of 9.6 kbps, and each frame lasts around 1 ms. The small VoIP packet size generally occupies only a single tile resource assigned for packet transmission. A straightforward calculation shows that in a 5-MHz carrier, more than 1,000 VoIP users can be accommodated solely from the tile resource perspective, which fundamentally lifts the tile resource as a potential bottleneck when evaluating VoIP capacity. Through the HARQ operation, the tile resource can be traded off against transmission power resource.

14.4.2 Flexible RL Frame Durations for Extended Coverage

To improve RF transmission efficiency, UMB, like EVDO, employs HARQ on both the forward and reverse links. The HARQ process consists of up to six transmissions, each lasting a single frame duration for most packet formats. For VoIP applications, however, continuous network coverage and reliable packet transmission are essential as the real-time nature of the service precludes the use of application layer retransmissions, which usually incurs long delays. Due to the limited power of transmission in the mobile terminal, under poor RF conditions, reverse link performance can become a matter of concern. To address this, UMB designs a special set of packet formats that use extended frame structures, in which each HARQ transmission lasts three frames instead of one frame. This effectively reduces the transmission power requirements to 1/3 for the same amount of information, greatly improving the VoIP network coverage. The trade-off is that more tile resources will be used for extended frame packet transmissions, which is another example of the tile-versus-power resource trade-off.

14.4.3 Persistent Assignment to Minimize Signaling Overhead

The UMB air interface packet transmissions are determined by the BTS scheduler for both the forward and reverse links. In each open frame, the scheduler decides which subset of users will transmit or receive packets, together with the tile allocations, power allocations, and packet format selections. The scheduling decisions are notified to the target users via the forward link control channel. For general packet data applications that can tolerate relatively long latency and large jitter, the scheduling decision can be made on a packet-to-packet basis, optimizing the aggregate spectral efficiency by analyzing RF dynamics. On the other hand, VoIP packets have a much more stringent delay requirement, and the VoIP traffic is composed of a stream of small but frequent packets. If each VoIP packet has to be explicitly scheduled and the scheduling decision has to be explicitly notified to the user on the forward control channel, this can lead to an excessively high signaling overhead when the number of VoIP users increases, degrading the overall RF performance. As the packets arrive periodically during a talk burst, it is much more efficient to preallocate a sequence of RF resource to the user throughout the talk burst as long as the RF condition of the mobile does not change significantly. This is the idea behind the persistent assignment specified in UMB. The resource assignment decisions by the scheduler comprise up two categories:

1. Nonpersistent assignment, which applies to a single-packet transmission. This type of assignment is highly suited to nonperiodical bursty data applications that have relatively loose delay requirements and allows the scheduler to select the best user for packet transmission.
2. Persistent assignment, which applies to all subsequent packet transmissions unless overridden by new assignments or implicitly nullified by packet transmission failures. This type of assignment is well suited to real-time applications such as VoIP, which have recognized packet arrival patterns and relatively stable packet sizes.

With persistent assignment, the scheduler signaling overhead is greatly reduced. For example, for a talk burst composed of 100 VoIP packets, nonpersistent assignment would cause up to 100 assignment messages to be sent to the mobile, and persistent assignment would, in most cases, need only one assignment and one possible deassignment message.

14.4.4 Seamless Mobility Support for VoIP

In UMB, the mobile terminal still maintains an active set with the network as in a CDMA system. The active set represents the top set of the sectors radiating strong signals to the mobile. The terminal transmits CDMA pilot and some other feedback channel information to the active set members in a dedicated CDMA control segment allocated in both the frequency and time domains. The embedded CDMA operation allows the terminal to communicate with multiple BTSs and determines which RF connectivity has the best quality for data communications as the packet transmissions in UMB operate in simplex mode on both forward and reverse links. The terminal receives packets from only one of the sectors within the active set, that is, the forward link serving sector (FLSS), and the terminal transmits packets to only one of the sectors within the active set, namely, the reverse link serving sector (RLSS). Typically, the FLSS and RLSS are the sectors with the best forward and reverse channel qualities, respectively. At any moment of time, the FLSS and RLSS can be the same, or they can be different if a significant link imbalance occurs. When the user moves across different cell or sector boundaries, the FLSS and RLSS must change to match the user's new RF conditions. If the user is engaged in an active VoIP call during this period, the process of changing FLSS or RLSS must be done quickly and reliably to ensure voice continuity.

In UMB, the terminal determines the desired FLSS (DFLSS) based on the strength of the forward pilot signal of each sector within the active set and the desired RLSS (DRLSS) based on the reverse pilot quality indications fed back by each sector within the active set. Since packet transmission on the forward link requires feedback support from the reverse link channels, and vice versa, the new sector serving in one direction must have reasonable channel quality in the other direction to maintain acceptable performance of the feedback channels. To that end, UMB specifies the following criterion to determine switching for the serving sector:

> The reverse pilot quality indicated by the FLSS must not be poorer than the best reverse pilot quality indicated by the active set by more than a preconfigured margin. Otherwise the terminal must find a different DFLSS that complies with this requirement. The rule applies to RLSS and DRLSS as well.

Once the terminal has determined a new serving sector, it sends a hand-off indication to the new sector. Meanwhile, all the feedback channels for data transmissions still communicate to the existing serving sector, so that it can continue to serve the user during the hand-off. Once the new serving sector detects the

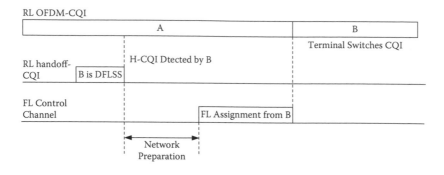

Figure 14.6   Serving sector switch timeline in UMB.

hand-off request, it coordinates with the existing serving sector and other network components to rearrange the data path for the user. When it is ready to serve the user, the new serving sector notifies the mobile on the forward control channel. The mobile then switches the data transmission and reception to the new serving sector. This ensures that data transmission interrupts due to the hand-off are minimized. Figure 14.6 illustrates the process.

14.4.5 Other Performance Improvement Techniques for VoIP in UMB

14.4.5.1 Spatial Time Transmit Diversity

Besides receiving diversity as employed in the EVDO system, UMB also introduces spatial time transmit diversity (STTD) on the forward link to further improve performance for users located at outer reaches of the network coverage. STTD is a technique to transmit multiple transformed versions of a data stream across multiple antennae to improve the reliability of data transmission. The data transmitted on each antenna is a certain combination of multiple temporal stream data points based on a preconfigured combination matrix. The transmission antennae are widely spaced to provide spatial channel diversity.

For example, [4] specifies what can be used as a combination matrix for STTD with two transmission antennae.

14.4.5.2 Spatial Diversity Multiple Access

When two users are at different locations, such as when the propagation paths between the BTS and the users are sufficiently separated, SDMA can be used to improve the RF capacity. That is, the BTS can transmit different data to the users on the same channel resource using differently steered antenna beams that cause little interference between data streams. The aggregate spectral efficiency can thus be increased multiplefold.

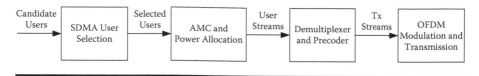

Figure 14.7   SDMA operation for forward link in UMB.

In SDMA operations, a steered antenna beam is created by transmitting the data over multiple antennae with different phases. The antennae are placed close together to create a narrow beam pointing in the desired direction when fed with a certain combination of antenna gains and phases. The available set of steered beams is specified by a predefined or downloadable set of precoding matrices. Figure 14.7 illustrates the SDMA operation on the forward link traffic channel. The terminal measures the difference between channel qualities of different beams. It reports the channel qualities, together with the preferred precoding matrix and the selected beam index. Based on feedback from the terminal, the BTS selects a set of users to be superposed on the same channel resource. Adaptive modulation and coding (AMC) and power allocation are performed for each selected user, based on the respective channel condition. Precoding is applied based on the beam index reported by the terminal within the precoding matrix, and the beam-formed data streams are sent to the transmitting antennae.

14.4.5.3 Fractional Frequency Reuse

Like the CDMA and EVDO systems, UMB systems operate with universal frequency reuse; that is, each sector in the network is configured to operate in the same carrier frequency. Since the traffic loading in different sectors is expected to fluctuate, the interference between sectors is dynamic as well. In general, the OFDMA system capacity is other-cell interference limited; hence, the system's performance can be improved if the data transmissions from different sectors are coordinated in the frequency domain to minimize cochannel interference. Fractional frequency reuse (FFR) is a technique used to achieve this goal.

When FFR is enabled, the entire RF resource is partitioned into multiple resource sets, defined as a set of two-dimensional resource pairs: subzone in frequency and interlace in time. Figure 14.8 depicts the concept of the resource set. The resource set partition is broadcast to the sector, and each terminal reports the RF conditions observed in each resource set. This enables the BTS to preferentially allocate the resource sets to different users. For example, a resource set can be preferentially allocated to users under good RF conditions by every BTS since these users generally require small channel transmission power and hence cause minimal interference to other cells. Some other resource sets can be preferred allocations to

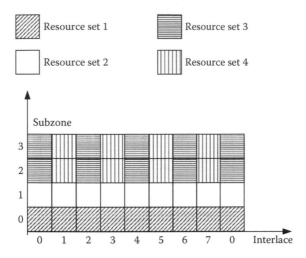

Figure 14.8   Resource set concept in UMB.

users close to cell boundaries. From the feedback given by the terminal, the BTS is able to sense which resource set is experiencing a high interference level from cells adjacent to the user's location and avoid allocating the same resource set to the user at that moment. In this manner, data transmissions to and from different BTSs are effectively coordinated to minimize the cross-cell interference among adjacent cells, especially when the traffic condition does not demand full band transmission simultaneously across all cells. For real network deployments, the traffic situations typically vary across cells and rarely become fully loaded at the same time; hence, the FFR is able to improve the overall performance of the system significantly.

A recent performance study [13] has shown that the OFDMA-based UMB system is able to support over 200 Erlang of VoIP calls within the 5-MHz spectrum. Compared to EVDO RevB, it offers yet another significant improvement on VoIP capacity.

14.5 Summary and Conclusion

In this chapter, we presented the challenges of supporting VoIP service in wireless mobile data networks and the various techniques that can be used to meet these challenges. With careful design and optimization, the 3G/4G mobile data networks are able to provide VoIP service with not only a similar service quality as in the existing 2G circuit voice network but also a significantly higher RF capacity for the application. By supporting voice service using VoIP techniques, the 3G/4G mobile data networks offer an integrated solution that has high spectral efficiency and high mobility for multimedia applications.

As a final note, while the discussions are based on 1x EVDO and UMB systems, most of the principles and many of the techniques also apply to high-speed packet applications and LTE networks since the technologies of the two wireless networks are quite similar.

References

1. 3GPP2 C.S0024-0_v4.0, *cdma2000 High Rate Packet Data Air Interface Specification*, version 4.0. October 2002.
2. 3GPP2 C.S0024-A_v1.0, *cdma2000 High Rate Packet Data Air Interface Specification*, version 1.0. March 2004.
3. 3GPP2 C.S0024-B_v2.0, *cdma2000 High Rate Packet Data Air Interface Specification*, version 2.0. March 2007.
4. 3GPP2 C.S0084–001.0, *Physical Layer for Ultra Mobile Broadband (UMB) Air Interface Specification*, version 2.0. August 2007.
5. 3GPP2 C.S0084–002.0, *Medium Access Control Layer for Ultra Mobile Broadband (UMB) Air Interface Specification*, version 2.0. July 2007.
6. 3GPP2 C.S0014-A_1.0, *Enhanced Variable Rate Codec, Speech Service Option 3 for Wideband Spread Spectrum Digital Systems*. April 2004.
7. 3GPP2 C.S0014-B_1.0, *Enhanced Variable Rate Codec, Speech Service Option 3 and 68 for Wideband Spread Spectrum Digital Systems*, version 1.0. May 2006.
8. RFC 3095, *Robust Header Compression (ROHC): Framework and Four Profiles: RTP, UDP, ESP, and Uncompressed*. July 2001.
9. Zhang, Q., Performance of robust header compression for VoIP in 13 EVDO system. GLOBECOM '06.
10. Chen, P., Da, R., Mooney, C., Yang, Y., Zhang, Q., Zhu, L., and Zou, J., Quality of service support in 13 EV-DO revision A systems. *Bell Labs Tech. J.*, 11(4), 169–184.
11. Soriaga, J., Hou, J., and Smee, J., Network performance of EV-DO CDMA reverse link with interference cancellation. GLOBECOM '06.
12. Bi, Q., Chen, P., Yang, Y., and Zhang, Q., An analysis of VoIP service using 13 EV-DO revision A system. *IEEE JSAC*, 24(1), 36–45, 2006.
13. Bi, Q., Vitebsky, S., Yuan, Y., Yang, Y., and Zhang, Q., Performance and capacity of cellular OFDMA systems with voice over IP traffic. *IEEE Trans. Veh. Technol.*, in press.

Chapter 15

IP Multimedia Subsystem (IMS)

Mojca Volk, Mitja Stular, Janez Bester,
Andrej Kos, and Saso Tomazic

Contents

15.1 Introduction

Communications in a wider sense have become a vital element of modern society and are about serving users with capabilities and services that meet their demands in a flexible and personalized fashion. Therefore, the focus of new-age telecommunications solutions has changed from technologies, functionalities, protocols, and methods to users, services, personalization, and contents, and it is the aim of telecommunication systems to provide adequate service environments.

Global evolution and convergence of communication systems have substantially redefined the concept of design, operation, management, and usage of networks and services, called next-generation network (NGN) concept [1–3]. Throughout the years, in pursuit of this concept, new technologies have emerged, and real-world experiences have identified suitable methods and enablers to provide the anticipated network and service convergence in a uniform and mobile manner. As a result, various reference architectures and proposals have emerged that reflect contemporary research and development trends in different fields of telecommunication systems and apply the latest technologies available [3–7]. One such standardized reference architecture is the Internet Protocol (IP) Multimedia Subsystem (IMS).

IMS represents a heterogeneous composition of proven telecommunication, Internet-oriented, and real-time multimedia technologies to establish a standardized global overlay platform that enables provisioning of highly converged multimedia services to users, regardless of their location, time, or mode of accessibility. This in turn requires extensive overall system transformation that addresses nearly all aspects of system structure, its capabilities and functions, as well as service provisioning. Openness, standardization, and user orientation of such systems are reflected in global interconnection between, until now, different and separate communications and content solutions. Hence, such systems are capable of providing various types of users with a flexible, personalized, and always available home service environment.

With respect to existing telecommunication solutions, IMS represents a convergent core packet data mobile service solution as an enhancement to the Universal Mobile Telecommunications System (UMTS) technology for third-generation (3G) mobile networks. But, over time it has become a widely adopted universal reference core architecture that reuses different types of existing access technologies available in both fixed and mobile domains.

In the prospects of global convergence, mobility no longer represents characterization of system architecture, operation, and the associated services but indicates the wireless nature of access to services and the possibility of being mobile using wireless or wire-line technologies. There are different types of wireless access technologies being promoted that vary in range limits, capacities, and level of mobility. Most prominent are unlicensed mobile access (UMA), interworking wireless local-area network (I-WLAN), and general packet radio service (GPRS). In this respect, IMS serves as a core service solution that enables global mobility as a service, sold and provided to end users, but is not limited in terms of mobile technologies and wireless or wired access.

15.2 Fundamentals of the NGN Concept

A packet-based network with the ability to provide telecommunications services based on various broadband QoS (quality-of-service)-enabled transport technologies is defined as the NGN, in which service-related functions are independent from transport-related technologies, and general mobility promotes ubiquitous home service environment.

The architecture is divided into functionally separated strata characterized by the following requirements: Service provisioning with the appropriate control and management mechanisms is transparently decoupled and independent from packet-based transport functionalities providing broadband capabilities available via multiple last-mile technologies. Clear separation of transport, control, and service functionalities introduces the possibility of mixing real-time and non-real-time service provisioning, thus enabling a complex multimedia service portfolio through a distributed and upgradable future-proof converged communications domain.

The system provides a wide range of multimedia services based on various service technologies building a flexible service development environment that exploits interface mechanisms to provide open interconnection capabilities to both legacy systems and outside service environments. Generally, the aim is to establish a personalized home service environment with support for generalized mobility and unrestricted access by users to different service and content providers while merging fixed and mobile domains into converged communications systems, supported with Internet-oriented principles.

15.3 IMS as an Enhancement to Mobile Domain for 3G UMTS Networks

In the past decades, interest in the mobile communications services and Internet has increased considerably and surpassed service consumption in other domains. This way, the mobile domain evolution is further accelerated to meet user demands and accomplish standardized and open principles of next-generation communication paradigms.

Today, packet-based services can be accessed via second-generation (2G) and 3G wireless networks. There are, however, constraints when providing multimedia and multisession services. Another issue is the demand for continuous maintenance of QoS throughout the session with respect to possibly different QoS demands from endpoints inside a single session. The addressed issues can be circumvented via a standardized approach, offered through IMS. It is therefore reasonable to assume emergence of IP-based broadband packet networks and mobile networks to offer universal mobile multimedia services, also known as the all-IP vision.

Considering mobile domain evolution, the common migration path is as follows. Starting with the 2G mobile network, the circuit-switched Global System for Mobile Communications (GSM) has been upgraded, with GPRS technology introducing the packet-switched domain. Known as the 2.5G solution, it provides enhanced mobile data capabilities. Furthermore, the GSM/GPRS network is upgraded and supplemented with a new air interface introducing high capacity and bit rates to enable a genuine 3G multimedia service environment, the UMTS. Two prominent technologies in this respect are Enhanced Data Rates for GSM Evolution (EDGE) for upgraded GPRS solutions and high-speed downlink packet access (HSDPA) for UMTS. In pursuit of an all-IP environment, UMTS architecture is further evolving. The IMS has been introduced as an enhancement above circuit-switched and packet-switched domains.

The IMS, presented by the Third Generation Partnership Project (3GPP), represents an evolved and standardized NGN concept with more detailed definition that implements new and already proven technologies, protocols, and principles, thus providing fundamental recommendations and detailed guidelines for network development and design. From an evolutionary point of view, IMS represents an enhancement to the mobile domain for 3G UMTS networks. Nevertheless, the concept has been widely adopted in both fixed and mobile domains and represents an access-agnostic core control and application subsystem that is applied as an enhancement to various environments (e.g., fixed and mobile telecommunications networks, broadcasting systems, cable systems). IMS is anticipated as the convergent core platform that assumes access agnosticism and global user and service mobility. Since IMS characteristics greatly reflect its mobile origin, the European Telecommunications Standards Institute (ETSI) TISPAN has presented an enhanced concept that also meets fixed domain requirements, known as ETSI TISPAN NGN [8].

A complete solution that provides support for IMS services consists of user equipment (UE), IP connectivity access networks (IP-CANs), and the specific functional entities of the core IMS domain. Due to evolutionary reasons, the most typical IP-CAN example is a GPRS core network inside a packed-switched mobile system with GSM/EDGE Radio Access Network/UMTS Terrestrial Radio Access Network (GERAN/UTRAN). Another two IP-CAN options are newly adopted, unlicensed mobile access (UMA) for mobile users and I-WLAN for nomadic users, which enable access over the unlicensed spectrum by reusing existent WLAN

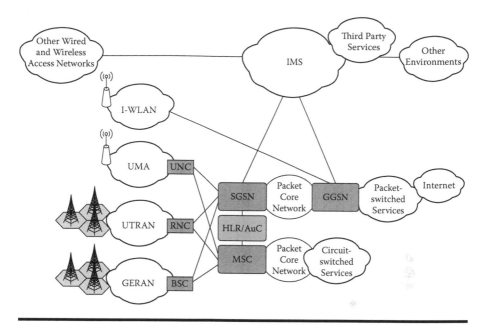

Figure 15.1 Evolved mobile domain. UMTS network consists of circuit-switched, packet-switched, and IP Multimedia Subsystem (IMS) in its core segment. Existent circuit-switched (GERAN/UTRAN access network and mobile switching center as a core entity) and packet-switched (GERAN/UTRAN access network and SGSN [service GPRS support node] and GGSN [gateway GPRS support node] as core entities) segments are retained, while IMS segment is introduced in parallel. Typical packet access segment technologies for the mobile domain involve GPRS technology with a GERAN/UTRAN radio part, unlicensed mobile access (UMA), and interworking wireless local-area network (I-WLAN). Most important, 2G/2.5G entities are shown that are used to access IMS services, that is, the radio network controller (RNC), base station carrier (BSC), UMA network controller (UNC), SGSN, and GGSN. The most common mobile solution exploits an existing packet-switched segment to access the IMS segment with its services.

equipment and interconnecting via the GPRS core network. The environment is represented in Figure 15.1.

15.4 IMS Architecture

Following NGN principles, the IMS architecture [2,5,9,10] consists of three strata as indicated in Figure 15.2:

■ Transport stratum, providing IP packet backbone and heterogeneous access network domain

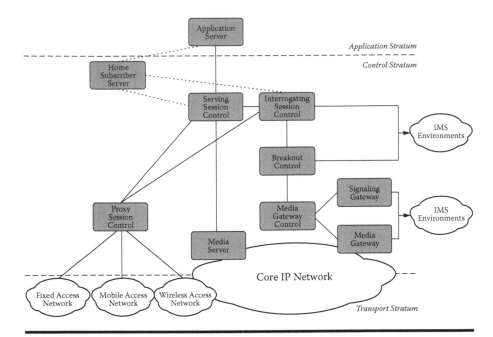

Figure 15.2 **IMS architecture outline. Architecture is composed of three strata and thus separates access and transport of media, call control and signaling, and application domain. Several entities are defined to perform session control, service provisioning, media handling, and interworking functions. IMS represents an access-independent core subsystem, applicable to various communication environments. To accomplish communication services, it requires interconnection to other subsystems or other communication environments.**

- Control stratum, enabling core functionalities for signaling, routing, interworking, and service control purposes
- Application stratum, providing an integrated service development, delivery, and execution environment

Any type of telecommunication system regardless of its evolution stage encompasses a portfolio of basic services required to accomplish essential communications (i.e., session control, service control, media handling, and interworking functions). In IMS, the respective entities are assigned to perform these functionalities. The next section presents a brief description of their capabilities and roles.

15.4.1 Call Session Control Function

Session control entities represent the key control stratum functionalities that provide control points for user authentication, session routing, and service control.

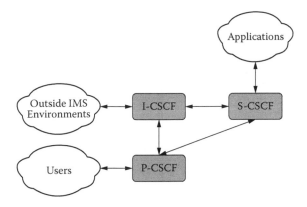

Figure 15.3 Call session control function (CSCF). In IMS, three functionally differ-ent implementations occur. Serving CSCF (S-CSCF) provides control functionalities for user authentication, session routing, and service control. Proxy CSCF (P-CSCF) represents the first point of contact in IMS from the user side; interrogating CSCF (I-CSCF) represents the first point of contact from other IMS environments.

They represent a packet-switched equivalent to the mobile switching center (MSC) entity of the circuit-switched mobile domain. There are, however, three function-ally different implementations that occur simultaneously in a complete IMS sys-tem: serving call session control function (S-CSCF), proxy CSCF (P-CSCF), and interrogating CSCF (I-CSCF) (see Figure 15.3).

15.4.1.1 Serving CSCF

The serving session control entity provides session, service, and charging control for a user. The S-CSCF, residing in the home network, is assigned to a user on the registration procedure, regardless of current location of the user (inside home net-work or in a visited network) and is located inside the signaling path of any further communication for this user.

The S-CSCF represents a single point of service evocation and triggering toward the application stratum via the dedicated IMS service control (ISC) interface for a given user based on user and service information acquired from the central user server (home subscriber server [HSS] entity).

15.4.1.2 Proxy CSCF

The proxy session control entity resides at the edge of the IMS network and repre-sents the first point of contact for requests, originating or terminating at the user side. In the most general scenario, a user in a visited network contacts the P-CSCF entity of the visited network, which forwards session signaling to the user's dedicated S-CSCF entity residing inside the home network.

15.4.1.3 Interrogating CSCF

The I-CSCF also resides at the edge of the IMS network and represents the first point of contact from outside IMS environments (i.e., registration requests and incoming requests toward native or visiting users inside the home IMS network). The P-CSCF is responsible for S-CSCF entity assignment in registration procedures. Also, it optionally implements the internetwork gateway for network topology, configuration, and capability hiding toward other environments.

15.4.2 Application Server (AS)

Application stratum consists of different types of application server (AS) entities that provide functionalities to complete various multimedia services. In IMS, ASs occur in three forms: SIP (Session Initiation Protocol) AS, IP multimedia service switching function AS (IM-SSF AS), and open service access service capability server AS (OSA SCS AS) (see Figure 15.4).

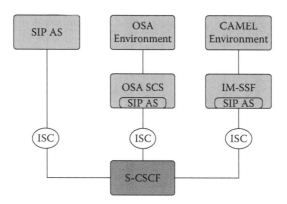

Figure 15.4 IMS application server options. Application server entities are considered the most important functionalities that provide value-added end-user multimedia services. Apart from native SIP-based services, executed within SIP application servers (SIP ASs), services can be developed and deployed in different service development environments, such as Parlay/Parlay X, connected through application programming interfaces (APIs) ASs (open service access service capability server application server, OSA SCS AS). Also, access to third-party service environments could be provided via gateway application server implementations (IP multimedia service switching function application server, IM-SSF AS, for CAMEL environment). All types of AS entities are from an underlying system point of view treated uniformly as SIP AS entities via the IMS service control (ISC) interface.

15.4.2.1 SIP Application Server

The SIP AS represents the most general AS type that hosts and executes service logic to provide end-user applications. It is based on SIP technology and is referred to as the service AS. Typical new-generation services are deployed on SIP AS entities (e.g., instant messaging, presence services, push-to-anything).

15.4.2.2 IP Multimedia Service Switching Function AS

In pursuit of openness of the system, apart from native ASs inside the operator's environment, IMS users benefit from applications provided by other environments or outside ASs via dedicated gateway servers. These servers induce appropriate technology conversion and protocol mappings to accomplish complete interconnection. An IM-SSF AS with a gateway toward the CAMEL (customized applications for mobile network enhanced logic) service environment is defined for IMS.

15.4.2.3 Open Service Access Service Capability Server AS

If different interface technologies are implemented that hide underlying telecommunication complexities and enable flexible service development and deployment, an environment is established in which multiple programming tools are available. In this way, service creation can be brought closer to Internet-oriented principles and computer programming paradigms while service deployment is greatly accelerated. OSA SCS AS is an AS type that implements standardized application programming interfaces (APIs) that provide interface technology for Parlay/Parlay X environments.

Nevertheless, from underlying an IMS system point of view all types of ASs are treated uniformly as SIP ASs via dedicated an ISC interface.

The goal of the application stratum with its ASs is to establish a service delivery platform (SDP) within or above the IMS system. The SDP is an integrated environment in which service development, deployment, operation, management, and charging are enabled in a converged and controllable way. A heterogeneous range of telecommunication services and content provided through the SDP are agnostic toward underlying telecommunication technologies, while the SDP environment itself aims at openness in a secure and robust manner.

15.4.3 Home Subscriber Server

The HSS represents a master database and an extended authentication, authorization, and accounting (AAA) server inside an IMS network (Figure 15.5). In this entity, information is stored that represents a vital compound of session and service control processes: user identification, addressing and numbering information, security information (e.g., access control information for authentication and authorization purposes), intersystem user location information, user profiles, and service

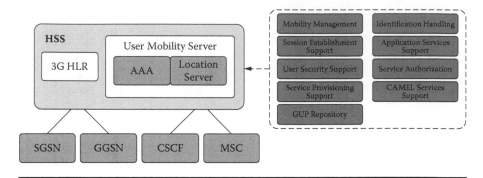

Figure 15.5 Generic HSS structure, basic interconnections, and capabilities. HSS serves to support session control and management and is a shared entity for all subsystems inside a heterogeneous communications environment (e.g., circuit-switched, packet-switched, and IMS domains of a migrated 3G mobile system).

profiles (see Figure 15.6). The HSS also generates security information required for authentication procedures, integrity control, and ciphering.

From a mobile domain point of view, a subset of HSS represents the equivalent to the home location register (HLR) and authentication center (AuC) and serves the packet-switched domain (services for the serving GPRS support node [SGSN] and gateway GPRS support node [GGSN]) and circuit-switched domain (services for MSC server) in parallel to authentication, service profile, and location information services for the IMS domain (services for CSCF).

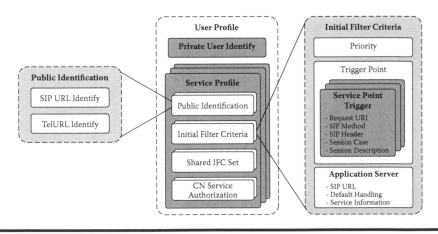

Figure 15.6 User and service profiles. User profiles represent a collection of services data and user-related data, that is, user identities and service profiles (subscribed services, their configuration, personalization information, etc.).

15.4.4 Gateway Control Functionalities and Gateway Functionalities

Gateway control functionalities and gateway functionalities on the control stratum of the IMS network represent a group of entities providing interconnection capabilities toward outside non-IMS environments (e.g., public switched telephone network [PSTN], Integrated Services Digital Network [ISDN], GSM). The media gateway control function (MGCF) controls the operation of media gateways (MGWs) and signaling gateways (SGWs) and in cooperation with the breakout gateway control function (BGCF) selects the appropriate gateways to complete the communications toward the outer environment. Also, the MGCF routes incoming requests to appropriate S-CSCF entities based on routing information.

The MGWs and SGWs perform media and signaling conversion between the IMS domain and other non-IMS domains, respectively.

15.4.5 Media Server

The multimedia resource function controller (MRFC) and multimedia resource function processor (MRFP) are entities that implement the media server inside the IMS network. The media server hosts resources that provide media-handling capabilities (e.g., sourcing, mixing, and processing of media streams and floor control) to support provisioning of services such as conferencing, multimedia announcements, and so on.

15.5 Important Principles in IMS

15.5.1 Interfaces and Protocols in IMS

IMS anticipates implementation of Internet-oriented protocols and interface technologies to establish a standardized and open service environment. Two principal protocols are defined.

The Diameter protocol is strictly used in transactions involving user and application data handling or authentication, authorization, and accounting procedures. For all other types of internal sessions and in communications toward other environments, the SIP is in use. It is appropriately extended to meet IMS requirements. In addition, some legacy protocols can be used with respect to eventual remaining existent solutions inside the IMS system or as an interconnection requirement toward other (legacy, NGN, or other) environments (e.g., H.323, SS7, MGCP/Megaco/H.248, SIGTRAN).

Mutual interactions and mechanisms among IMS entities are precisely defined. The detailed definitions are provided through so-called reference points that specify the interface and the functions that belong to it. In IMS, several reference points between entities are defined based on SIP and Diameter protocol.

15.5.2 User and Service Identities

End users and services inside IMS are identified in different ways through interdependent private and public identities and profiles.

Every IMS user has one or more IP multimedia private identities (IMPIs) assigned. The IMPI is a unique, permanent, and global identity stored in the HSS, and it is used within the home network to identify the user's subscription but not the user. It is generally used for accounting purposes through mechanisms for registration, authorization, administration, and accounting.

Every IMS user also has one or more IP multimedia public identities (IMPUs) that is used for incoming communication requests (SIP routing in the direction toward the user). The IMPU takes the form of an SIP URI (Internet naming) or TelURI (telephone numbering).

Each IMS user can register into the IMS network from multiple locations or terminals simultaneously. Therefore, each private identity (IMPI) with a unique IP address corresponds to one or more public identities (IMPUs).

User profiles are stored in the HSS. They represent a collection of service data and user-related data (i.e., user identities and service profiles, e.g., subscribed services, their configuration, personalization information). Each IMPU corresponds to exactly one service profile, while the service profile could corespond with several IMPUs simultaneously.

15.5.3 Registration, Basic Communication, and Service Control

The most important entities that enable the user to enter the IMS network, communicate, and use services are S-, P-, and I-CSCF; HSS; and AS (Figure 15.7). While the AS provides service logic invocation and execution, the CSCF is engaged in routing and service control based on user and service information stored in the HSS. These entities represent the base of the IMS core network and are fundamental for any deployment of the IMS environment.

User registration into the IMS system is generally treated separately from attachment to the IP access network. It is assumed that the IP-CAN bearer is already established prior to IMS registration. In the end-user terminal, an element named the IM services identify module (ISIM) is incorporated that represents a container of parameters and functionalities required for the registration procedure (e.g., IMPI, IMPU, home network domain name). Registration and a basic communication example for an IMS user are represented in Figure 15.8. The most general case is assumed where the user is located in visited networks.

Service control in IMS is provided through interactions between S-CSCF and AS via the ISC interface. There are several possible ways of service control based on different mechanisms.

A local copy of the user profile is saved inside the S-CSCF and contains the service profile with defined filter criteria. Based on priority and information regarding

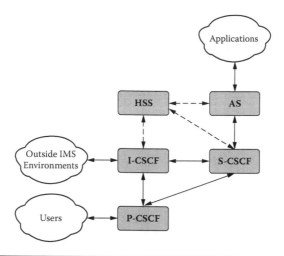

Figure 15.7   Basic IMS entities. While the AS provides service logic invocation and execution, the S-CSCF is engaged in routing and service control based on user and service information stored in the HSS. P-CSCF and I-CSCF provide access functionalities.

when a service should be provided to a user (service trigger points, filter criteria), the request is forwarded to a specified AS. Based on SIP request filtering services such as black/white lists, IP private branch exchange (PBX) services, monitoring, prepaid services, and so on are provided.

Some services could be provided via direct session establishment between a user and an AS or service, accessible via public address. This type of service involves messaging, voice portals, conference services, presence services, and so on.

The AS can also be the initiating side of the session. Service examples in this case are conference services, missed calls services, and the like.

Generally, the IMS assumes advanced multimedia real-time services that require a complex combination of different service control mechanisms for their operation.

15.5.4 Security

Security issues in IMS are considered completely separate from security issues regarding the packet transport domain. In this respect, there are two separate security associations established to ensure safe and controlled communication and service provisioning to every user.

There are several security viewpoints in IMS. Authentication for the end user in IMS is provided by the S-CSCF based on the user profile. To provide confidentiality protection, communication between the UE and the P-CSCF is encrypted. To protect SIP signaling messages between the UE and the P-CSCF, the involved parties exchange integrity keys. Data are verified for origin based on integrity key;

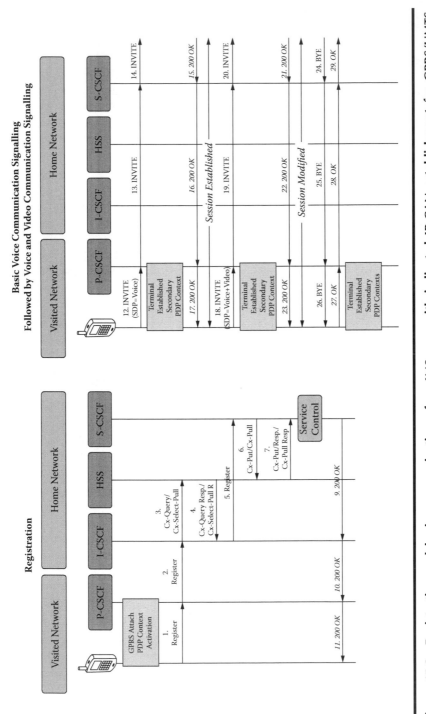

Figure 15.8 Registration and basic communication of an IMS user with indicated IP-CAN establishment for GPRS/UMTS. Message flow is SIP based, procedures involve basic IMS entities, and PDP (Packet Data Protocol) context activation events are indicated.

verification is also used to detect if data have been tampered with or duplicated. The engaged security mechanisms are based on a 3GPP standardized IMS-AKA (authentication and key agreement) procedure.

Due to business sensitivity issues, network topology hiding is available that includes hiding of the number of S-CSCF elements, their capabilities, and the capabilities of the network. For this purpose, topology hiding internetwork gateway (THIG) functionality is implemented within I-CSCF.

15.5.5 Charging

Charging and billing represent key system functionalities and are provided on packet core, subsystem, and application levels. The basic functions of the charging infrastructure involve charging information generation and charging data collection, provided by network elements that are forwarded to appropriate charging and billing systems. Two charging architectures are defined for the IMS subsystem: offline charging and online charging.

Offline charging functions generate call detail records (CDRs) and transfer them to the billing system after the resource or the service usage process has been completed. Billing in this case is of the non-real-time type.

When online charging is engaged, additional usage authorization is required and is provided by the online charging system (OCS). Billing in this case is provided in real time; therefore, continuous interaction between the billing system and usage control is anticipated.

Regardless of architecture type, charging can be based on one of the two charging principles. In event-based charging, case charging is applied to SIP transactions between the user and the IMS system; messages, divert services, and content download services are charged in this way. Session-based charging assumes charging of SIP sessions. Voice calls, IMS session, and calls to voice mail are services for session-based charging.

The mode of charging provided for the IMS depends on the involved entities and the charging type. ASs, media servers, and S-CSCF support both online and offline types, while other entities provide for only offline charging. Interactions for charging purposes inside the IMS are based on the Diameter protocol.

15.6 Mobile Domain Characteristics of IMS and Impact of Mobile Access Network on IMS Functionalities and Operation

When considering IMS in view of the mobile domain, several parallels are evident. First, the general aim of NGNs is to provide global mobility to both users and services. This requires implementation of several mobile principles. Second, the mobile origin itself is substantially reflected in the functional composition of the architecture and entities. Third, if mobile access technology is engaged, the IMS entities and their procedures are partly affected. The first is presented in this section.

In general, connections to IP-CAN and to IMS are considered independent. When a user accesses the IMS core network, it utilizes services provided by IP-CAN for packet-based communication. In the IMS network, the P-CSCF entity serves as an entry point from the user domain. Therefore, prior to connection to the IMS core network, the user must obtain an IP-CAN connection to acquire its IP addresses and learn the P-CSCF address. Since IMS is composed of an independent IP backbone that is implemented separately from the access network backbone for security purposes, this presents two separate authentication and registration procedures for IP-CAN followed by procedures for IMS.

Examining prominent access technologies in the mobile domain engaged to access IMS core network and services, there are basically two distinctive groups: GPRS access network with GERAN/UTRAN/UMA radio segment for mobile solutions and I-WLAN via GPRS for nomadic solutions (Figure 15.9). All of them exploit the existing packet-switched access segment of typical GSM/GPRS networks and vary in radio segment. While GERAN and UTRAN remain as legacy access solutions, UMA and I-WLAN are new.

UMA defines a new radio segment in addition to GERAN/UTRAN while employing already available broadband access connections. Even though treated as competition, UMA in fact represents a complementary solution for mobile and converged operators, foremost addressing indoor usage reusing existing indoor Wi-Fi networks, and thus enhances coverage and range issues of UMTS networks. I-WLAN also stands aside as a complement to UMA that exploits Wi-Fi via available broadband connectivity, foremost for fixed and nomadic users. In this case, the solution does not employ GPRS access but is based on direct SIP-based tunneling connectivity toward core IMS.

The principle of access technology independent of IMS is important. In this way, any specific requirements for mobile access should be dealt with by the access network alone. Nevertheless, there are some access-dependent particularities that affect IMS functionalities as well. Some most evident are presented in the following section for a common mobile IMS solution that uses a GPRS access network with the GERAN/UTRAN/UMA radio segment as an IP-CAN.

15.6.1 GPRS-Specific Concepts of IMS

Within the GPRS access network, IP-CAN bearers are provided via the Packet Data Protocol (PDP) context activation procedure (Figure 15.10 [1]). For IMS-related signaling, a dedicated PDP context or a general-purpose PDP context is established. P-CSCF discovery takes place after the GPRS attachment procedure and after or as part of the PDP context activation procedure for IMS signaling.

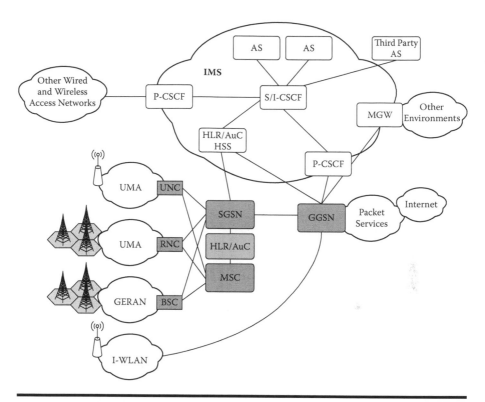

Figure 15.9   A detailed schema of introducing IMS into a mobile system. Existent circuit-switched (GERAN/UTRAN access network and mobile switching center as a core entity) and packet-switched (GERAN/UTRAN access network and SGSN and GGSN as core entities) segments are retained while the IMS segment is introduced in parallel. The most common mobile solution exploits an existing packet-switched segment to access IMS and its services.

When these procedures are completed, the user is granted basic access connectivity with acquired IP addresses.

If one or more IP addresses change at any point, the user must reregister for IMS services. In the event of loss of coverage, some types of PDP contexts are affected. In this case, the P-CSCF indicates modification and reacts accordingly.

Apart from basic packet connectivity, additional features are available to support QoS and security. UE has the ability to request prioritized handling over the radio interface for IMS signaling. Signaling connections are restricted to only specified destinations (i.e., the assigned P-CSCF and to the DNS (domain name server) and DHCP (Dynamic Host Configuration Protocol) servers inside the IMS operator domain. An IMS signaling flag is also available for determination of rules and restrictions that apply to PDP context at the GGSN in addition to applied local GPRS policy.

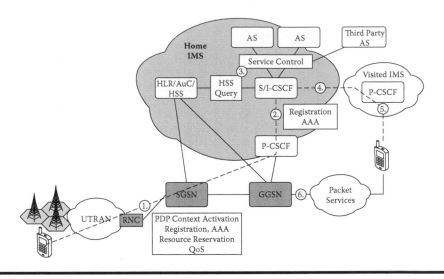

Figure 15.10 Basic communication procedures for IMS services accessed via GPRS technology using the UTRAN interface (example). Before connecting to the IMS core network, the user equipment acquires access network connectivity (attachment to GPRS access network, PDP context activation, AAA procedures, and management of quality of service [1]). These procedures serve to establish a packet-based bearer and provide user equipment with IP addresses and the P-CSCF address, required for the user equipment to enter the IMS core network. Once connected to the access network, the user enters the IMS domain starting with the registration procedure by which the user is authenticated and authorized according to its profile [2]. Afterward, services are provided to the user based on service control mechanisms [3]. For this example, IMS communication between two users is represented where one user is located in the visited IMS network. To complete the communication, appropriate interconnection mechanisms are applied [4,10,11].

This way, the packet-based access network connectivity is ensured, and the user is granted the access to the IMS core network starting with the registration procedure as specified by the IMS regardless of the chosen access network technology.

15.7 Conclusions and Prospects

The IMS as the next step in the evolution of NGNs extends beyond the definition of an upgraded and optimized solution, implementing newly available technologies, protocols, and services. It introduces a substantially remodeled concept in which call control, transport, and access segments build an enabling infrastructure, on top of which a value-added service delivery platform is positioned. Implemented as SIP and Diameter-based architecture, it exploits the initial horizontal nature of the mechanisms and procedures while pursuing controlled, secure, and managed operation and open interconnection.

The core IMS network presents a challenge to telecommunication operators and service providers when deciding strategies for next-generation solutions. Even though most of the services available to end users could be achieved by means of alternative solutions eluding IMS implementation, particularly NGN, VoIP (voice-over Internet Protocol), and fixed mobile convergence (FMC), such strategies present problems of standardization, openness, universality, and limitations regarding the use of different technologies and service design. These addressed issues are in favor of implementing IMS. Also, standardized seamless mobility and the principle of home environment as the essential next-generation characteristics could today be provided in the environment presented by IMS.

The concept of mobility is no longer limited to the access domain but has become one of the most important services in converged environments. From this point of view, IMS represents a corresponding and optimized solution due to its mobile origin, which is considerably reflected in architecture that implements basic mobility-enabling infrastructure. An integral IMS-based solution encompasses access-agnostic and secure call control and service delivery infrastructure for convergent fixed-mobile communications, for which mobility is an inherent characteristic, available to underlying technologies and overlying service delivery platforms.

Currently, there are several convergent initiatives evaluating available technologies to provide different scenarios to address mobile voice and multimedia service markets worldwide. The choice of available technologies is broad and differs in stage of evolution, ranging from GSM/GPRS to UMTS and alternative solutions (e.g., UMA and I-WLAN). If optimally combined, service availability is extended and access agnosticism is addressed. With ever more global ubiquity and mobility of end-user services, convergence is under way, diminishing the fixed-wireless–nomadic-mobile differentiation. In this respect, interworking is a major issue.

With these characteristics, IMS provides an open and flexible converged service environment that enhances mobile communication systems with Internet-oriented paradigms and broadband capabilities to provide "full multimedia" 3G and 4G capabilities and beyond.

There are, however, some issues remaining in 3G systems. There is a 4G initiative that aims at resolving these to provide customized personal mobile multimedia services based on integrated broadband wireless and mobile access solutions. The main emphasis will be on hybrid communications with ubiquitous high-bandwidth multimedia and full-motion data and video services. Important issues, such as interworking at high speed, various UE, alternative access technologies with increased capacity, and withdrawal of remaining circuit-switched segments are to be resolved. The 4G system addresses primarily the access segment; nevertheless, it represents a step closer to global mobility and provisioning of the home environment based on complete service portability. In this respect, the IMS as core infrastructure represents a viable and possible choice.

References

1. Third Generation Partnership Project, *General Universal Mobile Telecommunications System (UMTS) Architecture*, 3GPP TS 23.101 (Release 6), V6.0.0 (2004-12). http://www.3gpp.org/ftp/Specs/html-info/23101.htm (accessed July 3, 2006).
2. Third Generation Partnership Project, *Network Architecture*, 3GPP TS 23.002, Stage 2 (Release 7), V7.1.0 (2006-03). http://www.3gpp.org/ftp/Specs/html-info/23002.htm (accessed July 3, 2006).
3. Huber, J.F., Mobile next-generation networks. *IEEE Multimedia*, 11(1), 72–83, 2004.
4. inCode Advisors, *UMA's Role within Mobile Network Evolution,* White paper. January 2006. http://www.incodewireless.com/ (accessed August 20, 2006).
5. Koukoulidis, V., and Shahl, M., The IP multimedia domain: Service architecture for the delivery of voice, data, and next generation multimedia applications. *Multimedia Tools Appl.*, 28(1), 203–220, 2006.
6. Nielsen, T. T., and Jacobsen, R.H., Opportunities for IP in communications beyond 3G. *Wireless Pers. Commun.*, 33(3–4), 243–259, 2005.
7. Bannister, J., Mather, P., and Coope, S., *Convergence Technologies for 3G Networks: IP, UMTS, EGPRS and ATM.* West Sussex, UK: Wiley, 2004.
8. Lee, C.-S., and Knight, D., Realization of the next-generation network. *IEEE Commun. Mag.*, 43(10), 34–41, 2005.
9. Third Generation Partnership Project, *IP Multimedia Subsystem (IMS),* 3GPP TS 23.228, Stage 2 (Release 7), V7.4.0 (2006-06). http://www.3gpp.org/ftp/Specs/html-info/23228.htm (accessed July 3, 2006).
10. Poikselkä, M., Mayer, G., Khartabil, H., and Niemi, A., *The IMS: IP Multimedia Concepts and Services in the Mobile Domain.* West Sussex, UK: Wiley, 2004.
11. Knightson, K., Morita, N., and Towle, T., NGN architecture: Generic principles, functional architecture, and implementation. *IEEE Commun. Mag.*, 43(10), 49–56, 2005.

Links

1. Third Generation Partnership Project home page:http://www.3gpp.org
2. ETSI home page:http://portal.etsi.org/Portal_Common/home.aspwww.incodewireless.com
3. IEEE home page:http://www.ieee.org
4. IMS Forum home page:http://www.imsforum.org
5. Fraunhofer Institute for Open Communication Systems home page:http://www.fokus.fraunhofer.de/bereichsseiten/testbeds/ims_playground/index.php?lang=de
6. Nokia home page:http://www.nokia.com
7. Ericsson home page:http://www.ericsson.com
8. Siemens home page:http://www.siemens.com

Chapter 16

Location-Based Services

B. Falchuk, D. Famolari, and S. Loeb

Contents

16.1 Introduction

Location-based services (LBSs) use the power of mobile networks to locate users and provide services specific to their current location. Imagine the following:

- You are about to call your boss, but when you highlight her name in your address book, you see that she is currently in a different time zone where it is the middle of the night. You decide to leave her a voice mail instead.
- You are in a crowded theater, and your phone automatically silences itself for all but the most urgent calls.
- You are wandering through the Museum of Natural History. As you pass each piece, exhibit-specific audio and video information is delivered to your mobile device.
- You are at the airport, and your phone beeps to tell you that an old friend you have not seen in years is sitting at Gate 7 and looking for someone to join her for dinner.

All of these scenarios (see Figure 16.1) are made possible by location awareness and presence (e.g., see [1])—two concepts now making their way into mobile devices and promising to enhance everything from social networking to marketing and advertising. *Location awareness* enables a device (and the person carrying it) to be geographically located, while *presence* allows a user to know the status of another user.

Mobile operators have been talking about LBSs since the end of the 1990s but have yet to find ways to successfully commercialize them on large scales. An

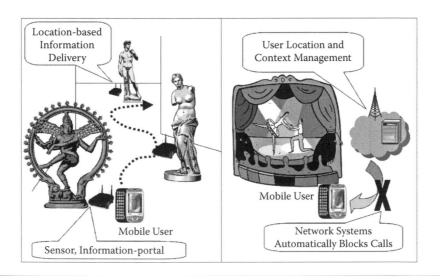

Figure 16.1   Two conceptual notions of LBS: museum (left) and theater (right).

archetypal example of such earlier attempts includes services in which advertisers and individuals seeking social networking send information invites via Bluetooth. These services failed to gain much traction because users either did not like or did not trust them. Privacy and confidentiality of location information are important issues that must be addressed before consumers will be willing to adopt LBSs. With the greater adoption of Web-based and cellular services, however, the general public appears more willing now to accept certain LBSs.

As discussed in the following section, many technologies can provide position fixes for mobile terminals. All vary depending on whether the network or the edge device initiates the query and the degree to which each party participates in the positioning. Some technologies, such as the Global Positioning System (GPS), can operate completely independently and separately from the mobile operator. This enables third-party LBS application development that can more quickly accelerate deployment and adoption of LBSs than solutions that are tightly controlled by the carrier.

The inclusion of GPS receivers in mobile handsets could jump-start LBS. ABI Research predicted that by 2011, there would be 315 million GPS subscribers for LBSs, up from a mere 12 million in 2006 (see http://www.abiresearch.com). The mobile industry is now favoring applications, such as turn-by-turn directions and other navigation services, through which functionality typically associated with in-car GPS systems is brought to the mobile device (see http://www.fiercewireless. com for details on such industry trends).

16.2 Survey of Positioning Technologies

This section outlines a variety of positioning technologies that form the basis of many LBSs. For more details, refer to resources such as [2].

16.2.1 Global Positioning System

The GPS is perhaps the most well-known positioning technology. It relies on a system of 24 or more geosynchronous satellites to constantly broadcast reference signals to end devices, or GPS receivers. In addition to the satellites, GPS relies on monitoring stations placed on the ground at specific positions around the world; these stations control the operation and synchronization of the satellites by monitoring their orbits and tracking offsets between their internal clocks. To localize a GPS receiver in three dimensions, the receiver must be able to decode signals from four separate GPS satellites. The distribution of satellite orbits is therefore designed to ensure that every part of the earth is visible by at least four GPS satellites at any given time. GPS satellites transmit pilot codes that GPS receivers use to determine their range from the satellite. The satellites also transmit navigation information that helps the receiver determine the present position of the satellite as well as the satellite's clock offset. By using range measurements together with the locations of those satellites, a GPS receiver can determine its location to within an accuracy of a few tens of meters in the neighborhood. Various modifications and enhancements to the GPS system can improve the position accuracy to within 10 meters.

While widely available, GPS suffers from a few drawbacks, the principal of which is the inability to receive satellite signals indoors and in high-multipath environments. As a consequence, LBSs based on GPS positioning technologies may fail when there is no direct line of sight between the receiver and the GPS satellites. In addition, performing the initial synchronization of the GPS satellites can be time consuming and can result in a long time to first fixes (TTFFs). Depending on the nature of the LBS, this value can be critical to the usability and utility of the service. Another potential drawback of GPS for LBSs in the cellular context is the reliance on devices with GPS receivers. These receivers add bulk and drain battery resources. However, manufacturers are increasingly integrating GPS chips into cellular handsets.[1]

Enhancements to the GPS were designed to combat some of these drawbacks. In particular, carriers have begun implementing systems that assist the GPS system by broadcasting additional reference signals to handsets. This technique is referred to as Assisted-GPS, or A-GPS, and carriers broadcast pilot signals from their fixed infrastructure that the terminal combines with GPS information to make a more accurate determination of its location. A-GPS addresses the inability to receive GPS signals indoors or without direct line of sight and can significantly improve the accuracy of GPS position fixes in areas of high multipath such as urban areas.

16.2.2 Enhanced Observed Time Difference and Uplink Time Difference of Arrival

Enhanced observed time difference (E-OTD) and uplink time difference of arrival (U-TDoA) are positioning technologies that rely not on satellites but on the cellular infrastructure for reference information. With E-OTD, the framing structure of the Global System for Mobile Communications (GSM) frame is used as a reference point to determine the flight times of GSM signals transmitted from the base station. Since radio signals travel at the speed of light, a known quantity, flight times provide the terminal with the range from its current location to the base station. E-OTD, however, requires an additional element in the network to collect transmitted signals and compare them with those received by the terminal. This is the job of location measurement units, or LMUs. LMUs are scattered throughout a provider's network and provide an additional set of measurements that can be used to determine offsets and to account for synchronization errors between the terminal and the base stations. E-OTD requires a good deal of synchronization between the terminal and a set of nearby base stations, between the LMU and the same set of base stations, as well as between the terminal and the LMU. This level of coordination can lengthen the initial time to acquire a position fix. To alleviate this issue, GSM operators tend to synchronize their LMUs periodically with nearby base stations. E-OTD does not require the terminal to make any ranging transmission to support the localization process and relies on terminal reception only. Thus, a terminal may remain in the idle mode while assisting in the localization process.

The U-TDoA is similar in concept to the E-OTD; however, it relies on the terminal to transmit ranging messages so that those signal arrival times may be measured at different LMUs and base stations. Thus, the transmitting and receiving responsibilities of the base stations and terminals are reversed in U-TDoA. In GSM systems, a mobile terminal can transmit to only one base station at a time; therefore, LMUs are required to listen to the ranging transmission to correlate reception characteristics at different locations. In E-OTD, a single LMU is required; however, in U-TDoA, multiple LMUs must be in range of the terminal and base stations. Also, since localization requires terminal transmissions, terminals must be in the active mode to participate in the localization process. Last, dedicated software is required on the mobile terminals to support the localization process in E-OTD; since the terminal's only role in U-TDoA is the transmission of GSM data frames, no such software is required for U-TDoA positioning.

16.2.3 Observed Time Difference of Arrival

E-OTD and U-TDoA both rely on the GSM framing structure to derive timing estimates and ranges. Observed time difference of arrival (OTDoA), however, is designed for cellular systems based on code-division multiple access (CDMA). OTDoA follows the same concepts as those of E-OTD and U-TDoA, namely, to rely on transmitted signals to derive range estimates from known locations. OTDoA differs because CDMA systems do not have a strictly synchronized framing structure as GSM and CDMA terminals are also capable of transmitting to multiple CDMA base stations simultaneously. While the framing structure in CDMA is not synchronized, all CDMA base stations are synchronized and transmit a high-frequency pilot code that terminals can use to accurately determine range and timing offsets. A consequence of base station synchronization in CDMA is that LMUs are not needed. OTDoA operates in a similar fashion as E-OTD. Terminals decode pilot channel signals from nearby base stations to determine range estimates between their current location and the known locations of the base stations. Both systems provide roughly equivalent accuracy as well, and both systems require that the mobile terminal have dedicated software to assist in the localization process.

16.2.4 Ultrawideband

A few researchers and small companies (as well as the U.S. military) are looking at ultrawideband (UWB) as a promising location detection technology. UWB uses very short bursts of radio energy to perform precise ranging and synchronization measurements. The technology is extremely accurate (to within a few centimeters) and requires very little power to operate. The technology has showed strong promise for performing indoor localization.

16.2.5 Wi-Fi Positioning

Cellular and satellite positioning services generally perform worse indoors than they do outdoors. This limitation is principally due to the high-multipath environment indoors that can affect timing measurements. During the mid-1990s, new wireless networking standards, collectively known as Wi-Fi, began to gain traction in homes and businesses. Today, Wi-Fi systems are widely deployed and provide an additional component for determining position. Wi-Fi networks are defined as wireless local-area networks (WLANs) and are generally small in range, on the order of up to 100 meters in coverage. While the indoor environment is not well suited for positioning technologies based on time difference arrival, the WLAN community has focused its positioning efforts on received signal strength. These systems estimate a terminal's position by comparing its received signal strength against previously recorded values at known locations. These techniques require a good deal of training to develop signal maps that serve as the reference basis for the positioning technology. This training involves taking a number of signal strength measurements at known locations and storing them in a database. When fulfilling a positioning request, the terminal sends the network its received signal strength information. The network consults the database and uses estimation and approximation algorithms to determine the location that provides the least estimation error. Wi-Fi positioning techniques attempt to tackle the problems caused by the indoor radio environment. Some systems, such those commercially offered by Ekahau (see http://www.ekahau.com), offer accuracies on the order of a few meters. However, diligent training and maintenance are required to keep the signal maps current. Also, these signal maps can vary widely. Changes in the number of people present, the deployment of new access points, rearrangement of furniture, and differences in receiver sensitivities can all alter the signal maps and result in inaccurate readings.

16.3 Types of Location-Based Services

Due to the somewhat elusive definition of LBS in the academic literature, the scope of LBSs is large, and opinions differ on what exactly composes an LBS. This section lists some examples of LBSs along with brief descriptions; subsequent sections present in more detail the functional components required to make these services a reality; the references provide more specific information. The following information illustrates a wide range of LBSs; many more are possible:

- Navigation
  - Driving directions, such as sending turn-by-turn navigation directions to automobiles
  - Traffic alerts, such as sending alerts to automotive users approaching heavy traffic areas

- Information
  - Advertising
    - Coupons and sales, such as pushing information to mobile users about local stores and products, including redeemable coupons
  - Travel
    - Location-based travel and tourist information, for example, dialing a number may provide more historic information about a local monument
  - Search, such as searching for information that is relevant to a current location
- Tracking
  - Fleet tracking, such as keeping track of service vehicles and their locations
  - People tracking, for example, keeping tabs on family members, getting notices when they roam out of predefined areas
- Emergency assistance and calling for mobile users
- Communication services options
  - Using dual-mode (Wi-Fi and cellular) phones opportunistically according to prevailing local conditions [3]
  - Seamless session continuity between cellular and Wi-Fi networks
- Leisure and gaming
  - Finding "buddies," friends, and dates close to you at a given time
- Geocaching and geodashing, which are "hide-and-seek" type gaming using GPS-enabled mobile devices

16.4 LBS Ecosystem

Providing end-to-end mobile LBS requires the coordination of many systems. While to some extent each communication provider has its own infrastructure into which LBS functionality would be provisioned, there are certain largely agreed-on functional requirements that all such systems share. This section describes these high-level functions and puts them together into a simplified context. Figure 16.2 shows that positioning technologies (e.g., cellular, satellite based, Wi-Fi, or other) allow ongoing positioning of a mobile user. A core communication infrastructure hosts the application servers and other management servers (e.g., authentication, accounting; see Third Generation Partnership Project [3GPP] Internet Protocol [IP] Multimedia Subsystem (IMS) architectures, for example) that together make up LBS logic and deliver services to mobile users. The LBS may make use of other arbitrary information services in the core or on the Internet. Among other things, the geographic information systems (GISs) or geographic databases (such as Geo-DB) allow conversion of geographic coordinates, generation of addresses from coordinates, and layered map creation. A simple scenario is as follows: As a mobile

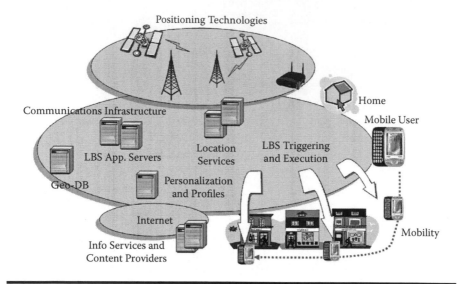

Figure 16.2   LBS ecosystem.

user migrates from home through an urban environment, for example, the user's location is updated in core servers, such as those providing cellular service. At each such update, LBS applications may be triggered; each service exploits the user's details, profiles, and location (perhaps filtered through a geographic database) and triggers some information delivery to the user (perhaps gathered from third-party Internet-based content providers).

In a prototypical case, the information relates to the stores currently near the user and personalizes the information to the specific user (perhaps by referring to a user-specific profile). See 3GPP for further rigorous LBS use cases through the various LBS infrastructure components.

16.5 Emerging Support in Standards Bodies

The notion of LBS is supported in several key geographic and wireless standards bodies (see [4] for details). The Open Mobile Alliance (OMA) comprises the world's leading communication providers; one of its foci is the Mobile Location Protocol (MLP), which is "an application-level protocol for obtaining the position of mobile stations ... independent of underlying network technology. The MLP serves as the interface between a Location Server and a Location Services Client."[2] The Internet Engineering Task Force (IETF) stewards a protocol called the Spatial Location Protocol (SLoP), which allows a client to "talk about" the spatial location of an Internet resource in a canonical way. The Open GIS Consortium (OGC), comprising over 200 companies and agencies, has defined a number of specifications that help with interoperability of LBSs. These standards bodies—and others—have a keen focus on supporting LBSs and have already provided valuable and relevant specifications.

16.6 Emerging Support in Middleware and Programming Tools

Typically, LBSs are delivered to, or executed on, mobile handsets. Today's predominant mobile operating systems (OSs) include

- Symbian
- Microsoft Windows Mobile
- Palm OS
- Linux

These operating systems have a spectrum of capabilities to support LBSs. Key to developing device-resident LBS applications for the mobile handsets is the availability of integrated development environments (IDEs). Micosoft's Visual Studio and Sun's Java Studio (and NetBeans Mobility Pack) are two such environments that provide major support for mobile application development. Note that to implement a successful LBS on a device a developer needs middleware and OS support not only for location application programming interfaces (APIs), for example, but also for other important aspects of intelligent networked services such as communication stacks (e.g., Transmission Control Protocol (TCP)/IP, Bluetooth), protocol stacks (e.g., Hypertext Transfer Protocol [HTTP]), Web services, and XML, to name only a few.

Software development kits (SDKs) are the underlying support, and APIs allow developers to quickly write source code that makes use of lower-level LBS enablers, shielding the developer from their implementation complexity. J2ME, for example, is a technical platform used for LBS implementations (especially on the client side). Figure 16.3 illustrates the architecture of the Java and J2ME stack. Standard device profiles and optional packages allow systematic cross-platform development of mobile Java applications. The figure shows that, for execution of Java on a mobile phone, the phone runs a Java Virtual Machine (JVM).

The JVM runs according to a so-called connected limited device configuration (CLDC); this means that the JVM and its APIs are tailored to run on a certain set of limited capability devices. The mobile information device profile (MIDP) sits atop the CLDC and provides add-on APIs for the developers that remain compliant with CLDC; for example, MIDP2.0 defines media, game, and messaging APIs for developers. Armed with these tools, the developer can more easily and effectively write cross-platform LBSs written in Java for mobile phones.

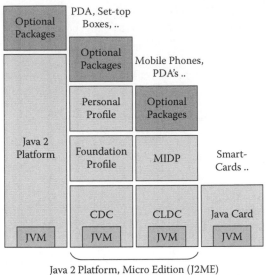

Figure 16.3 Java stack highlighting layers related to mobile computing.

16.7 Current Commercial Trends

This section outlines some commercial and grassroots trends by providing general descriptions and examples if possible. While the service providers mentioned here may come and go over time, the themes they represent should be valid for much longer. Note that we are not interested here in cases in which online services are simply made available as Wireless Application Protocol (WAP) pages[3]. Flash-point LBSs at the time of writing include mapping and directions, navigation, tourism, people and fleet tracking, trip planning, and real-time traffic information. These and others are outlined next.

16.7.1 People-Tracking and Personalized Services

Wireless people tracking refers to the technique of locating an individual user in terms of geospatial coordinates (see previous sections for techniques) and exploiting this information in some useful way. While fleet tracking via in-truck GPS receivers is a common practice, offering value-adding LBSs directly to mobile customers is seen by operators as a possible windfall of revenues. That is why people-tracking and related services are becoming very compelling—a *person* with a GPS receiver in hand not only can be tracked but also can be offered information and m-commerce services at any time for which the user will pay a service fee. For years, radio-frequency technologies have been used to ensure that criminals on probation remain inside the home; now, cheaper GPS technologies allow some jurisdictions to track parolees on a very fine level with GPS "bracelets." In the commercial world, one trend sees middle-class families with wireless telephony services paying for child- or family-tracking services from their providers. These services use location technologies such as GPS or cellular-based positioning to monitor if and when a child moves out of a specified region. If the child does break the region's virtual "barrier," the parent is notified. In 2006, Verizon Wireless began offering a similar service called Chaperone, by which the child is given a personal wireless phone that in turn enables several tracking modes for parents. In the same year, Sprint Nextel began offering a service called Family Locator that allows all family members to be tracked and monitored. These and other services typically require special equipment and mobile application software installation and incur extra premiums on monthly bills.

Social networking applications such as dating and meet-up services have experienced something of a renaissance at the time of writing. Today's dating services can take advantage of ubiquitous Internet access, instant messaging (IM), and cellular networks. Many dating services now incorporate location information. The company Meetmoi.com allows daters to update not only their personal profiles but also their locations. Once the server understands the user location, it can send the user a list of other singles (and their SMS [short message service] handles) that happen to be currently in the same geographic area. Messaging and meet-ups can then occur between the singles. Proxidating.com is a similar LBS dating service that relies on daters having Bluetooth receivers that come into range of each other and exchange compatibility profiles. Other more technically advanced dating services will be able to seamlessly infer singles' latitude and longitude coordinates through a location service provider and offer a variety of meet-up possibilities.

With respect to automotive telematics, service providers can mine GPS receivers inside vehicles to glean useful information, including vehicle location, trajectory, route, and so on and can then provide the driver with conditions that the driver should expect to encounter in the coming minutes (e.g., based on the car's trajectory). Such services are offered by major telecom and telematics providers. Adaptive route planning (navigation) is also a widespread LBS in which a vehicle's current location, destination, and road conditions in-between are constantly monitored to provide an optimal journey.

16.7.2 Opportunistic m-Commerce

LBSs allow service providers to find customers that are most likely to consume their services and to make the best use of where they are at a given moment in time. Loki.com is one of several services that deliver location-based information to a laptop independently from cellular network visibility. It achieves this by intermittently reporting the set of Wi-Fi access points that the laptop currently sees to its servers; the server then attempts to infer the user's location based on a database of known access points (correlated to geo-locations). After this succeeds, Loki.com is able to pass the location information along to third-party plug-in modules, which in turn provide various location-based information, such as weather, traffic, and shopping bargains, on a Web browser toolbar. Elsewhere, most of the main online search and map providers (e.g., Google, Yahoo!) hold important LBS information. Not surprisingly, all enable mobile users to transmit their location and do "local searches." The results are the set of services matching the search query that are within some distance of the searcher's current location (e.g., their zip code or address). Searching for a local "pizza restaurant" or for local "movie listings" are prototypical examples.

Service providers are also turning to online video games as an outlet for LBSs. Online games such as World of Warcraft and Second Life offer gamers rich three-dimensional (3D) environments in which their avatars interact with the environment and with other peers. Microsoft's 2006 purchase of Massive signals that in-game advertisements are deemed valuable; Massive specializes in technology that allows advertisers to inject dynamic logos and custom information onto virtual billboards in 3D (online) video games. Although no fine-grain (or one-to-one) billboard personalization has occurred at the time of writing, it is likely that eventually individual gamers will see virtual billboards that are targeted either directly to them or to small groups of players (who currently inhabit the same part of the world). While these billboards will begin as static imagery, they could also constitute codes and coupons that unlock discounts on real-world or in-game goods and services.

16.7.3 Recreational and Grassroots

As GPS receivers have shrunk and become highly portable and affordable, more and more interesting grassroots uses of geospatial information have been demonstrated. Not all of these constitute services per se, but all show tremendous promise. *Geocaching* [5] is a social game played by a large number of users. Not unlike orienteering or hide-and-seek, the game is about navigating to and finding treasures located in various disparate regions. Once found by a user, the user gets to relocate parts of the treasure to some new position and challenge others to navigate to it. Although played largely as open source and free, there are obvious advertising possibilities for this sport. *Geodashing* is a game in which users armed with GPS receivers race to pass through a set of geospatial coordinates that together form a so-called dash. Finally, other grassroots use of GPS information and geospatial information includes GPS way-pointing and GPS drawings. In the former, users document interesting locations through text annotations, photos, and other links (e.g., to services); together, these can be used by other users as a sort of guidebook to local information. GPS drawing [6] is an experimental phenomenon among conceptual artists who, with logging enabled on their GPS receivers, trace out (sometimes enormous) shapes on the earth and upload the logs to a server where they can be displayed.[4] *Geotagging* is the process by which imagery or photographs are tagged with geospatial coordinates. This is a specialization of photo tagging, given mass market appeal most notably by Flickr.com and Zoomr.com.

There are several LBS possibilities in which location information is related to IM. A "where are my buddies" service for IM clients has gained in popularity. This service is available, for example, as an open source plug-in to AOL's AIM and revolves around IM users receiving notification whenever one of their buddies registers a location that is less than some threshold distance away from their own. Similar services—with names like Buddy Beacon and Boost Loopt—are being offered by major telecom providers. Such a scenario is sometimes generally referred to as context-sensitive computing or location-aware computing [7]. Research has delivered several wireless context-based systems; [8] describes a prototypical one. In context-sensitive systems, the computing platform gathers and understands aspects of the mobile user's surroundings (i.e., context), including but not limited to the user's location, activities, photographs taken recently, messages, and so on. By correlating and aggregating user-context information, a deep understanding of user needs can be inferred, and as a result deeply targeted services can be offered by an LBS provider.

16.8 Current and Future Challenges

LBS creation is an ongoing research issue. Traditional intelligent network communication services have been designed and created on graphical SDKs emphasizing events and flow control. The IMS and implementations of location enablers must now be programmatically integrated with services at creation time; thus, LBS creation tools have become more complex [9]. LBS SDKS have emerged from various vendors. It is thought that these tools must continue to improve to better support LBS development. Some of the salient issues include

- Making LBS portable across devices
- Separating concerns of LBS creation and provisioning

These issues and others are being considered by various standards and industry organizations (e.g., 3GPP, OMA, Europe-based Information Society Technologies at http://cordis.europa.eu/ist/).

While LBSs attempt to provide users with value based on their current location, other geospatial attributes (e.g., the user's speed or acceleration) are also important. Current research illustrates that as LBSs become more peer to peer (i.e., services beneficial to one mobile user are located on another mobile user's platform), these attributes should come into play (e.g., see [10]). Privacy is another flash point in LBS; it has several dimensions:

- Users do not want providers to abuse the location or context information that they may have logged about them.
- Users should be allowed to remain anonymous from service providers and other users if possible.
- Effective identity and consent management is essential.

Detailed studies of these and other related issues can be found in the literature [11,12].

16.9 Conclusion

LBSs have arrived, and ongoing improvements to underlying protocols, equipment, and middleware will only make them more effective and profitable for communication providers. Although some challenges exist, it is widely believed that soon almost all mobile users will rely on LBSs to get personalized information and service based on their current location and context.

Notes

1. In the United States, cellular service providers are required to meet so-called E-911 regulations that require them to locate a cellular phone to within a few hundred meters for emergency purposes. To meet this need, many providers have mandated that all their phones include GPS locator chips that provide location assistance in case of emergency. These locator chips are not fully functional GPS receivers and do not provide the capability to track users or support LBSs.
2. Quotation taken from http://www.openmobilealliance.org.
3. For example, one of the Web's most visited sites, eBay, makes its auction portal available via mobile devices. Such an initiative, however, does not exploit user location in any deep way.
4. See [6] for an architecture to support this in a systematic, large-scale fashion.

References

1. IETF RFC 3920, *Extensible Messaging and Presence Protocol (XMPP): Core.*
2. Kupper, A., *Location Based Services*. New York: Wiley, 2005.
3. Falchuk, B., Loeb, S., Famolari, D., Eiger, M., Elaoud, M., Komandur, K., and Shallcross, D., Intelligent network-centric admission control for multi-network environments. *Proc. IEEE Int. Symp. Consumer Electron. (ISCE'06)*, 296–301, 2006.
4. Mohapatra, D., and Suma, S., Survey of location based wireless services. *Proc. IEEE Int. Conf. Personal Wireless Commun. (ICPWC'2005)*, New Delhi, January 2005.

5. Geocaching—The Official Global GPS Cache Hunt Site. http://www.geocaching.com/

6. Falchuk, B., Web application supporting large-scale collaborative GPS art. *Proc. Web Technol., Appl. Serv.*, IASTED, Calgary, 2006.

7. Jones, Q., and Grandhi, S., P3 systems: Putting the place back into social networks. *IEEE Internet Comput.*, 9(5), 38–46, September 2005.

8. Koolwaaij, J., Tarlano, A., Luther, M., Nurmi, M., Mrohs, B., Battestini, A., and Vaidya, R., Context watcher—Sharing context information in everyday life. *Proc. Web Technol. Appl. Serv.*, IASTED, Calgary, 2006.

9. Telcordia Technologies, *Converged Application Server (CvAS)*. http://www.telcordia.com

10. Falchuk, B., and Marples, D., Ontology and application to improve dynamic bindings in mobile distributed systems. *Proc. 2nd Int. IEEE Wireless Internet Conf.*, Boston, 2006.

11. Kölsch, T., Fritsch, L., Kohlweiss, M., and Kesdogan, D., Privacy for profitable LBS. *Proc. 2nd Int. Conf. Security in Pervasive Comput.*, 168–178, 2005.

12. Stross, R., Cellphone as tracker: X marks your doubts. *New York Times*, November 19, 2006.

Links

1. Third Generation Partnership Project (3GPP). http://www.3gpp.org/

2. Java Platform Micro Edition, Sun Microsystems. http://java.sun.com/javame/index.jsp

3. Microsoft .NET. http://www.microsoft.com/net/default.mspx

4. AutoDesk Location Services (and Java SDK). http://www.autodesk.com

5. Open Mobile Alliance (OMA) Mobile Location Service Enabler V1.0. http://www.openmobilealliance.org/release_program/mls_v1_0.html

6. Internet Engineering Task Force. http://www.ietf.org

7. Open Geospatial Consortium (OGC). http://www.opengeospatial.org/

8. Microsoft Visual Studio. http://msdn.microsoft.com/vstudio

9. Java NetBeans Mobility Pack. http://developers.sun.com

Chapter 17

Authentication and Privacy in Wireless Systems

Thomas M. Chen and Nhut Nguyen

Contents

17.1 Introduction

Authentication and *privacy* refer to the problems of ensuring that communications take place only between the right parties without disclosure of information to unauthorized eavesdroppers. Radio communications are highly appealing for the convenience of mobility: the freedom from a fixed location. For this reason, wireless services have been growing rapidly. An ITU (International Telecommunication Union) study found more than 2 billion cellular phone subscribers in the world in 2005 and more than the 1.26 billion land phone lines. It has become easy to find Institute of Electrical and Electronics Engineers (IEEE) 802.11 wireless local-area networks (LANs) in residences, hotels, stores, and corporate sites. Smart phones and a variety of wireless messaging devices can send SMS (short message service) messages and e-mail and browse the World Wide Web through a number of wireless services. IEEE 802.16 "WiMax" is starting to offer broadband wireless services in the local loop.

The enthusiasm for wireless services makes it easy to ignore the inherent insecurity of the radio medium. The most obvious risk is that radio signals can be received by anyone within range of the transmitter. Radio communications are easy to intercept, possibly by someone who is beyond sight. In contrast, it is more difficult to intercept wired communications. An eavesdropper must physically access a wired medium and therefore tends to be more visible. In wireless communications, the loss of privacy (or confidentiality) is always a possibility and motivates the need to devise measures to protect privacy. The primary means used to protect privacy in wireless systems is cryptography, which is described in this chapter.

Another clear risk is impersonation; someone presents a false identity to attempt to access unauthorized services. For land-line phones, impersonation is a much smaller risk because phones are typically in an indoor environment under private ownership. It is implicitly assumed that the owner is responsible for physical access. The identity of a land-line phone user is associated with that fixed location. However, users in a wireless network are mobile, so their identities cannot be associated with a particular location. Instead, mobile users must carry their credentials (e.g., passwords) and present them to the network to verify their identities. It is important that authentication credentials are difficult to duplicate by someone else. Authentication of mobile user identities, also largely based on cryptography, is another topic covered in this chapter.

17.2 Cryptography Basics

Due to the nature of radio signals, it is not feasible to prevent eavesdropping. It must be assumed that any radio transmission might be overheard by an unwanted third party. Encryption is a mathematical process for transforming the original data (called *plaintext*) into a form (*ciphertext*) that makes sense only to the intended receiver but no sense to an eavesdropper. Traditionally, cryptography is done on the basis of secret keys shared by both parties, which is the reason it is called *symmetric cryptography*. In the 1970s, a different approach called asymmetric cryptography was invented by which the transmitter uses a public key and the receiver uses a secret key.

In symmetric cryptography, the secret key known only to the transmitter and receiver is used for both encryption and decryption (Figure 17.1) [1]. It is usually assumed that an eavesdropper may have knowledge of the encryption (and the reverse decryption) algorithm, and publicly known encryption algorithms are often used. Thus, the strength of the system depends on the difficulty for an eavesdropper to discover the key. *Cryptanalysis* refers to the process of breaking a cipher by discovering the secret key or the original plaintext (or both) based on at least observations of the ciphertext. Hence, strong encryption algorithms should leave no statistical structure in the ciphertext that could give clues about the plaintext or encryption key. In addition, strong algorithms should be resistant to brute force attacks where every possible key is tried. Therefore, longer encryption keys are better than short ones because a brute force attack would have to go through a larger number of

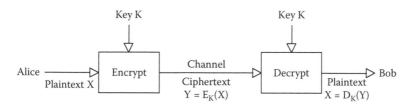

Figure 17.1   Symmetric or secret key cryptography.

possible keys. For example, doubling the key length from n bits to $2n$ bits would mean an increase from 2^n possible keys to $(2^n)^2$ possible keys.

A large number of encryption algorithms are known, but one of the most widely used is the Data Encryption Standard (DES) adopted by the U.S. government in 1976 [2]. It was based on an earlier Lucifer algorithm developed by IBM. DES encrypts plaintext as separate 64-bit blocks using a 56-bit key. Basically, blocks go through 16 cycles, each cycle performing substitution and permutation operations. Each cycle can be viewed as "scrambling" the text more. Although the logical operations in each cycle are similar, each cycle depends on a different 48-bit subkey derived from the 56-bit key.

In 2001, the U.S. government adopted Rijndael as the Advanced Encryption Standard (AES), replacing DES [3]. It operates on data blocks of 128 bits. There are 9, 11, or 13 cycles for keys of 128, 192, and 256 bits. Each cycle or "round" involves substitution, shift, mixing, and XOR (exclusive OR) operations. Although the logical operations are identical in each cycle, each cycle depends on a different subkey derived from the key.

A common problem for all symmetric key encryption algorithms is the need for every pair of communicating parties to share a secret key, protected against eavesdropping, before they have a secure channel. Surprisingly, there is a protocol invented by Diffie and Hellman (called the Diffie–Hellman key exchange) that allows two parties to share a secret number confidentially over an unsecure channel [4]. Alternatively, all parties could share secret keys with a trusted third-party key distribution center. When two parties need to communicate, they can both fetch a secret key securely from the key distribution center to use for the duration of their session. However, this scheme does not scale well with the number of users. As the population increases, the number of keys needed will increase exponentially, and the load on a key distribution center will grow similarly.

In 1976, Diffie and Hellman proposed public key or asymmetric cryptography (Figure 17.2) [4]. In public key cryptography, each party knows his or her own private key, and there is no need to share this key with anyone else. Everyone knows his or her public key, but the relationship between the public and private keys is designed to make it computationally difficult to discover the private key from the

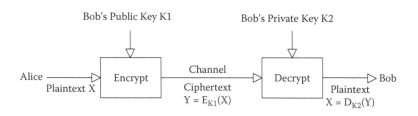

Figure 17.2 Asymmetric or public key cryptography.

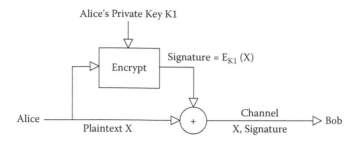

Figure 17.3 A digital signature.

public key. Anyone can encrypt a message with the recipient's public key, but only the recipient with the private key can decrypt the message.

In 1977, RSA was the first practical public key cryptosystem; it was invented by Rivest, Shamir, and Adelman and patented in 1983 [5]. The security of RSA basically depends on the difficulty of factoring a very large number into two prime numbers.

A public key cryptosystem such as RSA comes in handy for digital signatures. A digital signature attached to a message serves the same purpose as a handwritten signature on a physical document. It verifies the originator of the message. This is possible because RSA happens to have the property that the public and private keys are interchangeable. Thus, a digital signature could be created if Alice encrypts a message with her private key and attaches this signature to the message (Figure 17.3). Bob can decrypt the signature with Alice's public key and compare the decrypted message with the received message. A match would imply that the signature could have come only from Alice because only Alice has the private key corresponding to her public key. It also verifies that the message was not modified during transmission. In practice, digital signatures are produced in a more efficient way by using a hash function such as message digest algorithm-5 (MD5) to produce a hash or message digest of the message first before encryption. A message could be very long, whereas a message digest is a short, fixed length, so it is much more efficient to encrypt a message digest. The recipient can decrypt the digital signature to recover the message digest. This is compared with a hash of the received message. A match implies that the message was not altered and came from the sender.

An association between users and their public keys is often handled by certificates. A certificate is essentially a digital document binding together a user's identity and public key, verified by a digital signature of a trusted third party called a certificate authority.

Finally, secret keys are useful in authentication protocols. Suppose that two parties want to verify each other's identity, and they share a secret key. Each party can prove its identity by demonstrating possession of the secret key, without revealing the secret key itself (which could then be stolen), through a basic challenge–response protocol (Figure 17.4). Alice issues an unpredictable (random) challenge

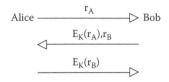

Figure 17.4   Challenge–response authentication with a shared secret key.

to Bob. Bob demonstrates his knowledge of the secret key by encrypting the challenge. Bob can verify the identity of Alice in the same way with another challenge.

The protocol works with public keys as well (Figure 17.5). Alice generates a challenge and encrypts it with Bob's public key K1. Bob can decrypt it with his private key and then encrypt the challenge with Alice's public key K2. Alice can decrypt this to recover the original challenge, which verifies the identity of Bob because only Bob could have decrypted the first message. Obviously, Bob can authenticate the identity of Alice similarly with a separate challenge.

17.3 Privacy and Authentication in IEEE 802.11 Wireless LANs

IEEE 802.11 wireless LANs have been commercially successful. Approved in 1999, IEEE 802.11a specified a data rate of 25–54 Mbps in the 5-GHz band, and 802.11b specified a data rate of 6–11 Mbps data rate in the 2.4-GHz band. Another flavor 802.11g, backward compatible with 802.11b, but with data rate increased to 25–54 Mbps, was approved in 2003. Currently, an 802.11n standard is being drafted for a data rate of 200–540 Mbps.

Security has been a long-standing weakness of 802.11 wireless LANs. Traditionally, authentication and privacy have depended on preshared secret keys. Security has been recognized as an important problem, and there are ongoing efforts to improve the standards.

17.3.1 Privacy

The original security scheme included in IEEE 802.11 wireless LAN standards was WEP (wired equivalent privacy), approved in 1999. It was recognized that

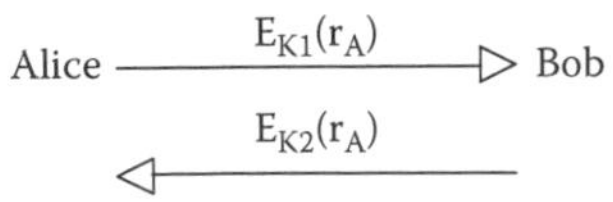

$$\text{Alice} \xrightarrow{\quad E_{K1}(r_A) \quad} \text{Bob}$$
$$\xleftarrow{\quad E_{K2}(r_A) \quad}$$

Figure 17.5   Challenge–response with public keys.

wireless networks are inherently vulnerable to eavesdropping. WEP was intended to provide privacy comparable to traditional LANs by using the RC4 algorithm to encrypt data packets sent out from an access point or wireless network card. RC4 is a stream cipher that generates a pseudorandom bit stream called a key stream that is combined with the plaintext using bit-by-bit XOR to produce the ciphertext. The key stream must be pseudorandom because the decryptor generates the same key stream to XOR with the ciphertext to recover the plaintext.

In WEP, the RC4 cipher uses a seed constructed from a secret key (preshared password) and a 24-bit initialization vector (IV) to encrypt each packet. The IV prevents repetition in the key stream from packet to packet when the same secret key is used. The IV is prepended to each packet and sent in plaintext to synchronize the decryptor at the receiving host. Unfortunately, the 24-bit IV length means that there is a 50% chance of seeing a repeated IV after 5,000 packets. In 2001, it was shown that a passive eavesdropper could recover the RC4 key after observing a sufficient amount of traffic (a few million frames). Alternatively, an attacker could send packets on the wireless LAN to stimulate reply packets. These attacks were implemented in software such as Aircrack and WEPCrack. By 2003, a number of shortcomings in WEP were widely recognized.

An interim replacement for WEP called Wi-Fi Protected Access (WPA) was published by the Wi-Fi Alliance and deployed commercially in 2003. IEEE 802.11i, also known as WPA2, was approved in 2004. The key establishment procedure and authentication are the same in WPA and WPA2. These depend on another standard, IEEE 802.1X, and an authentication server to share a secret 256-bit pairwise master key (PMK). In the first step, the access point asks for identification information from the host, such as user name and MAC (media-access control) address. The access point receives this information and forwards it to the authentication server. The authentication server can use any number of authentication protocols to verify the identity of the host. During the authentication process, the access point simply passes messages between the host and authentication server.

After a PMK is established between a host and access point, the PMK is used to derive 128-bit pairwise transient keys (PTKs). There are four PTKs that are mixed by an algorithm, along with MAC addresses and nonces from the host and access point, to finally generate per-packet encryption keys. Different from WEP, packets are encrypted by AES used in counter mode.

17.3.2 Authentication

IEEE 802.11 specified two types of authentication: open system authentication (OSA) and shared key authentication (SKA). A host first learns the name or service set identifier (SSID) of networks within its range. After choosing a network to join, it issues an authentication request message specifying the authentication scheme. The access point can accept or reject the requested authentication scheme in its response.

By default, an access point uses OSA, which essentially provides no authentication of hosts. In SKA, authentication depends on a host knowing a preshared secret key (the method of exchanging secret keys was not specified in 802.11). Knowledge of the secret key is demonstrated through a challenge–response protocol (Figure 17.4). The access point sends a random 128-byte challenge to the host. The host encrypts the challenge using WEP (discussed below) with the secret key and an IV of its choosing. The host sends the encrypted challenge and IV to the access point for decryption. If the access point can recover the secret key, it verifies that the host knows the secret key and allows access to the wireless LAN.

Problems with the authentication protocol are known. In actuality, the challenge–response protocol verifies only that the host is one in a group of hosts who all know the same key. IEEE 802.11 did not specify a way for hosts to obtain unique secret keys. Thus, an access point cannot authenticate the exact identity of a host. Also, authentication is unidirectional; there is no provision for hosts to verify the identity of an access point.

17.4 Privacy and Authentication in IEEE 802.16 Wireless Metropolitan-Area Networks (WiMax and Mobile WiMax)

Privacy and authentication in WiMax and Mobile WiMax networks are protected by the privacy sublayer of the MAC layer in the protocol stack [6]. The privacy sublayer in the 802.16 protocol stack defines security procedures that use X.509 certificates as the main mechanism for subscriber station (SS) authentication and other cryptography-based security functions [7].

17.4.1 Privacy

Symmetric encryption is used to protect privacy of data exchanged between an SS and a base station (BS). Messages are encrypted using traffic encryption keys (TEKs) generated by the BS and sent to the SS using the Privacy and Key Management (PKM) protocol. PKM exchanges are authenticated using the HMAC-SHA1 algorithm. TEKs are encrypted using the 3DES encryption algorithm. Keying information to protect PKM exchanges is derived from a shared authentication key (AK) established between the SS and the BS using public key cryptography as part of the authorization process.

In WiMax standards, the cryptographic algorithm specified for message encryption is DES with cipher block chaining (DES-CBC). Mobile WiMax enhances encryption strength with the introduction of the AES in Counter with CBC-MAC (AES-CCM) algorithm.

17.4.2 Authentication

The BS authenticates an SS using the PKM with the certificate of the SS as a credential and as part of the authorization process. In the early 802.16 standards, the SS did not authenticate the BS. The authorization process starts when the SS sends an authentication information message (informative) and then an authorization request message to the BS. The authorization request message includes the X.509 certificate of the SS along with other security information. The BS verifies the certificate to decide if the SS is authorized to access network resources and then uses the public key contained in the certificate to establish and encrypt an AK using the RSA algorithm. The encrypted AK is sent back to the SS along with other security information. The correct use of the AK to derive other keying information (e.g., the key encryption keys, KEKs, used to encrypt the TEKs) establishes the authenticated identity of the SS to the BS.

The privacy sublayer of the early 802.16 standards had a few deficiencies, noticeably the lack of BS authentication and the cryptographically weak DES algorithm based on 56-bit keys used for message encryption. These deficiencies were addressed in the newer 802.16e (Mobile WiMax) standards with the PKMv2, protocol, which includes the mutual authentication based on the Extensive Authentication Protocol (EAP) and stronger encryption algorithms such as AES [6].

17.5 Privacy and Authentication in Global System for Mobile Communications Networks

17.5.1 Privacy

In a Global System for Mobile Communications (GSM) network, the main mechanisms used to protect privacy are cryptography-based ciphering and temporary identity information [8].

17.5.1.1 Ciphering

To protect the privacy of signaling and media information being transmitted over the air interface, a symmetric ciphering process is used. This privacy protection mechanism is active when the ciphering feature for a mobile subscriber is activated by the network.

The network uses a random number RAND and the subscriber-specific secret key Ki to generate a ciphering key Kc using the cryptographic A8 algorithm. Each subscriber is assigned a unique secret key Ki that is stored both in the subscriber identity module (SIM) card attached to a mobile station and in the authentication center (AuC) of the network. The random number RAND is then sent to the mobile station.

The mobile station uses the received random number RAND and the subscriber-specific secret key Ki stored in the SIM card to generate a ciphering key Kc using the same cryptographic A8 algorithm. Note that since the inputs (Ki and RAND) and the key generation algorithm (A8) are the same, the generated ciphering key Kc is identical on both the mobile station and the network (hence symmetric ciphering.)

The generated ciphering key Kc and the frame number of the information frame being transmitted over the air interface are used as input to the ciphering algorithm A5 to generate an encryption bit mask S2. The length of the bit mask is the same as the information frame length.

The mobile station encrypts the information frame to be transmitted by exclusive-ORing (XOR) the information frame with the bit mask S2 before sending it out on the air interface.

On the receiving end (the network), a decryption bit mask S2 is generated using the same inputs (Kc and frame number) and cryptographic algorithm A5. Since the inputs and the algorithm are the same, the bit masks generated by the mobile station and the network are identical.

The network decrypts the received frame by XORing it with the decryption bit mask S2 to get the original information frame.

The same process is used on the reverse direction; that is, the network encrypts and the mobile station decrypts, but with a different bit mask S1. The bit mask S1 is generated by using the same A5 algorithm with the frame number and the ciphering key Kc as the inputs (Figure 17.6).

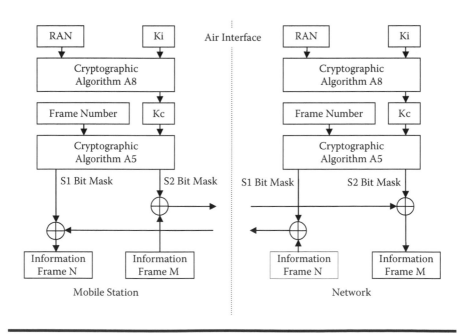

Figure 17.6 Ciphering for privacy in GSM networks.

17.5.1.2 Temporary Identity Information

To further protect the privacy of subscriber identity, a security mechanism that utilizes temporary identity information is used. On successful registration to the network, a subscriber is assigned an international temporary mobile station identifier (TMSI) by the VLR (visitors location registry) of the serving network. In subsequent transactions between the mobile station and the serving network, the subscriber is identified by this TMSI instead of the permanent and private international mobile station identification (IMSI). The mapping between a TMSI and the IMSI of a subscriber is known and valid only in the serving network. An attacker who captures TMSI information that is exchanged over the air interface cannot derive the subscriber's identity from that information.

17.5.2 Authentication

Authentication of a mobile subscriber in a GSM network is performed using a challenge–response protocol [8]. The network sends a random number RAND as a challenge to the mobile station. The mobile station uses the received random number RAND and the secret key Ki stored in the SIM card as input to the cryptographic algorithm A3 to generate an authentication response and sends it back to the network.

The network compares the received response with the expected response (SRES) for the challenge RAND. The expected response was generated using the same cryptographic algorithm A3 and the same random number RAND sent to the mobile station.

If the received response matches the expected response, the mobile station is authenticated, and the network allows the mobile station to continue. Otherwise, the authentication process fails, and the mobile station requests are rejected by the network.

Generating the random number RAND, the ciphering key Kc, and the expected response SRES for many mobile stations by using cryptographic algorithms is time and resource consuming. Thus, generating these values by the network that serves thousands of subscribers in real time is not practical. Instead, a prefabricated approach that utilizes the concept of an authentication triplet is used in GSM networks.

An authentication triplet contains the random number RAND, the ciphering key Kc, and the expected response SRES. In a GSM network, authentication triplets are generated by a network element named the AuC. Physically, the AuC usually coresides with the HLR (home location register). A subscriber record in an HLR/AuC may contain a number of precalculated authentication triplets to be used for that subscriber.

The network, specifically the mobile switching center (MSC)/VLR, uses one authentication triplet for each authentication procedure and the subsequent encryption of the session. When the network authenticates a mobile subscriber, an authentication triplet must be retrieved from the HLR/AuC by the MSC/VLR and used. To reduce traffic between network elements, authentication triplets are usually retrieved by the MSC/VLR in bulk in advance. Also, if the MSC/VLR detects that the number of triplets for a subscriber is below a set threshold, it can initiate a procedure to contact the HLR to replenish the authentication triplets so that it does not have to contact the HRL/AuC when it needs to authenticate the mobile subscriber.

Note that the authentication process described is one-way authentication only. During the authentication process, only the mobile subscriber is authenticated by the network, but the network is not authenticated by the mobile subscriber.

17.6 Privacy and Authentication in Second-Generation Code-Division Multiple Access Networks

Privacy and authentication procedures in second-generation (2G) code-division multiple access (CDMA) networks are based on the cellular authentication and voice encryption (CAVE) algorithm and are specified in the Interim Standard 41 (IS-41) standards of the Telecommunication Industry Association (TIA) [9].

17.6.1 Privacy

With CDMA technology, information to be transmitted over the air interface is spread out using a pseudonoise (PN) code. Since all users use the same radio-frequency spectrum over the air interface in CDMA networks, this code is needed to identify a particular pair of users of the radio-frequency spectrum. Thus, the CDMA technology has an inherent privacy mechanism built in: The receiver can get the transmitted information only if it knows the code used to spread the original information.

However, if the spreading code is based solely on public information that a perpetrator can obtain, this advantage may disappear. Anyone who knows the code can monitor the air interface and decode the transmitted information with the right equipment.

In CDMA networks, privacy of information transmitted over the air interface is protected by combining the built-in privacy mechanism of the CDMA technology with the use of cryptographic algorithms and encryption [9].

17.6.1.1 Voice Privacy

Voice privacy in CMDA networks is protected by using a secret private long code mask (PLCM). A PLCM is derived from an intermediate value named the voice privacy mask (VPM), which is generated using the CAVE algorithm and secondary key information called shared secret data (SSD). The SSD itself is also generated by using the CAVE algorithm and a secret primary key named A-Key that is known only to the mobile station and the AuC in the network. An SSD is a 128-bit integer value whose first half (SSD_A) is used for authentication, and the second half (SSD_B) is used for privacy protection.

The voice privacy feature of a CMDA network can be activated or deactivated. If voice privacy is activated, a PCLM is used to change the characteristics of the spreading code used to spread voice information before being transmitted over the air interface. In other words, transmitted voice information is scrambled with a secret PCLM, which is used to alter the characteristics of the spreading code. Unless the PCLM (which itself is a secret) and the method to alter the characteristics of the spreading code are known, it is extremely difficult to despread a signal transmitted over the CDMA air interface to obtain the original voice information.

17.6.1.2 Signaling Privacy

To protect signaling privacy (e.g., to whom a person is calling), the symmetric encryption process is used to encrypt signaling messages. An encryption key called the cellular message encryption algorithm (CMEA) is generated by both the MS (mobile station) and the network using a common SSD_B value as input to the CAVE algorithm. The signaling messages are then encrypted and decrypted by the mobile station and the network using this CMEA key and the CMEA.

A TMSI is also used to further protect subscriber identity in CDMA networks in a similar manner as in GSM networks. A TMSI is assigned to the mobile station after successful authentication and is used in subsequent transactions between the mobile station and the network to conceal the identity of a subscriber.

17.6.2 Authentication

The challenge–response process and cryptographic algorithms are used for authentication in CDMA networks. A secondary key SSD_A and a RAND number are fed into the CAVE algorithm to generate both response and expected response [9].

The network sends the RAND number as the challenge to the mobile station to be authenticated. The mobile station feeds the received RAND number and the stored SSD_A value to the CAVE algorithm to generate an authentication response AUTHR (18 bits) and sends it to the network as the response. The network uses the same RAND number and the SSD_A value for the mobile station as input to the CAVE algorithm to compute an expected response. The network then compares the AUTHR received from the mobile station against the computed expected response to determine the authenticity of the subscriber.

There are two procedures available to the network for authentication purposes: global challenge and unique challenge. With global challenge, all mobile stations are challenged with the same RAND number, which is broadcast over a broadcast channel. In the unique challenge procedure, a specific RAND number is used for each mobile station that is requesting access.

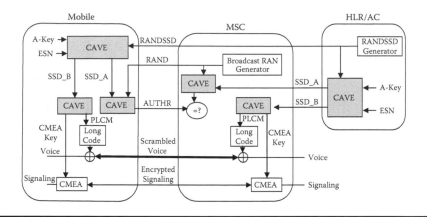

Figure 17.7 **Authentication and privacy in 2G CDMA networks.**

For additional security, the network may also use a call history COUNT register. The COUNT register is a 6-bit counter that counts the times the mobile station has made calls and is maintained in both the mobile station and the network. To be authenticated, the value of the COUNT register in the mobile station must match that of the COUNT register for that mobile station maintained in the network.

As described, privacy and authentication mechanisms used in CDMA networks rely on secondary key information SSD which is derived from the primary key A-key using the CAVE algorithm. To keep SSD information in the mobile station and the network in sync and to provide further security by changing the common secondary key information SSD, the network may initiate an SSD update procedure to change the SSD value on the mobile station. The network sends a RANDSSD number to the mobile station to start the SSD update procedure. The mobile station uses the received RANDSSD number, its equipment serial number (ESN) and the primary key A-Key as inputs to the CAVE algorithm to generate a new SSD value. To prevent illegal SSD updates by a rogue BTS (base transceiver station), the mobile station authenticates the network before updating SSD with the new value. The mobile station sends a RANDBS number to the network as a challenge. The network uses this RANDBS to calculate an AUTHBS response to the challenge and sends it back to the mobile station. The mobile station verifies the received response to authenticate the network. It updates the SSD with the new SSD value only if it has successfully authenticated the network (Figure 17.7).

17.7 Privacy and Authentication in Third-Generation Networks

Third-generation (3G) mobile networks are based on specifications developed by either the Third Generation Partnership Project (3GPP) or the Third Generation Partnership Project 2 (3GPP2) standardization bodies.

17.7.1 3GPP-Based 3G Networks

The 3GPP standards define the operation of 3G networks using the WCDMA (wideband CDMA) technology for radio access networks (RANs). For core networks, GSM-based standards specify circuit-switching (CS) domain operations while standards based on general packet radio service (GPRS) specify packet-switching (PS) domain operations. In newer releases of the specifications, the IMS, which is totally IP based, is specified for the core network operations.

Privacy and authentication mechanisms in 3GPP-based networks are largely based on security mechanisms developed in GSM standards but with many enhancements to address the identified shortcomings, noticeably the lack of mutual authentication, of GSM networks.

17.7.2 3GPP2-Based 3G Networks

The 3GPP2 took an evolutionary approach toward 3G networks. The core network and the RAN, respectively, evolved from the IS-41 and the IS-95 specifications of the 2G CDMA networks. The 3GPP2 defined the specifications for IS-2000, which specify the operations of cdma2000 1xRTT (radio transmission technology) and 3xRTT networks in revisions 0, A, and B. The 1xRTT specifications then further evolved into two 3G specifications: the cdma2000 1xEVDO (evolution data only) standards specified by IS-856/3GPP2 C.S0024, which define the enhancement for data applications only, while the cdma2000 1xEVDV (evolution data and voice) standards (defined in IS-2000/3GPP2 C.S001–0005 Revision C) specify enhancement for both data and voice applications.

The privacy and authentication mechanisms used in these 3GPP2 networks are largely based on an enhanced version of the ones used in the IS-41 standards, but newer releases of the specifications are adopting the mechanisms used in 3GPP-based 3G networks.

17.7.3 Privacy

17.7.3.1 3GPP Networks

In 3GPP networks, the privacy of information exchanged between a mobile subscriber and the network over the open air interface is also protected using a symmetric encryption process.

User voice traffic and certain signaling messages in dedicated wireless channels are protected with symmetric encryption using the Universal Mobile Telecommunications System (UMTS) encryption algorithm (UEA) with the session encryption key CK established during the authentication and key agreement (AKA) procedure described in Section 17.7.4.1. The algorithm also makes use of other information, such as the sequence number COUNT-C, the 1-bit direction information DIRECTION, the length of the key stream LENGTH, and the radio bearer identification BEARER as inputs to the f8 ciphering algorithm to further protect the privacy of exchanged information [10].

For user data traffic, the encryption procedures are set up by an IP-based security protocol. As of Release 7 of the 3GPP specifications, the IPSec (IP Security) protocol is used as the main protocol for privacy protection of data traffic in 3G networks. After a successful AKA procedure, the identity module of the mobile subscriber uses the agreed-on session key for encryption (CK) and the session key for integrity verification (IK) to set up IPSec tunnels with the network. Privacy (as well as integrity) of data exchanged between the mobile subscriber and the network are then protected by these IPSec tunnels using the session keys (CK and IK) and cryptographic algorithms specified by the IPSec standards [11].

17.7.3.2 3GPP2 Networks

In cdma2000 1 × RTT and 3 × RTT networks (i.e., up to Release B of IS-2000 specifications), the voice and signaling privacy protection process is similar to the one used in 2G CDMA networks. Symmetric encryption based on the CAVE algorithm is used to encrypt signaling traffic. Voice traffic is scrambled using the PLCM also derived from the CAVE algorithm. The encryption algorithm for signaling traffic encryption is enhanced with the new ECMEA (enhanced CMEA) algorithm. The enhanced subscriber privacy (ESP) requirements are also specified to protect the keys used for privacy protection [12]. Data traffic is protected by a symmetric encryption process that uses the ORYX ciphering algorithm.

In cdma2000 1 × EVDO networks, the Diffie–Hellman algorithm is used for session encryption key exchange instead of the CAVE algorithm. The encryption algorithm is also replaced with one that is based on the more secure AES encryption standards.

17.7.4 Authentication

17.7.4.1 3GPP-Based Networks

The authentication method used in 3GPP-based 3G networks is mostly an enhanced version of the one in GSM networks. This enhancement addresses one of the biggest security deficiencies in 2G networks: The network is not authenticated by the mobile subscriber during the authentication process. With this enhanced authentication method, mutual authentication between a mobile subscriber and the network is achieved. 3GPP has specified this enhanced authentication method as part of the AKA protocol to protect communications in 3G networks [10].

In this method, the home subscriber server (HSS) in a 3G network generates authentication vectors (AVs), which are an enhanced version of the authentication triplets used in GSM networks. The AV (a quintuplet) contains a random challenge RAND, a network authentication token AUTN, the expected response XRES, a session key for integrity check IK, and a session key for encryption CK. To calculate the AUTN, the HSS uses a cryptographic algorithm with a sequence number SQN that is kept in synch between the HSS and the identity module of the mobile subscriber and the long-term secret that it shares with the identity module as inputs. As in GSM networks, the network uses one AV for each authentication procedure.

The authentication process starts when the network sends the random challenge RAND and the network authentication token AUTN to the identity module of the mobile user. The identity module authenticates the network by verifying the received AUTN with the one calculated from the shared long-term secret and the sequence number SQN stored in the identity module. If the calculated AUTN and the received AUTN match, the network is authenticated. In this case, the identity module produces an authentication response RES using the shared long-term secret and a cryptographic algorithm and sends it back to the network.

The network verifies the received response to authenticate the mobile subscriber. If the received authentication response RES matches with the expected response XRES in the AV, the mobile subscriber is authenticated. The session key for integrity check IK and the session key for encryption CK are then used to establish IPSec tunnels between the mobile user equipment and the network for privacy and integrity protection of communications between the mobile user and the network [11].

17.7.4.2 3GPP2-Based 3G Networks

The AKA procedures for cdma 1xRTT and cdma 3xRTT (i.e., in 3GPP2 specifications Revision B and earlier) are based on the CAVE algorithm used in 2G CMDA networks as described.

For enhanced authentication in 3G networks, 3GPP2 defines the requirements for enhanced subscriber authentication (ESA) [12]. Since the AKA procedures defined by 3GPP specifications met most of these requirements, the 3GPP2 Revision C (cdma2000 1 × EVDV) specifications adopt the 3GPP AKA described as the basis for 3GPP2 AKA.

17.8 Challenges and Open Research Issues

Wireless networks have difficult security challenges due to two aspects: the openness of radio channels and the mobility of users. There is much security infrastructure in place for authentication and privacy based on well-known techniques in symmetric and asymmetric cryptography. Obviously, these security mechanisms are working successfully for the most part today.

One of the major difficulties in all wireless systems is key management. Keys should not be static because keys can be discovered by attackers given enough time.

Thus, there is a need to exchange and manage secret keys. While it can be done, exchanging keys over unsecure networks involves risks and complexity.

Some security techniques depend on public keys and a public key infrastructure (PKI) for certificates. It could be said that the PKI has been deployed more slowly than some expectations, and its success is an open question.

17.9 Summary

This chapter covered cryptographic techniques to enable privacy and authentication in the prevalent wireless networks, namely, IEEE 802.11 wireless LANs, 802.16 wireless metropolitan-area networks (MANs), 2G GSM and CDMA cellular networks, and 3G networks. Privacy is typically protected by symmetric encryption. Authentication is a more difficult problem and involves more elaborate procedures in wireless systems.

References

1. Garrett, P., *Making, Breaking Codes: An Introduction to Cryptology.* Upper Saddle River, NJ: Prentice Hall, 2001.
2. NBS (U.S. National Bureau of Standards), *Data Encryption Standard.* FIPS Publ. 46, January 1977.
3. NIST (National Institute of Standards and Technology), *Specification for the Advanced Encryption Standard AES.* FIPS Publ. 197, 2001.
4. Diffie, W., and Hellman, M., New directions in cryptography. *IEEE Trans. Inform. Theory*, IT-22, 644–654, November 1976.
5. Rivest, R., Shamir, A., and Adleman, L., A method for obtaining digital signatures and public-key cryptosystems. *Commun. ACM*, 21(2), 120–126, February 1978.
6. *Part 16: Air Interface for Fixed and Mobile Broadband Wireless Access Systems Amendment for Physical and Medium Access Control Layers for Combined Fixed and Mobile Operation in Licensed Bands*, IEEE Standard 802.16E-2005. December 2005.
7. Johnston, D., and Walker, J., Overview of IEEE 802.16 security. *IEEE Security Privacy*, 2(3), May 2004.
8. Mouly, M., and Pautet, M.-B., The GSM system for mobile communications. *Cell & Sys*, Palaiseau, France, 1992.
9. *Cellular Radiotelecommunications Intersystem Operations*, TIA IS-41D. December 1997.
10. *Third Generation Partnership Project; Technical Specification Group Services and System Aspects; 3G Security; Security Architecture*, Version 6.10.0, 3GPP TS 33.102. October 2006.
11. *Third Generation Partnership Project; Technical Specification Group Services and System Aspects; Access Security for IP-Based Services*, Version 6.6.0, 3GPP TS 33.203. October 2006.
12. *Enhanced Subscriber Authentication (ESA) and Enhanced Subscriber Privacy (ESP)*, Version 1, 3GPP2 S.R0032. December 2000.

Links

1. IEEE 802 standards. http://standards.ieee.org/getieee802/
2. WiMax Forum. http://wimaxforum.org/
3. 3GPP. http://www.3gpp.org/
4. 3GPP2. http://www.3gpp2.org/
5. GSM Association. http://www.gsmworld.com/
6. CDMA Development Group. http://www.cdg.org/

Chapter 18

Network Management Tools

Teresa Piliouras

Contents

18.1 Introduction

Network design and planning are complementary to network management because a well-designed network will be easier to manage. A network that is configured to collect network management data also provides a foundation for informed planning and new design decisions.

According to the International Organization for Standardization (ISO) Network Management Forum, network management should encompass these functions: configuration management, fault management, accounting management, performance management, and security management. Figures 18.1 through 18.6 provide an overview of the network management tasks that automated tools support as well as differences in the level of functionality between simple and advanced products.

• **Tasks:**	• **Implementation:**
– Gather information on current configuration – Use data to change configuration – Store data, maintain inventory, and produce reports including vendor contact information, lease line circuit numbers, quantity of spares, etc.	– Manual versus automated tools – Simple tool: Central storage of all network data – Advanced tool: Automatic gathering and storage of network data, automatic change of running configuration, etc.

Figure 18.1   Configuration management.

• **Tasks:**	• **Implementation:**
– Identify fault occurrence – Isolate cause – Correct, if possible	– Information gathering (event driven versus polling) – Simple tool: Identify problem(s), but not cause(s) – Advanced tool: Interpret network events, produce reports, and assist in fault correction and fault

Figure 18.2   Fault management.

• **Tasks:**	• **Implementation:**
– Gather data on network resource utilization – Set usage quotas – Bill users (may be based on a number of different approaches)	– Define appropriate metrics for system usage – Simple tool: Monitor metrics exceeding quotas – Advanced tool: Automate billing, assist in forecasting resource requirements

Figure 18.3   Accounting management.

• **Tasks:**	• **Implementation:**
– Gather data on network resource utilization – Analyze data in both real time and off-line modes – Perform simulation studies	– Define appropriate metrics for system performance – Simple tool: Graphically display network devices and links – Advanced tool: Set thresholds and error rates and then perform prescribe corrective actions. Collect historical data, which can in turn be used for simulation and other analysis

Figure 18.4   Performance management.

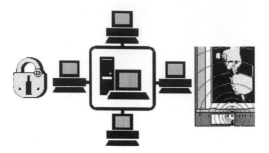

| • **Tasks:**
  – Identify sensitive information & its location
  – Secure and maintain access points to information
• **Note:**
  – This is not the same as operating system security or physical security! | • **Implementation:**
  – Locate access points (ftp, remote login, e-mail, etc.)
  – Secure access (encryption, packet filtering, etc.)
  – Identify potential or actual security breaches |

Figure 18.5 Security management.

18.2 Network Management Protocols

To perform network management tasks, data must be collected from every device in the network. This includes such information as the device name, device version, number of device interfaces, device parameters, and device status. To process and consolidate this information across multiple devices and platforms, network management protocols are needed. Network management protocols provide a uniform way of accessing standard metrics for network devices made by any manufacturer. The major network management protocols include Common Management Information Service Element/Common Management Information Protocol (CMISE/CMIP) and SNMP (Simple Network Management Protocol) and Remote

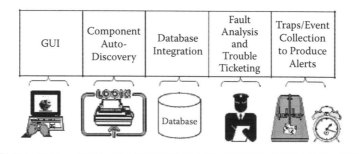

GUI	Component Auto-Discovery	Database Integration	Fault Analysis and Trouble Ticketing	Traps/Event Collection to Produce Alerts

Figure 18.6 Network management tool functionality.

Networking Monitoring (RMON). These open standards have largely supplanted the proprietary, single-vendor network management protocols used in legacy mainframe environments.

CMISE/CMIP[1] provides a common network management protocol for all network devices conforming to the ISO reference model. It is designed to be a total network management solution. However, it requires a high degree of overhead and is difficult to implement. Therefore, its use is largely limited to telephone companies and service providers with very demanding, complex, and large-scale network management tasks and the highly skilled staff required to perform these tasks.

SNMP and RMON are complementary and are very widely implemented by device manufacturers. For this reason, this discussion focuses on SNMP and RMON because these are the protocols of choice in most organizations, large or small. The major SNMP standards are documented in a series of Requests for Comments (RFCs) put out by the IETF (Internet Engineering Task Force).[2]

At the core of SNMP is the definition of the management information base (MIB). The MIB provides a precise specification of all the network management information that will be collected over the network for each network device (also known as a network element). Within the MIB, SNMP represents network resources as objects. Each object relates to a specific variable associated with a particular aspect of a managed object (e.g., a printer object might consist of an interface object, status object, etc.), represented by either a scalar (i.e., a single value) or a table (containing multiple entries, as shown in the example given in Figure 18.7). The MIB defines how each variable is named, coded, and interpreted. The collection of all managed objects in the network is referred to collectively as the MIB.[3]

The MIB conforms to the structure of management information (SMI) for Internets based on the Transmission Control Protocol/Internet Protocol (TCP/IP). This SMI in turn is modeled after the SMI of the OSI (Open Systems Interconnection). While the SMI is equivalent for both SNMP and OSI environments, the actual objects defined in the MIB are different. SMI conformance is important because it

Sample Routing Table

Destination	Next
2	3
3	4
3	2

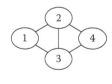

Note: Each table entry must be assigned an OID (Object ID) for retrieval purposes.

Figure 18.7   Sample routing table.

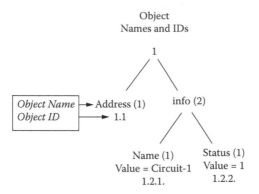

Figure 18.8 Example of object naming and OID.

means that the MIB is capable of functioning in both current and future SNMP environments. In fact, the Internet SMI and the MIB are completely independent of any specific network management protocol, including SNMP.

MIBs are tree-like, hierarchical, structured data schemas defined according to the ISO Abstract Syntax Notation One. This syntax defines each MIB component as a node labeled by an object identifier and a short text description. Figure 18.8 illustrates a generic object and its associated identification (OID) and name. Figure 18.9

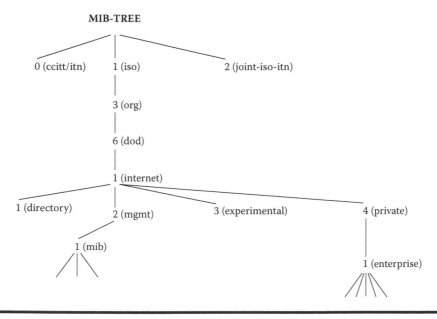

Figure 18.9 Overview of IETF management information base (MIB). (Source: IETF.)

presents an overview of the entire IETF MIB structure, with the Internet in the middle (and of which SNMP is only a small part).

The SNMP MIB repository contains four types of attributes:

1. Management attributes
2. Private attributes
3. Experimental attributes
4. Directory attributes

MIB-I contains a limited list of objects dealing with IP internet working routing variables. MIB-II extends this to provide support for a variety of media types, network devices, and SNMP statistics and is not limited to TCP/IP.

The management attributes are the same for all agents supporting SNMP. Thus, SNMP managers may work with agents from various manufacturers. To offer more functionality, vendors may populate the private attributes with proprietary, value-added data.

Most real MIB implementations employ extensions, which provide additional information storage for various network management and operational tasks. Such extensions might include [1]:

- Enhanced security information relating to access of the various managed objects, the network management system, and the MIB (so it can be manipulated).
- Extended data for user and user group administration. Users and user groups can themselves be managed MIB objects that must be created, maintained, and eventually deleted.
- Configuration histories and profiles to provide data for reporting, backups, and alternate routing (as part of local-area network [LAN] fault and performance management functions).
- Trouble-tracking data to help expedite resolution of network problems. The augmented MIB is used to translate event reports into trouble tickets, assign work codes to staff, recognize and categorize, relate problems, escalate problems by severity, and close trouble tickets after eliminating the problems and associated causes.
- Extended set of performance indicators to support advanced performance management (relating to resource utilization, threshold violations, trends, bandwidth utilization between interconnected LANs, etc.).

RMON is a very well-known and popular example of an SNMP MIB extension. Using a network management station, the utilization of the MIB and its extensions should be carefully monitored. This involves regular inspection of SNMP command frequencies, the number and types of traps being generated, information retrieval frequencies initiated by GetNextRequest (the significance of this SNMP command is described in Section 18.2.3), and the proportion of positive to negative poll responses.

In addition to the MIB, SNMP defines the following components:

■ Management agents
■ Management station
■ Network management protocol

18.2.1 Management Agents

Each agent possesses its own MIB view, which includes the Internet standard MIB and, typically, other extensions. However, the agent's MIB does not have to implement every group of defined variables in the formal IETF MIB specification. The management agent contains only the portion of the SNMP MIB monitoring and control data that is relevant to a particular device as implemented by the device manufacturer. This means, for example, that gateways need not support objects applicable only to hosts and vice versa. This eliminates unnecessary overhead, facilitating SNMP implementation in smaller LAN components that have little excess capacity.

Agents respond to action and information requests or provide unsolicited information to the network management station. An agent performs two basic functions:

1. *MIB variable inspection.* The agent is responsible for examining the values of counters, thresholds, states, and other parameters.
2. *MIB variable alteration.* The agent is also responsible for resetting counters, thresholds, and so on. For example, it is possible to reboot a node by setting a variable.

An agent MIB implementation can be hosted on several types of platforms, including [1]:

■ Object-oriented databases
■ Relational databases
■ Flat-file databases
■ Proprietary format databases
■ Firmware

Thus, MIB information is distributed within agents. A typical configuration might include, at the agent level, a disk-based relational database or a combination of PROM (programmable read only memory) with static object attributes and RAM (random-access memory) with dynamically changing information.

18.2.2 The Management Station

The network management station consists of:

■ Network interface and a monitor for viewing network information
■ Network management applications
■ Database of managed network entities

Managers execute network manager station (NMS) applications and often provide a graphical user interface that depicts a network map of agents. Typically, the manager also archives MIB data for trend analysis.

This is implemented in two different ways:

1. Each agent's MIB entries are copied into a dedicated MIB segment.
2. MIB entries are copied into a common area for immediate correlation and analysis.

The manager provides presentation and database services, which must be implemented carefully to ensure the appropriate features, functions, and performance are present to support network management. Some issues involved in this implementation are summarized as follows [1]:

- Use of object-oriented databases
  - Advantages:
    - They are naturally well suited to network management because the MIB itself is formally described as a set of abstract data types in an object-oriented hierarchy.
    - They have the ability to model interface behavior via stored methods.
  - Disadvantages:
    - This technique is not yet mature for product implementations.
    - There are no standards yet for query and manipulation languages.
    - The MIB object class hierarchy is broad and shallow in the inheritance tree.
    - Performance characteristics are not yet well documented.
- Use of relational databases
  - Advantages:
    - The technology is mature, stable, and well supported.
    - There is a standard access language (structured query language).
    - There are translators for translating ER (entity-relationship) models into relational schema.
    - There are many application choices and vendors.
  - Disadvantages:
    - They are not well suited to storing OO (object-oriented) models.
    - Performance is highly dependent on database tuning and the application.
    - Use of other databases

Flat-file databases and other proprietary formats can be tailored very specifically to MIBs and can be optimized for performance. However, this may lead to network management design and implementation that is more complex and time-consuming to develop, maintain, and change.

18.2.3 Network Management Protocol

The network management protocol defines the messaging format and conventions by which management stations and agents communicate. Objects (in the form of read-only OIDs or read–write OIDs, also known as Traps) are exchanged between an SNMP manager and agent through messages. SNMP uses the OSI level 4 transport layer User Datagram Protocol (UDP) for message transmission, as shown in Figure 18.10. The network management protocol defines the SNMP message types, which are sent as single packets, as illustrated in Figure 18.11 and summarized as:

- *GetResponse.* This is a poll initiated by a management station to request information from a management agent.
- *GetNextRequest.* This is a poll initiated by a management station to request the next item in the management agent's MIB table. This message must be sent repeatedly to obtain all the information contained in the agent's MIB because items can be retrieved only one at a time.
- *GetNextResponse.* This is the agent's message reply to the management station's GetRequest.
- *SetRequest.* This is a command sent by a management station to an agent to set an item parameter. This command is used to perform device configuration management actions.

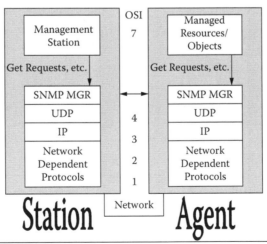

Figure 18.10 SNMP architecture.

SNMP Message:

SNMP Version #	Community Name	SNMP PDU

> • **Community Name:**
> – This is sent in plaintext and contains an unsecured password or IP address (which could be based on a domain name or some other name assigned to a group, department, etc.).
> • **Five Valid SNMP Message Types, or Protocal Data Units (PDUs):**
> – GetRequest
> – GetResponse
> – GetNextRequest
> – GetNextResponse
> – SetRequest
> – Trap
> • **IETF RFC 1157 Defines PDU Conventions**

Figure 18.11 SNMP message format and types.

■ *Trap.* This is an unsolicited message sent by an SNMP management agent. It is event driven because the agent does not need to be asked or polled by the management station to send the trap. Seven traps are defined in MIB-2. The first six traps are defined by SNMP, and the last is a vendor-specific message type:

1. System cold start
2. System warm start
3. Link down
4. Link up
5. Authentication failure
6. Exterior Gateway Protocol (EGP) neighbor loss
7. Enterprise specific (potentially unlimited in number)

The SNMP does not support the concept of management-station-to-management-station communication. It supports only the manager and agent interaction. However, it does not set a limit on the number of management stations or agents that can coexist on the network.

SNMPv1, as its name implies, is a simple and easy protocol to implement. For this reason, it is widely supported by device manufacturers. However, a number of drawbacks with SNMPv1 led to the creation of a new protocol, SNMPv2. The SNMPv1 drawbacks included:

■ It is officially standardized only for TCP/IP networks.
■ It is inefficient for large table retrievals due to the implementation of GetNextRequest.

■ The password security provided by the cleartext community string in the SNMP message can be easily broken.

■ This version supports only a limited number of MIB entries and message types, as described.

■ It cannot process bundled message requests, so each request must be sent separately, leading to increased network overhead and traffic.

■ EGP MIB-2 cannot detect routing loops. The EGP is designed to detect failure of a device to acquire or reach its exterior gateway neighbors.

SNMPv2 was designed to address the deficiencies in SNMPv1, particularly with respect to security and functionality. Because it is a separate protocol, distinct from SNMPv1, SNMPv1 and SNMPv2 network management stations and agents are incompatible and cannot interact directly on the same network. If this is needed, proxy agents are used to translate between SNMPv1 and SNMPv2, as shown in Figure 18.12.

SNMP proxy agent software permits an SNMP manager to monitor and control network elements that are otherwise not addressable using SNMP. For example, a network manager might wish to migrate to SNMP-based network management, but there are devices on the network that use a proprietary network management scheme. An SNMP proxy manages both device types in their native mode by acting as a protocol converter to translate the SNMP manager's commands into the

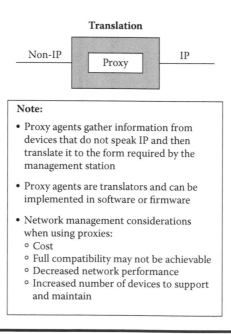

Figure 18.12　Proxy agent translation.

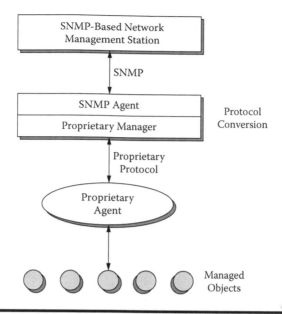

Figure 18.13   Proxy agent integration with SNMP.

proprietary scheme. This strategy facilitates migration from a proprietary environment to an open one. This is illustrated in Figure 18.13.

The SNMPv2 SMI provides extensions to allow the capture of new data types and more object-related documentation. SNMPv2 also supports two new message formats or protocol data units (PDUs):

1. *GetBulkRequest.* This is a poll initiated by a management station to request and retrieve blocks of data from tables in the management agent's MIB table.
2. *InformRequest.* This message enables a manager to send trap information to another manager. SNMPv2 also defines a manager-to-manager MIB to support distributed management architecture.

SNMPv2 also added the following security capabilities:

■ MD5 processing capabilities were added to authenticate and time-stamp a message profile and its origin. MD5 is a message digest authentication algorithm developed by RSA Incorporated. MD5 computes a digest of the message that is used by the receiver to verify the contents of the message. The message can also be encrypted using the Data Encryption Standard (DES) as part of another processing step. A message recipient must then perform these actions: (1) Decrypt the message, (2) check the time stamp to verify the message is recent and is the latest message sent from the originator, (3) process the profile, and (4) verify the authentication. Using an MD5 message digest imposes at least a 10% increase in message processing time; using DES doubles the required message processing time. Thus, the added security of SNMPv2 is not without substantial cost and processing overhead.

■ A mechanism for time-stamping loosely synchronized workstations was provided.

As with SNMPv1, the SNMPv2 security provisions lacked the strength needed for mission-critical networks. They both failed to adequately address the security of MIB data transmission and access control. Access is based on a *shared* password (i.e., the community string) that could be altered or intercepted during transmission. This could lead to an unauthorized user gaining read-only or read–write access to network devices.

Thus, these protocols provide only limited control over message privacy and content and authorization and control of device access, remote configuration, and administration. SNMPv3 was designed to address these serious security limitations by providing mechanisms for authentication, privacy, authorization, view-based access control (to limit what different user groups could see and do), and standards-based remote configuration. SNMPv3 does not attempt to deal with denial-of-service and traffic analysis attacks [2]. While SNMPv1 and SNMPv3 are widely used within the industry, the market for SNMPv2 has not emerged because it lacks the simplicity of SNMPv1 and the security of SNMPv3.

RMON refers to the IETF standards defined in RFCs 1757, 1513, and 2021. It defines an interface between an RMON agent and an RMON management application. Generally, RMON agents are used to monitor and troubleshoot LANs. Although RMON is not designed for wide-area networks (WANs), some vendors offer WAN RMON products for analyzing T1/E1 and T2/E2 lines to a central site. RMON extends the SNMP MIB to define additional MIBs for network management. RMON has two versions, RMON 1 and RMON2, as shown in Figure 18.14. RMON 1 specifies several additional MIBs, outlined as [3]:

1. *Statistics*. Statistics are used to collect real-time, current network statistics (percentage broadcast traffic, percentage multicast traffic, percentage utilization, percentage of errors, percentage of collisions, etc.).
2. *History*. This refers to the collection of statistics over time.
3. *Alarm*. This MIB defines predetermined thresholds that may trigger an event when an alarm condition occurs.
4. *Host*. This MIB tracks individual host statistics, such as packets in/out, multicast/broadcast, bytes in/out, errors, and so on.
5. *hostTopN*. This MIB defines data collection on the N (statistically) most active hosts, tracking information such as packets in/out, multicast/broadcast, bytes in/out, errors, and so on. Typically, N is set to 10.

Name:	ISO.org.dod.internet.mgmt.mib-2.rmon...			
OID:	1 .3 .6 .1	.2	.1	.16 .[..]

RMON-1 (RCF-1757 & 1513)	RMON-2 (RCF-2021)
.1 Statistics	.11 ProtocolDir
.2 History	.12 ProtocolDist
.3 Alarm	.13 AddressMap
.4 Hosts	.14 nlHosts
.5 HostTopN	.15 nlMatrix
.6 Matrix	.16 alHost
.7 Filter	.17 alMatrix
.8 Capture	.18 usrHistory
.9 Events	.19 probeConfig
.10 TokenRing (RFC-1513)	

Figure 18.14   RMON nomenclature overview.

6. *Matrix.* This MIB defines host-to-host conversation statistics, such as packet, byte, and error counts.

7. *Filters.* This MIB defines data relating to packet structure and content matching. It defines bit pattern matching at the data link, IP header, and UDP header level.

8. *Capture.* This MIB defines the collection of data that is uploaded to a management station for subsequent detailed packet analysis.

9. *Event.* This MIB defines data specifying reactions/actions to predetermined conditions or thresholds.

10. *Tokenring.* This MIB defines token-ring-related RMON extensions and includes token ring station for status and statistics data, for order data, and for configuration data and token ring source routing (for utilization data).

The first three MIBs—Host, HostTopN, and Matrix—are directly involved with monitoring traffic flow and net flow switching. They provide the most granularity in traffic flow information. The filter, packet capture, and event MIBs are associated with the most processing and are intended for in-depth data analysis.

RMON 2, as the name implies, is the second version of RMON. It defines even richer data sets for network analysis by providing these additional MIBs:

11. *ProtocolDir.* This defines a master list of supported protocols for each probe. It is used by the network management system to determine which protocols (IP, Novell, DECnet, Appletalk, Custom Protocol, etc.) the probe is able to decode.

12. *ProtocolDist.* This MIB defines segment protocol statistics, providing aggregate statistics on the traffic generated by each protocol on a per-segment basis.

13. *AddressMap.* This MIB is used to map network and MAC addresses to physical ports.

14. *nlHost.* This is used to define host in/out statistics based on a network-layer address.
15. *nlMatrix.* This is used to define host-to-host network layer statistics capturing traffic data between host pairs at the network layer.
16. *alHost.* This is used to define host in/out statistics based on application-layer addresses.
17. *alMatrix.* This defines traffic statistics between host-to-host pairs based on an application-layer address.
18. *usrHistory.* This is used for the same basic purpose as the RMON1 history group. It is used for data logging and collecting historical statistics that are user defined.
19. *probeConfig.* This is used to define standardized probe configuration parameters, including capabilities (supported RMON groups), software revision level, hardware revision level, current date and time settings, reset control (i.e., running, warm boot, cold boot), download file name, IP address of Trivial File Transfer Protocol (TFTP) download server, and download actions (e.g., to PROM or to RAM).

To summarize, RMON MIBs must be implemented by a device manufacturer so data can be collected by RMON (hardware/software) monitoring devices known as *probes.* A RMON software agent gathers the information for presentation on a network-monitoring console with a graphical user interface, as shown in Figure 18.15.

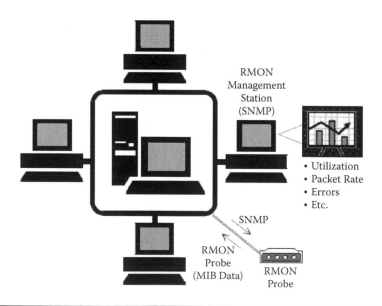

Figure 18.15 Network monitoring using RMON.

18.3 Summary

This chapter discussed the need for automated tools to support complex network design, planning, and network management tasks. It also reviewed the major network management protocols, with particular emphasis on SNMP and RMON because these protocols are widely used and implemented in vendor products.

Notes

1. CMISE is part of a family of protocols that define the exchange of management information between entities (manager/agent). The manager/agent model was copied in SNMP but is implemented with less functionality. CMISE defines a user interface and corresponding services to be provided for each network component through the Common Management Information Service (CMIS) protocol. CMIP specifies the PDU format and implementation of CMIS services through the CMIP.

2. The IETF, a large, open, international body composed of researchers, vendors, network designers, and others, is concerned with the development, evolution, and "smooth" operation of the Internet. The IETF maintains an RFC Editor that publishes and archives all RFCs. For the most recent RFC information on IETF protocols and standards, refer to http://www.rfc-editor.org/overview.html.

3. The actual IETF MIB designation depends on the protocol, the protocol version, and the context. For example, the original SNMP MIB (also known as MIB-I), defined in 1988 by RFC 1066, included variables needed to manage the TCP/IP stack. It was updated by RFC 1213. The updated version, designated MIB-II, adds new variables. The SNMPv2 MIB is defined by RFC 3418 etc.

References

1. Perkins, D., *A Consolidated Overview of Version 3 of the Simple Network Management Protocol [SNMPv3], Network Working Group — Internet.* February 15, 2004. ftp://ftp.rfc-editor.org/in-notes/internet-drafts/draftperkins-snmpv3-overview-00.txt

2. Terplan, K., *Effective Management of Local Area Networks,* 2nd ed. New York: McGraw-Hill, 1996.

3. NetScout. http://www.netscout.com/products/

Chapter 19

Blueprints and Guidelines for Telecommunication Service Providers

Kornel Terplan and Paul Hoffmann

Contents

19.1 Introduction

The time for proprietary solutions is over. Telecommunication service providers (TSPs) can no longer afford to maintain numerous and different support, documentation, and management systems. Standards bodies and industry associations can provide help in streamlining processes and organizational structures. Three applicable solutions—the enhanced telecommunications operations map (eTOM), the telecommunications management network (TMN), and the control objectives for information and related technology (CobiT)—are addressed in this section.

19.2 The Enhanced Telecommunications Operations Map

The eTOM business process framework serves as the starting point for development and integration of business and operations support systems (OSSs), and it helps drive the TeleManagement Forum (TMF) members' work in developing next-generation OSS (NGOSS) solutions. In the case of service providers (SPs), it provides a neutral reference point as they consider internal process reengineering needs, partnerships, alliances, and general working agreements with other providers. For suppliers, the eTOM framework outlines potential boundaries of software components and the required functions, inputs, and outputs that must be supported by products.

The eTOM framework includes the enterprise processes required by SPs. It is not an SP business model, however; it does not address strategic issues or questions of who an SP's target customers should be, what market segments the provider should serve, and so forth. A business process framework represents one part of the strategic business model and plan of an SP.

Figure 19.1 shows the highest conceptual view of the eTOM business process framework (level 0 processes). This view provides an overall context-differentiating strategy and life-cycle processes from operations processes in two large groupings, shown as two separate boxes. It also differentiates key functional areas in five horizontal layers and illustrates the interacting internal and external entities within a given enterprise.

Figure 19.2 shows the level 0 view of level 1 processes. This detail is needed to position and analyze business processes. Seven vertical process groups—representing the end-to-end processes required to support customers and manage a business—are presented. The eTOM focuses on the core customer operation processes of fulfillment, assurance, and billing (FAB). Operations support and readiness is differentiated from FAB real-time processes to increase the focus on enabling support and automation in FAB (i.e., online, immediate support of customers). The strategy-and-commit vertical and the two life-cycle management verticals are also differentiated because, unlike operations processes, they do not directly support the customer, and they adhere to different business time cycles.

The horizontal process groupings shown in Figure 19.2 distinguish functional operations processes and other types of business functional processes (e.g., marketing vs. selling and service development vs. service configuration). The functional processes shown at left (within the strategy-and-commit, infrastructure life-cycle management, and product life-cycle management vertical process groupings) enable, support, and direct work in the operations verticals.

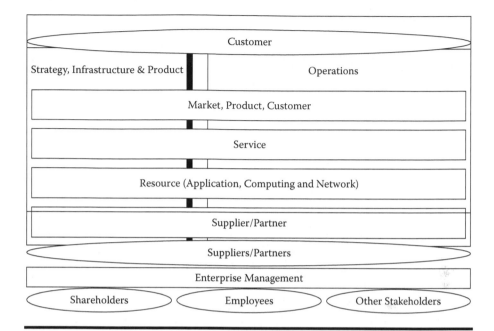

Figure 19.1 eTOM business process framework: Level 0 processes.

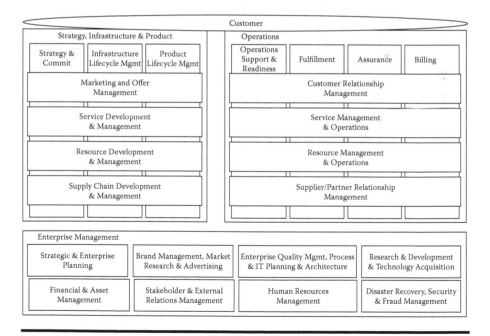

Figure 19.2 eTOM business process framework: Level 1 processes (Level 0 view).

In summary, eTOM involves a number of improvements relative to the earlier TOM:

■ It expands the scope of operations to all enterprise processes.
■ It distinctly identifies marketing processes due to their heightened importance in today's e-business world.
■ It identifies enterprise management processes so that everyone involved is aware of his or her critical tasks, thereby enabling process framework acceptance across the enterprise.
■ It moves FAB to the high-level framework to emphasize that customer priority processes are the focus of the enterprise.
■ It defines an operations support and readiness vertical process grouping applicable to all functional layers other than enterprise management. To integrate e-business and make customer self-management a reality, there must be an understanding within the enterprise regarding the processes needed to allow increasing amounts of direct, online customer operations support and customer self-management.
■ It recognizes three enterprise process groupings that are distinctly different from operations processes: strategy and commit, infrastructure life-cycle management, and product life-cycle management.
■ It recognizes the different cycle times of strategy and life-cycle management processes and the need to separate these processes from those involving customer priority operations, in which automation is most critical. It does so by decoupling the strategy-and-commit process and the two life-cycle management processes from the day-to-day, minute-to-minute cycle times of customer operations processes.
■ It moves from a customer care or service orientation to a customer relationship management (CRM) orientation that emphasizes customer self-management and control, increasing the value customers contribute to the enterprise and the use of information to customize and personalize individual customer needs. Elements are added to this customer operations functional layer to represent better selling processes and to integrate marketing fulfillment within CRM.
■ It acknowledges the need to manage resources across technologies (i.e., application, computing, and network) by integrating the network and systems management functional process into the resource management and operations area. Also, it situates IT management in this functional layer as opposed to an outbound process layer.

19.3 The Telecommunications Management Network

The TMN is a special network implemented to help manage the overall telecommunications network for a TSP. As such, it interfaces with one or more individual networks at several points to exchange information. It is logically separate from the networks it manages, and it may be physically separate as well. However, a TMN may use part of the telecommunications network for its own communications.

The TMN effort is chartered by the Telecommunications Standardization Sector of the International Telecommunication Union. Its development began in 1988 and has concentrated primarily on the network's overall architecture, using the Synchronous Digital Hierarchy (the international version of the North American Synchronous Optical Network [SONET]) technology as a target. However, TMN techniques are applicable to a broad range of technologies and services.

TMN is an extension of the Open Systems Interconnection (OSI) standardization process. It attempts to standardize some of the functionality, and many of the interfaces, of managed networks. When a TMN is fully implemented, the result will be a higher level of integration. TMNs are usually described as involving three different types of architectures:

1. The functional architecture describes the appropriate distribution of functionality within TMN, in the sense of allowing for the creation of function blocks from which a TMN of any complexity can be implemented. Requirements for TMN-recommended interface specifications are based on the definitions of function blocks and the reference points between them.
2. The information architecture, based on an object-oriented approach, provides the rationale for application of OSI systems management principles to TMN principles. The OSI principles are mapped to the TMN principles and, as necessary, expanded to fit the TMN environment.
3. The physical architecture describes interfaces that can actually be implemented, together with examples of the physical components that make up the TMN.

Management functions are grouped into the five areas identified as part of the OSI model. Examples are:

1. Fault management (alarm surveillance, testing, and problem administration)
2. Configuration management (provisioning and rating)
3. Performance management (monitoring of service quality and traffic control)
4. Security management (management of access and authentication)
5. Accounting management (rating and billing)

The management requirements that helped shape the TMN specifications address planning, provisioning, installing, maintaining, operating, and administering communication networks and services. TMN specifications use standard Common Management Information Protocol (CMIP) application services when appropriate. However, one of the key concepts of the TMN specifications is their introduction of "technology-independent" management, which is based on an abstract view of managed network elements. This abstract view and a single communication interface to allow diverse equipment to be managed. Thus, TMN-managed networks can consist of both TMN-conforming and nonconforming devices.

TMN specifications define an intended direction, but many smaller details involved in the process must be determined. Published TMN specifications address the overall architecture, the generic information model, management services and functions, management and transmission protocols, and an alarm surveillance function. Future areas of focus will be the service layer, traffic (i.e., congestion), and network-level management.

The relatively slow pace of TMN specification development has not prevented companies from recognizing the benefits of the TMN approach to management. The TMF is incorporating TMN into its specifications, and many companies are beginning to build, or specify, management systems and components that comply with TMN principles. Systems that comply with these principles reduce costs and improve services due to several reasons:

- Standard interfaces and objects make it possible to rapidly and economically deploy new services.
- Distributed management intelligence minimizes management reaction time to network events.
- Mediation makes it possible to handle similar devices in an identical manner, leading to more generic operation systems and vendor independence.
- Mediation allows management and transparent upgrading of existing device inventories.
- Distributed management functions increase scalability, isolate and contain network faults, and reduce network management traffic and the load on operations systems.

Many of the benefits that accrue from the TMN principles are due directly to the distributed architecture and its mediation function (MF). The TMN architecture addresses communication networks and services as collections of cooperating systems. By managing individual systems, TMN has a coordinated effect on the overall network.

This coordination can be illustrated through a simple example. Within an enterprise, one operations system may deal with network element inventory, another may deal with traffic planning, and several element managers may deal with network elements of various types. When a customer requires a circuit of a specific bandwidth and quality, all these systems must be coordinated to meet the customer's needs. The TMN architecture not only facilitates this effort but also allows for this function to be distributed among several systems. Such a distribution allows a TMN-based system to handle global networks by enabling workloads to be spread across multiple systems.

This ability to subdivide and distribute the total management effort requires clear definitions of functions, interfaces, and the information model. These topics, defined in the TMN specifications, are outlined subsequently.

The TMN architecture identifies specific functions and their interfaces. These functions are what allow a TMN to perform its management activities. The architecture also provides flexibility in terms of building a management system by allowing certain functions to be combined within a physical entity. The function blocks described subsequently, along with typical methods of their physical realization, are defined according to the TMN specifications (Terplan, 2001).

19.3.1 Operations Systems Function

The operations systems function (OSF), in the form of a TMN-compliant management system or set of management applications, monitors, coordinates, and controls TMN entities. The OSF makes it possible for general activities such as management of performance, faults, configuration, accounting, and security to be performed. In addition, specific capabilities in regard to planning of operations, administration, maintenance, and provisioning of communication networks and systems should be available. These capabilities are realized through an operations system that can be implemented in many different ways. An example would be a descending abstraction system (e.g., business, service, and network) in which the overall business needs of the enterprise are met by coordinating the underlying services. In turn, individual services are realized through coordination of network resources.

19.3.2 Workstation Function

The workstation function (WSF) provides the TMN information, such as access control, topological map displays, and graphical interfaces, to the user. These functions are realized through a workstation.

19.3.3 Mediation Function

The MF acts on information passing between an OSF and a network element function (NEF) or Q adapter function (QAF) to ensure that the data produced through the MF complies with the needs and capabilities of the receiver. MFs can store, adapt, filter, threshold, and condense information. In addition to providing the abstract view necessary to treat dissimilar elements in a similar manner, MFs may also help local management in regard to their associated NEFs (in other words, MFs may include element managers).

The MF is realized through a mediation device. Mediation can be implemented as a hierarchy of cascaded devices, using standard interfaces. This cascading of mediation devices and the various interconnections to network elements provide TMNs with a great deal of flexibility, as well as allowing for future design of new equipment to support a higher level of processing within the network element without the need to redesign an existing TMN.

19.3.4 Q Adapter Function

The QAF function is used to connect non-TMN-compliant NEFs to the TMN environment and is realized in a Q adapter, which allows legacy devices (i.e., those that do not support TMN management protocols, including Simple Network Management Protocol [SNMP] devices) to be accommodated within a TMN. A Q adapter typically performs interface conversion functions (i.e., it acts as a proxy).

19.3.5 Network Element Function

The NEF function is realized through the network elements themselves, which can present a TMN-compliant or noncompliant interface. Such elements include physical elements (switches), logical elements (virtual circuit connections), and services (operations systems software applications). Figure 19.3 illustrates the functions occurring within a TMN environment. The portions falling outside the TMN environment are not subject to standardization. For example, the human interface portion of the WSF is not specified in the TMN standard.

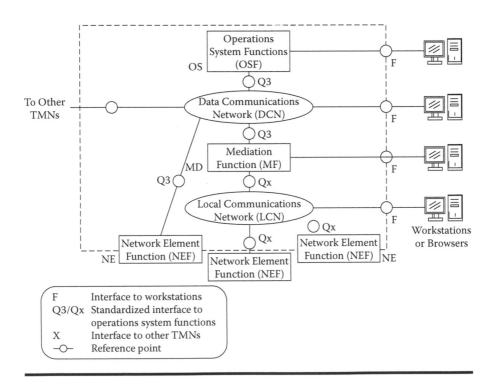

Figure 19.3 Functions within the TMN architecture.

Within the TMN specification are well-defined reference points identifying the characteristics of the interfaces between function blocks. The reference points identify the information that passes between the function blocks. The function blocks exchange information using the data communications function (DCF). The DCF may perform routing, relaying, and internetworking actions at OSI layers 1 to 3 (i.e., physical, data link, and network, respectively) or their equivalents. These functions are performed in the data communications network. Figure 19.3 also shows the reference points (F, G, M, Qx, Q3, and X) defined in the TMN specifications. These reference points are characterized by the information shared between their endpoints, and they can be further explained as follows:

■ Is the interface between a workstation and an operations system or a mediation device.
■ Is the interface between a workstation and a human user. The specification of this interface is outside the scope of TMN.
■ Is the interface between a Q adapter and a non-TMN-compliant network element. This interface, not specified in TMN, is actually one of the most important, given that today's networks consist primarily of devices that do not comply with the TMN standard.
■ Qx refers to the interface between a Q adapter and a mediation device, a TMN-compliant element and a mediation device, or between two mediation devices.
■ Q3 refers to the interface between a TMN-compliant element and an operations system, a Q adapter and a mediation device, a mediation device and an operations system, or between two operations systems.
■ Is the interface between operations systems in different TMNs. The operations system outside the X interface may be part of either a TMN or a non-TMN environment. This interface may require increased security relative to the level required by Q interfaces. In addition, access limitations may be imposed.

At present, only the Q3 interface has been specified to any detailed degree. The definition outlined includes the management protocol of the Q3 (CMIP), alarm surveillance capabilities, and operations in the generic model used to describe the network. *Alarm surveillance* refers to a set of functions that enable monitoring and polling the network concerning alarm-related events or conditions.

The TMN information model, which focuses on the management protocol object classes required to manage such a network, includes an abstraction of the management aspects of network resources and related support management activities. Information about objects is exchanged across TMN-standard interfaces.

The TMN specifications make up a generic information model that is technology independent. This independence allows management of diverse equipment in a common manner through an abstract view of network elements. This concept is vital if a TMN is to achieve its goals. The generic information model also serves as a basis for defining technology-specific object classes. These classes support a technology-independent view while enabling more precise management.

For example, a switch used to perform common management activities, such as provisioning or performance gathering, could be defined according to TMN specifications. In addition, this generic switch definition could be extended to cover the peculiarities of a particular vendor's switch. The extended definition could be used for such specialized activities as controlling the execution of diagnostic routines. TMN generic modeling techniques can be used by resource providers or management system providers to define their own objects.

The TMN specification includes an information model that is common to managed communication networks. This model can be used to generically define the resources, actions, and events that exist in a particular network.

The TMN architecture is an excellent means of visualizing network and service management solutions. Figure 19.4 shows a simple solution [1] representing a typical network management system (NMS) layer integrating various element management systems (EMSs). Additional management applications help support the service management layer (SML) and business management layer (BML).

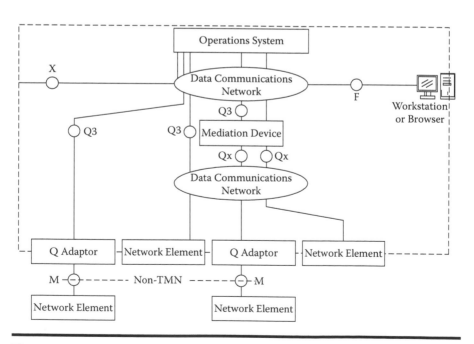

Figure 19.4   Sample TMN physical architecture and interfaces.

The visualization technology may be used to support various regional NMSs. On top of these NMSs, service-oriented messages, events, and alarms may be extracted and displayed in various service centers.

19.4 Control Objectives for Information and Related Technology

CobiT is an information technology (IT) governance, control framework, and maturity model. Its purpose is to ensure that IT resources are aligned with an enterprise's business objectives so that services and information, when delivered, meet quality, fiduciary, and security needs. It is also intended to provide a mechanism to balance IT risks and returns. CobiT defines 34 significant processes, links 318 tasks and activities to them, and defines an internal control framework for them all.

CobiT can be used by business or IT management, but its origins are in auditing. It was developed by the Information Systems Audit and Control Association, which is an international organization based in the United States. More recently, the IT Governance Institute has made some contributions.

CobiT is often introduced in an enterprise via the audit route. As a result, IT managers often view CobiT as a threat to their positions rather than as a useful and powerful framework for communicating effectiveness and value to their companies.

CobiT processes and control objectives are segmented into four domains:

1. Planning and organization (PO)
   PO1 Define a strategic IT plan
   PO2 Define the information architecture
   PO3 Determine the technological direction
   PO4 Define the IT organization and relationships
   PO5 Manage the IT investment
   PO6 Communicate management aims and directions
   PO7 Manage human resources
   PO9 Assess risks
   PO10 Manage projects
   PO11 Manage quality
2. Acquisition and implementation (AI)
   AI1 Identify automated solutions
   AI2 Acquire and maintain application software
   AI3 Acquire and maintain technology infrastructure
   AI4 Develop and maintain IT procedures
   AI5 Install and accredit systems
   AI6 Manage changes

3. Delivery and support (DS)
    DS1 Define and manage service levels
    DS2 Manage third-party services
    DS3 Manage performance and capacity
    DS4 Ensure continuous service
    DS5 Ensure system security
    DS6 Identify and allocate cost
    DS7 Educate and train users
    DS8 Assist and advise customers
    DS9 Manage problems and incidents
    DS10 Manage problems and incidents
    DS11 Manage data
    DS12 Manage facilities
    DS13 Manage operations
4. Monitoring (M)
    M1 Monitor the processes
    M2 Assess internal control adequacy
    M3 Obtain independent assurance
    M4 Provide for independent audit

CobiT is based on established frameworks, such as the Software Engineering Institute's capability maturity model, ISO 9000, and Infrastructure Library (ITIL). However, CobiT does not include control guidelines or practices, which are the next level of detail. Unlike ITIL, CobiT does not include process steps and tasks because it is a control framework rather than a process framework. CobiT focuses on what an enterprise needs to do, not how it needs to do it, and the target audience is auditors, senior business management, and senior IT management.

19.5 The ITIL Processes

The ITIL is based on delivering best-practice processes for IT service delivery and support, rather than defining a broad-based control framework. It focuses on the method. ITIL has a much narrower scope than CobiT because of its own focus on IT service management, but it defines a more comprehensive set of processes within that narrower field of service delivery and support. ITIL is more prescriptive about the tasks involved in those processes, and as such, its primary target audience is IT and service management.

Some enterprises have combined CobiT and ITIL to provide a more comprehensive IT governance and operations framework. The principles behind the CobiT and ITIL frameworks are consistent.

Auditors often use CobiT in combination with the ITIL self-assessment workbook to assess the service management environment. CobiT provides a set of key goal and performance indicators and critical success factors for each of its processes. These add value to ITIL because they establish the basis for managing.

The development processes of the two frameworks are not linked to each other, and both would benefit from closer collaboration. However, they are unlikely to contradict each other in any substantive way.

Reference

Terplan, K. *Operations Support System Essentials: Support System Solutions for Service Providers.* New York: John Wiley & Sons, 2001.

Chapter 20

Antenna Technology

James B. West

Contents

20.1 Introduction

This chapter is a brief overview of contemporary antenna types used in cellular, communication links, satellite communication, radar, and other microwave and millimeter wave systems. In this presentation, microwave is presumed to cover the frequency spectrum from 800 MHz to 94 GHz.

20.2 Fundamental Antenna Parameter Definitions

An *antenna* [1–5] is fundamentally a device that translates guided wave energy into radiating energy. Antenna physics, unlike radio-frequency (RF)/microwave transmission lines, simultaneously exhibits spatial and frequency (time) dependencies. Schantz provided an excellent phenomenological description of an antenna device; he characterized antenna functionality from multiple perspectives [1].

Electromagnetic radiation is the emission of energy from a device in the form of electromagnetic waves.

Radiation pattern is a graphical or mathematical description of the radiation properties of an antenna as a function of space coordinates. The standard (r, θ, φ) spherical coordinate system is typically used, as shown Figure 20.1. Radiation properties include radiation intensity (watts/solid angle), radiation density (watts m^2), gain, directivity, radiated phase, and polarization parameters, all of which are discussed in subsequent paragraphs. Radiated power, rather than electric field, is commonly used at microwave frequencies since radiated power is readily measured with contemporary antenna metrology equipment.

As an example, a *power radiation pattern* is an expression of the variation of received or transmitted power density as quantified at a constant radius about the

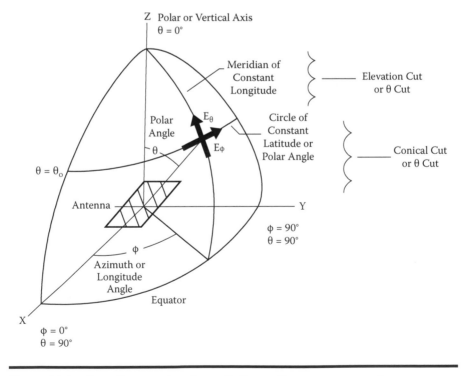

Figure 20.1   Coordinate system for radiation pattern measurement.

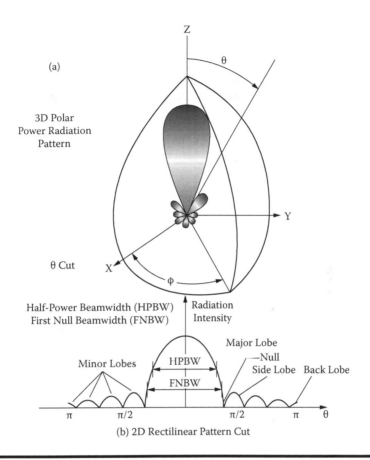

(a)

3D Polar
Power Radiation
Pattern

θ Cut

Half-Power Beamwidth (HPBW)
First Null Beamwidth (FNBW)

Radiation
Intensity

Major Lobe

—Null

Minor Lobes   HPBW   Side Lobe   Back Lobe

FNBW

π   π/2   π/2   π   θ

(b) 2D Rectilinear Pattern Cut

Figure 20.2   Antenna parameters.

antenna. Power patterns are often graphically depicted as normalized to their main beam peak.

Radiation pattern cuts are two-dimensional (2D) cross sections of the three-dimensional (3D) power radiation pattern. A cut in the plane of the theta unit vector is the *theta cut*, while a pattern cut in the phi unit vector direction is the *phi cut*, as shown in Figure 20.2. The *principal plane cuts* are 2D, orthogonal antenna cuts taken in the E field and H field planes of the power radiation patterns, as illustrated in Figure 20.3.

An isotropic radiator is a hypothetical lossless antenna having equal radiation in all directions. The 3D radiation pattern of an isotropic radiator is a sphere.

A directional antenna is an antenna that has a peak sensitivity that is a function of direction, as illustrated in Figures 20.2 and 20.3.

An omnidirectional antenna is an antenna that has essentially a nondirectional radiation pattern in one plane and a directional pattern in any orthogonal plane.

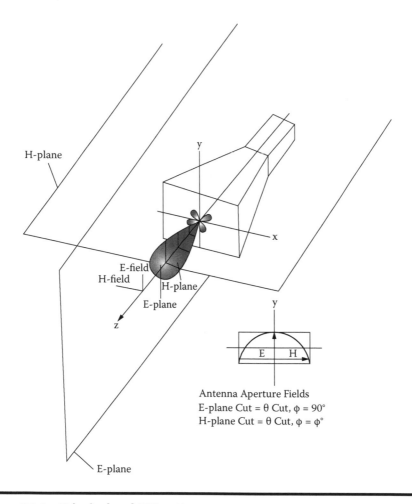

Figure 20.3   Principal and E/H pattern cuts.

Radiation pattern lobes are portions of the radiation pattern bound by regions of lesser radiation intensity. Typical directional antennas have one major *main lobe* that contains the antenna's radiation peak and several *minor lobes*, or *side lobes*, which are any lobes other than the major lobe. A *back lobe* is a radiation lobe that is located approximately 180° from the main lobe. The *side lobe level* is a measure of the power intensity of a minor lobe, usually referenced in decibels to the main lobe peak of the power radiation pattern. The *front-to-back ratio* is a measure of the power of the main beam to that of the back lobe. A *pattern null* is an angular position in the radiation pattern at which the power radiation pattern is at a minimum. The parameters are depicted in Figure 20.2.

For field regions, the 3D space around an antenna is divided into three regions. The region closest to the antenna is called the *reactive near field*, the intermediate region is the *radiating near-field (Fresnel)* region, and the farthest region is the *far-field (Fraunhofer)* region. The reactive near-field region is where reactive fields dominate. The Fresnel region is where radiation fields dominate, but the angular radiation pattern variation about the antenna is a function of distance away from the antenna.

The far field is the region where the angular variation of the radiation pattern is essentially independent of distance. Various criteria have been used to quantify the boundaries of these radiation regions. The commonly accepted boundaries between the field regions are

$$\text{Reactive near field/Fresnel boundary: } R_{\text{RNP}} = 0.62 \cdot \sqrt{\frac{D^3}{\lambda}} \qquad (20.1)$$

$$\text{Fresnel/far-field boundary: } R = \frac{2D^2}{\lambda}, \qquad (20.2)$$

where D is the largest dimension of the antenna, and λ is the wavelength in free space.

The far-field formulation assumes a 22.5° phase error across a circular aperture designed for a −25-dB side-lobe level, which creates a ±1.0-dB error at the 25-dB side-lobe level [6]. A more stringent far-field criterion is applicable to lower side-lobe levels [7]. Antennas operate in the far field for typical communication systems and radar applications.

The directivity of an antenna is the ratio of the radiation intensity in a given direction to that of the radiation intensity of the antenna averaged over all directions. This ratio is usually expressed in decibels. The term *directivity* most often implicitly refers to the *maximum directivity*.

Directivity is a measure of an antenna's ability to focus power density (transmitting) or to preferentially receive an incoming wave's power density as a function of spatial coordinates. A passive, lossless antenna does not amplify its input signal due to conservation of energy considerations; rather, it redistributes input energy as a function of spatial coordinates. Consider an isotropic radiator and a highly directive antenna such as a parabolic reflector. Assume that each antenna is lossless, is impedance and polarization matched, and has the same power at the input terminal. The matched isotropic radiator couples the same total input power into free space as that of the parabolic reflector, but the power density as a function of spatial coordinates is drastically different. The parabola focuses the majority of the total input power into a very narrow spatial sector and is commensurately more "sensitive" in this region. In contrast, the isotropic radiator uniformly distributes the input power over all directions and lacks the directional sensitivity. These concepts are illustrated in Figure 20.4.

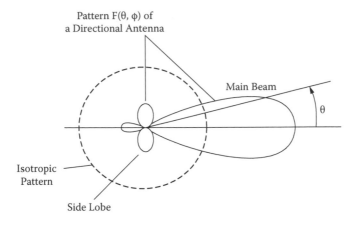

Figure 20.4   Comparison of two-dimensional (2D) isotropic and directional pattern cuts.

Directivity does not account for resistive loss mechanisms, polarization mismatch, or impedance mismatch factors in an antenna. As noted, a high value of directivity means that a high percentage of antenna input power is focused in a small angular region. By way of example, parabolic reflector antennas used on radio astronomy can have directivities exceeding 80 dB. An isotropic radiator, in contrast, has a directivity of 0 dB (1.0 numeric) since it radiates equally in all directions. A half-wave resonant dipole radiating in free space has a theoretical directivity of 2.14 dB.

Half-power beam width is the angle between two directions in the maximum lobe of the power radiation pattern where the directivity is one half the peak directivity, or 3 dB down, as illustrated in Figure 20.2. The half-power beam width is a measure of the spatial selectivity of an antenna. Low-beam-width antennas are very sensitive over a small angular region. Half-power beam width and directivity are inversely proportional for antennas in which the major portion of the radiation resides within the main beam.

For power gain, the gain of an antenna is the ratio of the radiation intensity in a given direction to the total input power accepted by the antenna input port. This ratio is usually expressed in decibels above an isotropic radiator (*dBi* for linear polarization and *dBic* for circular polarization).

Gain is related to directivity through the *radiation efficiency* parameter. Radiation efficiency is a measure of the insertion dissipative power losses internal to the antenna:

$$G = \eta_{cd} \times D, \tag{20.3}$$

where G is the power gain, η_{cd} is the antenna efficiency due to conductor and dielectric dissipative losses, and D is the directivity of the antenna.

Power gain usually implicitly refers to the peak power gain. The gain parameter includes the effects of dissipative losses but does not include the effects of polarization and impedance mismatches.

Antenna polarization refers to the electric field polarization properties of the propagating wave received by the antenna. Wave polarization is a description of the contour that the radiating electric field vector traces as the wave propagates through space. The most general wave polarization is *elliptical*. *Circular* and *linear polarizations* are special cases of elliptical polarization.

Examples of elliptical, circular, and linear polarized propagating electric field vectors are illustrated in Figure 20.5. *Vertical and horizontal polarizations* are sometimes used as well and loosely refer to the orientation of the linear polarized electric field vector at the plane of the antenna. The elliptically polarized wave is described by its *polarization ellipse*, which is the planar projection of the contour that its electric field vector sweeps out as the wave propagates through space. The orientation of the polarization ellipse is the *tilt angle*, and the ratio of the *major diameter* to the *minor diameter* of the ellipse is the *axial ratio*. The direction of rotation of the radiated field is expressed as either left handed or right handed as the wave propagates from a reference point. Left-hand circular polarization (LHCP) and right-hand circular polarization (RHCP) are commonly used in contemporary microwave systems.

In most applications, it is desirable to have the receiving antenna's polarization properties match those of the transmitting antennas and vice versa. If an antenna is polarized differently than the wave it is attempting to receive, then power received by the antenna will be less than the maximum, and the effect is quantified by a *polarization mismatch loss*, which is defined by

$$\text{PML} = |\boldsymbol{\rho}\text{r} \cdot \boldsymbol{\rho}\text{t}|2 = |\cos \psi p|2, \tag{20.4}$$

where ρr is the receive polarization unit vector, ρt is the transmit polarization unit vector, and ψp is the angle between the two unit vectors. This formulation is applicable to elliptical, circular, and linear polarizations.

Some commonly encountered polarization mismatch losses are depicted in Figure 20.6. It is apparent from the figure that a pure RHCP wave is completely isolated from an LHCP wave. This property is exploited in polarization diversity systems, for example, satellite communication systems, and cellular radio systems.

The *axial ratio*, usually expressed in decibels, is a measure of the ellipticity of a polarized wave. A 0-dB axial ratio describes pure circular polarization. A 6-dB axial ratio is typically used as a rule of thumb for the maximum axial ratio in which an elliptical wave is considered circular. The axial ratio does not describe the sense of the propagating wave. The axial ratio of an antenna is a 3D function of spatial coordinates.

An arbitrary elliptically polarized wave can be mathematically decomposed to an LHCP component and an RHCP component. Consider two circular antennas of opposite polarization sense, one in the transmit mode and the other in the receiving

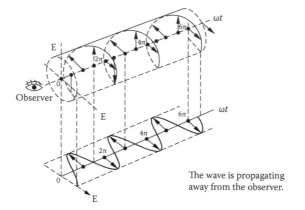

The wave is propagating
away from the observer.

(a) Elliptically Polarized Radiating Wave

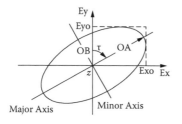

Rotation of a plane electromagnetic
wave and its polarization ellipse at
$Z = 0$ as a function of time.

(b) Polarization Ellipse

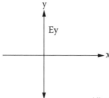

Exo = Eyo, 90° phase shift
between Exo and Eyo

(c) Circular Polarization

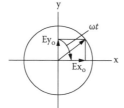

(d) Linear Polarization

Figure 20.5 Antenna polarization.

		Wave Polarization					
		Vertical ↑	Horizontal →	Right Hand Circular	Left Hand Circular	45° Right Linear ↗	45° Left Linear ↘
Wave Polarization	Vertical ↑	0 dB	∝	3 dB	3 dB	3 dB	3 dB
	Horizontal →	∝	0 dB	3 dB	3 dB	3 dB	3 dB
	Right Hand Circular	3 dB	3 dB	0 dB	∝	3 dB	3 dB
	Left Hand Circular	3 dB	3 dB	∝	0 dB	3 dB	3 dB
	45° Right Linear ↗	3 dB	3 dB	3 dB	3 dB	0 dB	∝
	45° Left Linear ↘	3 dB	3 dB	3 dB	3 dB	∝	3 dB

NOTE: Direction of propagation is into the page.

Figure 20.6   Polarization loss between receive and transmit antennas.

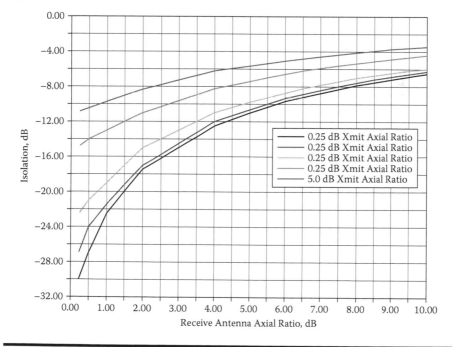

Figure 20.7   Receive antenna axial ratio (dB).

mode. As each antenna deviates from pure circular polarization toward elliptical polarization, the orthogonal (undesired) polarization components of each antenna become increasingly influential, which in turn reduces the isolation between the antennas. Similarly, a deviation from pure circular polarizations between receiving and transmitting antennas of the same polarization sense manifests itself as an increased loss. Figure 20.7 quantifies the minimum polarization isolation given the axial ratio and polarization sense of each antenna [8].

Radiation efficiency accounts for losses at the input terminal of the antenna and the losses within the structure. Efficiency due to conductor and dielectric losses is defined as

$$\eta_{cd} = \eta_c \times \eta_d, \tag{20.5}$$

where η_c is the efficiency due to conduction loss, and η_d is the efficiency due to dielectric loss.

Conduction loss is a function of the conductivity of the metal used. The dielectric loss is typically specified in terms of the loss tangent parameter and can be anisotropic. Tabulations of both parameters are available in the literature [9]. Conduction and dielectric loss efficiencies can also be determined experimentally.

The *input impedance* of an antenna is the impedance at the input terminal reference plane. *Impedance* is the ratio of the voltage to the current at the terminal plane, which is related to ratio of the electric and magnetic fields at the terminal plane. An antenna can be thought of as a *mode translator* of electromagnetic waves. Guided waves propagating through the input transmission line from the input generator to the antenna input terminal are transformed into unbounded (radiation) electromagnetic waves. Thus, the input port of an antenna radiating into free space can be thought of as the impedance with real and imaginary components that are a function of frequency.

The real or resistive component of the input impedance is required for the transfer of real power from the input generator into the radiated electrical and magnetic fields to initiate an average power flow out of the antenna. This resistive component is called the *radiation resistance*. The reactive near fields in the immediate vicinity of the antenna structure, that is, the reactive near-field region, are manifested as a reactance at the antenna input terminal. A *matched impedance condition* over the desired frequency range is required to optimally transfer input generator power into radiated waves. Both the real and imaginary components of the input impedance can vary with frequency, further complicating the impedance-matching problem. Standard transmission line and broadband circuit-matching techniques can be applied to ensure the optimal transfer of energy between the input transmission line and the antenna's input terminal [10,11].

Impedance mismatch loss is a measure of the amount of power from the input generator that passes through the antenna's input terminals. The input reflection coefficient is a measure of the amount of power that is reflected from the antenna input back into the generator. Mismatch loss (MML) is not a dissipative, or resistive, loss but rather a loss due to reflections of guided electromagnetic waves back to the input generator.

The impedance mismatch loss is defined as

$$\text{MML} = (1 - |\Gamma|^2), \qquad (2.6)$$

where Γ is the voltage reflection coefficient at the antenna input terminal.

The voltage reflection coefficient is usually specified and measured as either a return loss or a voltage standing wave ratio (VSWR):

$$\text{Return loss} = 10\log(|\Gamma|^2) \qquad (20.7)$$

and

$$\text{VSWR} = \frac{1 + |\Gamma|}{1 - |\Gamma|} \qquad (20.8)$$

The overall antenna efficiency is minimally related to the previously mentioned parameters of conductor/dielectric losses, polarization loss, and impedance mismatch factors. In addition, there are other efficiency parameters that are specific to certain classes of antennas. These additional terms include

- Illumination efficiencies (reflector and lens antennas)
- Spillover efficiencies (reflector and lens antennas)
- Blockage efficiencies (reflector antennas)
- Random phase and amplitude errors (arrays, lens, apertures, and reflector antennas)
- Aperture taper efficiencies (arrays, lens, apertures, and reflector antennas)
- Others

The effective area is the ratio of the available power at the antenna input terminals due to a polarization-matched plane wave incident on the antenna from a given direction to the power density of the same plane wave incident on the antenna. Effective area is a measure of an antenna's power-capturing properties under plane wave illumination. The effective area of an aperture antenna is related to its physical area by its *aperture efficiency*:

$$\eta_{ap} = \frac{A_{em}}{A_p} = \frac{\text{Maximum effective area}}{\text{Physical area}}. \qquad (20.9)$$

The effective area is related to its directivity by

$$A_{em} = \frac{\lambda^2}{4\pi} \times D_0 \eta_{cd} (1 - |\Gamma^2|) \times |\rho r \cdot \rho t|^2. \qquad (20.10)$$

The effective area parameter is useful in first-order array and aperture antenna directivity calculations.

Bandwidth is a frequency range in which a particular electrical parameter of an antenna conforms to a specified performance. Examples of parameters include gain, beam width, side-lobe level, efficiency, axial ratio, input impedance, and beam direction or squint. Since all of the mentioned parameters have different frequency dependencies, it is impossible to uniquely specify a single bandwidth parameter.

Electrically small antennas feature dimensions that are small relative to their operating wavelengths. Conventional antenna structures typically operate most efficiently when their largest dimension is an appreciable fraction of a wavelength (i.e., $\lambda/2$ or $\lambda/4$), which is unacceptably large for many applications, particularly those in the high-frequency (HF) through lower ultrahigh-frequency (UHF) regions. It is therefore highly desirable to determine the maximum theoretical bandwidth of an antenna that encloses a given volume in space. The fundamental limits of electrically small antennas have been examined by several researchers [12–14]. Their findings can be summarized as follows: An antenna that can be enclosed within a sphere of radius r has maximum bandwidth when its geometric structure optimally fills the volume of the sphere. This implies that 3D radiating elements will exhibit larger bandwidths than thin-wire radiators, which had been verified within the research community through several examples [1].

The bandwidth of the antenna can be expressed by its quality factor, or Q, which is the ratio of reactive power to real power. Chu has shown that the Q of an electrically small antenna can be expressed as [13]

$$Q = \frac{1+2(kr)^2}{(kr)^3[1+(kr)^2]} \cong \frac{1}{(kr)^3} \text{ for } kr \ll 1, \tag{20.11}$$

where k is $2\pi/\lambda$, the wave propagation constant; λ is the free-space wavelength; and r is the radius of the enclosing sphere.

For antenna with $Q > 2$, the fractional bandwidth is related to the quality factor by the following expression:

$$\text{Fractional band width} = \frac{\Delta f}{fo} = \frac{1}{Q} \tag{20.12}$$

Equation 20.12 shows that lower values of Q translate to larger bandwidths.

Equation 20.13 is known as the fundamental limit on the electrical size of the antenna. Figure 20.8 illustrates this limit for 100% radiation efficiency (no dissipative loss) as a function of kr. Note that Q proportionately lowers with decreasing radiation efficiency. The classic resonant $\lambda/2$ resonant dipole and the Goubau electrically small multielement monopole are compared to the theoretical limit [15].

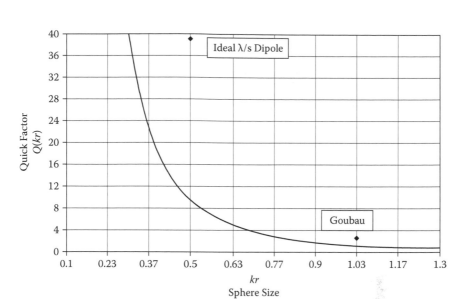

Figure 20.8   Fundamental Q limit for 100% radiation efficiency.

The Goubau antenna has traditionally been the standard for broadband electrically small antennas. Excellent discussions of bandwidth properties for several new antenna elements are available in the literature [16,17].

Reciprocity is a theorem of electromagnetics that requires receive and transmit properties to be identical for linear and reciprocal antenna structures. This concept is useful in antenna measurements since it is normally more convenient to measure the properties of an antenna under test (AUT) in the receive mode. Also, it can be easier to understand some antenna concepts through a receive mode formulation, while others are easier through a transmit mode interpretation.

The Friis transmission formula describes the coupling of electromagnetic energy between two antennas under far-field radiation conditions. The power received by the receiving antenna is related to the power transmitted through the transmitting antenna by the following expression:

$$\frac{P_r}{P_t} \eta_{cdr} \cdot \eta_{cdt} \cdot (1- |\Gamma_r|^2) \cdot (1- |\Gamma_r|^2) \cdot \left(\frac{\lambda}{4\pi}\right)^2 D_{or} \cdot D_{ot} \cdot \left|\bar{\rho}_r \cdot \bar{\rho}_t\right|^2, \qquad (20.13)$$

where θ and φ are the usual spherical coordinate angles, and the r and t subscripts refer to the receive and transmit modes, respectively.

Regarding antenna noise temperature, all objects with a physical temperature greater than 0 K radiate energy. Antenna noise temperature is an important parameter in radio astronomy and satellite communication systems since the antenna, ground, and sky background noise contribute to the total system noise. This system noise ultimately sets a limit to the system signal-to-noise ratio. Satellite system link performance is typically specified in terms of the ratio of system gain to the system noise temperature, which is called the **G/T** figure of merit. The noise temperature of an antenna is defined as follows:

$$T_A = \frac{\int_0^{2\pi} \int_0^{\pi} T_B(\theta, \phi) G(\theta, \phi) \sin\theta d\theta d\phi}{\int_0^{2\pi} \int_0^{\pi} G(\theta, \phi) \sin\theta d\theta d\phi}, \tag{20.14}$$

where

$$T_B = \varepsilon(\theta, \phi) T_m = (1 - |\Gamma|^2) T_m = \text{the equivalent brightness temperature,} \tag{20.15}$$

where $\varepsilon(\theta, \varphi)$ is the emissivity, T_m is the molecular (physical) temperature (K), and $G(\theta, \varphi)$ is the power gain of the antenna.

The brightness temperature is a function of frequency, polarization of the emitted radiation, and the atomic structure of the object. In the microwave frequency range, the ground has an equivalent temperature of about 300 K, and the sky temperature is about 5 K when observing toward the zenith (straight up, perpendicular to the ground) and between 100 and 150 K when looking toward the horizon.

20.3 Radiating Element Types

20.3.1 Wire Antennas

Wire antennas [18–20] are arguably the most common antenna type, and they are used extensively in contemporary microwave systems. The classic *dipole, monopole,* and *whip* antennas are used extensively throughout the microwave frequency bands. Broadband variations of the classic wire antenna include the *electrically thick monopole* and *dipole, biconical dipole, bow tie, coaxial dipole* and *monopole,* the *folded dipole, discone,* and *conical skirt monopole.*

Various techniques to reduce the size of this type of antenna for a given resonant frequency include foreshortening the antenna's electrical length and compensating with lumped impedance loading, top hat capacitive loading, dielectric material loading, helical winding of monopole element to retain electrical length, but with foreshortened effective height. An excellent discussion on electrically small antennas is found in [5].

20.3.2 Loop Antennas

Loop antennas [18,21] are used extensively in the HF through UHF band and have application at the L band and above as field probes. The circular loop is the most commonly used, but other geometric contours, such as rectangular, are used as well. Loops are typically classified as *electrically small*, for which the overall wire length (circumference multiplied by the number of turns) is less than 0.1 wavelength, and *electrically large*, for which the loop circumference is approximately 1 wavelength. Electrically small antennas suffer from low radiation efficiency but are used in portable radio receivers and pagers and as field probes for electromagnetic field strength measurements. The ubiquitous amplitude modulation (AM) portable radio antenna is a ferrite material-loaded multiturn loop antenna. *Electrically large* loops are commonly used as array antenna radiating elements. The radiation pattern in this case is in end fire toward the axis of the loop.

20.3.3 Slot Antennas

Slot antennas [22] are used extensively in aircraft, missile, and other applications where a physical low profile and ruggedness are required. Slot antennas are usually half-wave resonant and are fed by introducing an excitation electric field across their gap. *Cavity-backed slots* have application in the UHF bands. The annular slot antenna is a very low-profile structure that has a monopole wire-like radiation pattern. Circular polarization is possible with crossed slots fed in phase quadrature. Stripline-fed slots are frequently used in phased-array applications for the microwave bands [23].

Waveguide-fed slotted array antennas find extensive use in the microwave and millimeter wave bands [24]. Linear arrays of resonant slots are formed by machining slots along the length of standard waveguide transmission lines, and a collection of these waveguide linear array "sticks" is combined to form 2D planar phased arrays. The most common waveguide slots currently in use are the *edge slot* and the *longitudinal broad wall slot*. This array type is a cost-effective way to build high-efficiency, controlled side-lobe-level arrays. One-dimensional electrically scanned phased arrays can be realized by introducing phase shifters in the feed manifold that excite each linear waveguide array stick within the 2D aperture.

20.3.4 Helical Antennas

The helical antenna [20] has seen extensive use in the UHF through microwave frequencies, both for single radiating elements and as phased-array antenna elements. An excellent helical antenna discussion can be found in [20]. The helical antenna can operate either in the *axial (end-fire) mode* or in the *normal mode*. The axial mode results in a directional, circularly polarized pattern that can operate over a 2:1 frequency bandwidth.

The normal-mode helix has found wide application for broad-beam, cardioid-shaped radiation patterns with very good axial ratio performance. Kilgus developed bifilar and quadrafilar helical antennas for satellite signal reception applications [25,26]. Circular polarization is generated by exciting each quadrafilar element of the helix in phase quadrature. Helical antennas are also used as radiating elements for phased arrays and as feeds for parabolic reflector antennas [27].

20.3.5 Yagi–Uda Array

The Yagi–Uda [19] array is a common directional antenna for the VHF (very high-frequency) and UHF bands. End-fire radiation is realized by a complex mutual coupling (surface wave) mechanism between the driven radiation element, a reflector element, and one, or more, director elements. Dipole elements are most commonly used, but loops and printed dipole elements are used as well. Balanis described a detailed step-by-step design procedure for this type of antenna [3]. Method-of-moment computer simulations are often used to verify and optimize the design.

20.3.6 Frequency-Independent Antennas

Classic antenna radiator types, such as the wire dipole antenna, radiate efficiency when their physical dimensions are a certain fraction of the operating wavelength. As an example, the half-wave dipole is strictly resonant only at one frequency, and acceptable performance can be realized for finite bandwidths. The frequency-independent antenna [28] is based on designs that are specified in terms of geometric angles. Rumsey has shown that if an antenna shape can be completely specified by angles, then it could theoretically operate over an infinite bandwidth [29]. In practice, the upper operating frequency limit is dictated by the antenna feed structure, and the lower operating frequency is limited by the physical truncation of the antenna structure. The *planar spiral, planar slot spiral, conical spiral*, and the *cavity-back planar spiral* antennas all exploit this concept.

A closely related and very useful antenna is the *log-periodic antenna*. This type of antenna has an electrical periodicity that is a logarithm of frequency. Because the antenna shape cannot be completely specified by angles, it is not truly frequency independent but nevertheless is extremely broadband. The most common architecture is the log-periodic dipole array. Balanis documented a detailed design procedure for this structure [3].

A key of frequency antennas is that the electrically active region of the antenna is a subset of the entire antenna structure, and the active region migrates to different regions of the antenna as a function of frequency [30].

20.3.7 Aperture Antennas

Horn antennas are the most commonly used aperture antennas [18,31,32] in the microwave and millimeter wave frequency bands. There is an abundance of horn antenna design information available in the literature. *E plane* and *H plane sectoral horns* generate fan beam radiation, that is, a narrow beam width in one principal plane and a broad beam width in the orthogonal principal plane. The *pyramidal horn* provides a narrow beam in both principal planes. Sector and pyramidal horns are typically fed with rectangular waveguide and are linearly polarized. The *ridged horn* is a broadband variation of the pyramidal horn [33]. A *dual-ridge square aperture horn* in conjunction with an orthomode transducer (OMT) can be used to generate broadband, circularly polarized radiation patterns.

Conical horns are typically fed with circular waveguide. An important variation of the basic conical horn includes the *corrugated horn*, which is typically used as a feed for high-efficiency, electrically large reflector antenna systems [19]. The corrugations are used to extinguish field diffractions of the edge of the aperture plane that can lead to spurious radiation in the back-lobe and side-lobe regions. The *aperture-matched horn* also minimizes aperture edge diffraction by blending the aperture edge to make a gradual transition into free space. *Dual-mode conical horns* use a superposition of waveguide modes in the throat region to suppress beam width, control orthogonal beam widths, and minimize cross polarization. *Multimode horns* can be used in monopulse radar applications where sum and difference beam can be generated by means of mode-generating or mode-combining feed structure.

References

1. Schantz, H., *The Art and Science of Ultrawideband Antennas*. Norwood, MA: Artech House, 2005.
2. *IEEE Standard Definitions of Terms for Antennas,* IEEE Standard 145–1983. Institute of Electrical and Electronics Engineers, 1983.
3. Balanis, C.A., *Antenna Theory, Analysis, and Design*, 2nd ed. New York: Wiley, 1997.
4. Macnamara, T., *Handbook of Antennas for EMC*. Norwood, MA: Artech House, 1995.
5. *Antennas and Antenna Systems*, Catalog Number 200, Watkins-Johnson Company, September 1990, pp. 101–127.
6. Hollis, J.S. et al., *Microwave Antenna Measurements*. Atlanta, GA: Scientific-Atlanta, November 1985.
7. Hacker, P.S., and Schrank, H.E., Range distance requirements for measuring low and ultralow sidelobe antenna patterns. *IEEE Trans. Antennas Propagation*, AP-30(5), 956–966, 1982.
8. Offutt, W.B., and DeSize, L.K., Methods of polarization synthesis, in *Antenna Engineering Handbook*, 2nd ed., Johnson, R.C., and Jasik, H., Eds.. New York: McGraw-Hill, 1984.
9. Howard K., *Reference Data for Radio Engineers*, 6th ed. New York: Sams/ITT, 1979, pp. 4-21–4-23.
10. Bowman, D.F., Impedance matching and broadbanding, in *Antenna Engineering Handbook*, 2nd ed., Johnson, R.C., and Jasik, H., Eds., New York: McGraw-Hill, 1984.
11. Cuthbert, T.R., *Broadband Direct-Coupled and Matching RF Networks*. Greenwood, AR: TRCPEP, 1999.
12. Wheeler, H.A., Fundamental limitations of small antennas. *Proc. IEEE*, 1479–1483, December 1947.
13. Chu, L.J., Physical limitations of omnidirectional antennas. *J. Appl. Phys.*, 19, 1163–1175, December 1948.

14. Hansen, R.C., Fundamental limitations in antennas. *Proc. IEEE*, 69(2), 170–182, February 1981.
15. Goubau, G., Multi-element monopole antennas. *Proc. Workshop Elec. Small Antennas*, ECOM, 63–67, May 1976.
16. Best, S.R., The performance of an electrically small folded spherical helix antenna. *IEEE Int. Symp. Antennas Propagation*, 2002.
17. Best, S.R., On the performance of the Koch fractal and other bent wire monopoles. *IEEE Trans. Antennas Propagation*, 51(6), 1293–1300, June 2003.
18. Salati, O.M., Antenna chart for systems designers. *Elec. Eng.*, January 1968.
19. Kraus, J.D., *Antennas*, 2nd ed. New York: McGraw-Hill, 1988.
20. Tai, C.T., Dipoles and monopoles, in *Antenna Engineering Handbook*, 2nd ed., Johnson, R.C., and Jasik, H., Eds. New York: McGraw-Hill, 1984.
21. Fujimoto, K., Henderson, A., Hirasaw, K., and James, J.R., *Small Antennas*. New York: Wiley, 1987.
22. Blass, J., Slot antennas, in *Antenna Engineering Handbook*, 2nd ed., Johnson, R.C., and Jasik, H., Eds. New York: McGraw-Hill, 1984.
23. Mailloux, R.J., On the use of metallized cavities in printed slot arrays with dielectric substrates. *IEEE Trans. Antennas Propagation*, Ap-35(55), 477–487, 1987.
24. Elliot, R.S., *Antenna Theory and Design*. Englewood Cliffs, NJ: Prentice Hall, 1981.
25. Kilgus, C.C., Resonant quadrafilar helix, *IEEE Trans. Antennas Propagation*, Ap-23, 392–397, 1975.
26. Donn, C., Imbraie, W.A., and Wong, G.G., An S band phased array design for satellite application. *IEEE Int. Symp. Antennas Propagation*, 60–63, 1977.
27. Holland, J., Multiple feed antenna covers L, S, and C bands segments. *Microwave J.*, 82–85, October 1981.
28. Mayes, P.E., Frequency independent antennas, in *Antenna Handbook, Theory, Applications, and Design*, Lo, Y.T., and Lee, S.W., Eds. New York: Van Nostrand Reinhold, 1988.
29. Rumsey, V.H., *Frequency Independent Antennas*. New York: Academic Press, 1966.
30. DuHamel, R.H., Frequency independent antennas, in *Antenna Engineering Handbook*, 2nd ed., Johnson, R.C., and Jasik, H., Eds. New York: McGraw-Hill, 1984.
31. Love, A.W., *Electromagnetic Horn Antennas*. New York: IEEE Press, 1976.
32. Olver, A.D., Clarricoats, P.J.B., Kishk, A.A., and Shafai, L., *Microwave Horns and Feeds*. New York: IEEE Press, 1994.
33. Walton, K.L., and Sunberg, V.C., Broadband ridge horn design, *Microwave J.*, 96–101, March 1964.

Chapter 21

Diversity

Arogyaswami J. Paulraj

Contents

21.1 Introduction

Diversity is a commonly used technique in mobile radio systems to combat signal fading. The basic principle of diversity is: If several replicas of the same information-carrying signal are received over multiple channels with comparable strengths, which exhibit independent fading, then there is a good likelihood that at least one or more of these received signals will not be in a fade at any given instant in time, thus making it possible to deliver an adequate signal level to the receiver.

Without diversity techniques, in noise-limited conditions, the transmitter would have to deliver a much higher power level to protect the link during the short intervals when the channel is severely faded. In mobile radio, the power available on the reverse link is severely limited by the battery capacity of handheld subscriber units. Diversity methods play a crucial role in reducing transmit power needs. Also, cellular communication networks are mostly interference limited, and again, mitigation of channel fading through use of diversity can translate into reduced variability of carrier-to-interference ratio (C/I), which in turn means lower C/I margin and hence better reuse factors and higher system capacity.

The basic principles of diversity have been known since 1927 when the first experiments in space diversity were reported. There are many techniques for obtaining independently fading branches, and these can be subdivided into two main classes. The first are explicit techniques, by which explicit redundant signal transmission is used to exploit diversity channels. Use of dual-polarized signal transmission and reception in many point-to-point radios is an example of explicit diversity. Clearly, such redundant signal transmission involves a penalty in frequency spectrum or additional power.

In the second class are implicit diversity techniques: The signal is transmitted only once, but the decorrelating effects in the propagation medium such as multipaths are exploited to receive signals over multiple diversity channels. A good example of implicit diversity is the RAKE receiver in code-division multiple-access (CDMA) systems, which uses independent fading of resolvable multipaths to achieve diversity gain. Figure 21.1 illustrates the principle of diversity in which two independently fading signals are shown along with the selection diversity output signal that selects the stronger signal. The fades in the resulting signal have been substantially smoothed out while also yielding higher average power.

If antennas are used in transmit, they can be exploited for diversity. If the transmit channel is known, the antennas can be driven with complex conjugate channel weighting to cophase the signals at the receive antenna. If the forward channel is not known, we have several methods to convert space-selective fading at the transmit antennas to other forms of diversity exploitable in the receiver.

Exploiting diversity needs careful design of the communication link. In explicit diversity, multiple copies of the same signal are transmitted in channels using a frequency, time, or polarization dimension. At the receiver end, we need arrangements to receive the different diversity branches (this is true for both explicit and implicit

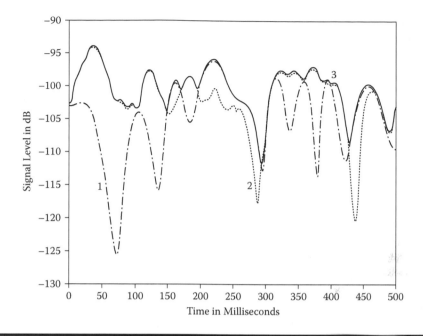

Figure 21.1 Example of diversity combining. Signals 1 and 2 are two independently fading signals. Signal 3 is the result of selecting the stronger signal.

diversity). The different diversity branches are then combined to reduce signal outage probability or bit error rate (BER).

In practice, the signals in the diversity branches may not show completely independent fading.

The envelope cross correlation p between these signals is a measure of their independence.

$$\rho = \frac{E\left[[r_1 - \bar{r}_1][r_2 - \bar{r}_2]\right]}{\sqrt{E|r_1 - \bar{r}_1|^2 \ E|r_2 - \bar{r}|^2}}$$

where r_1 and r_2 represent the instantaneous envelope levels of the normalized signals at the two receivers, and $\bar{r}_1$ and $\bar{r}_2$ are their respective means. It has been shown that a cross correlation of 0.7 [1] between signal envelopes is sufficient to provide a reasonable degree of diversity gain. Depending on the type of diversity employed, these diversity channels must be sufficiently *separated* along the appropriate diversity dimension. For spatial diversity, the antennas should be separated by more than the *coherence distance* to ensure a cross correlation of less than 0.7. Likewise, in frequency diversity, the frequency separation must be larger than the *coherence bandwidth,* and in time diversity the separation between channel reuse in time should be longer than the *coherence time*. These coherence factors in turn depend on the channel characteristics. The coherence distance, coherence bandwidth, and coherence time vary inversely with the angle spread, delay spread, and Doppler spread, respectively.

If the receiver has a number of diversity branches, it has to combine these branches to maximize the signal level. Several techniques have been studied for diversity combining. We describe three main techniques: selection combining, maximal ratio combining and equal gain combining.

Finally, we should note that diversity is primarily used to combat fading, and if the signal does not show significant fading in the first place (e.g., when there is a direct path component), diversity combining may not provide significant diversity gain. In the case of antenna diversity, array gain proportional to the number of antennas will still be available.

21.2 Diversity Schemes

There are several techniques for obtaining diversity branches, sometimes also known as diversity dimensions. The most important of these are discussed in the following sections.

21.2.1 Space Diversity

Space diversity has historically been the most common form of diversity in mobile radio base stations. It is easy to implement and does not require additional frequency spectrum resources. Space diversity is exploited on the reverse link at the base station receiver by spacing antennas apart to obtain sufficient decorrelation. The key for obtaining minimum uncorrelated fading of antenna outputs is adequate spacing of the antennas. The required spacing depends on the degree of multipath angle spread. For example, if the multipath signals arrive from all directions in the azimuth, as is usually the case at the mobile, antenna spacing (coherence distance) on the order of 0.5 to 0.8λ is quite adequate [2]. On the other hand, if the multipath angle spread is small, as in the case of base stations, the coherence distance is much larger. Also, empirical measurements show strong coupling between antenna height and spatial correlation. Larger antenna heights imply larger coherence distances.

Typically, 10 to 20λ separation is adequate to achieve $\rho = 0:7$ at base stations in suburban settings when the signals arrive from the broadside direction. The coherence distance can be three to four times larger for end-fire arrivals. The end-fire problem is averted in base stations with trisectored antennas as each sector needs to handle only signals arriving _60_ off the broadside. The coherence distance depends strongly on the terrain. Large multipath angle spread means smaller coherence distance.

Base stations normally use space diversity in the horizontal plane only. Separation in the vertical plane can also be used, and the necessary spacing depends on vertical multipath angle spread. This can be small for distant mobiles, making vertical plane diversity less attractive in most applications.

Space diversity is also exploitable at the transmitter. If the forward channel is known, it works much like receive space diversity. If it is not known, then space diversity can be transformed to another form of diversity exploitable at the receiver (see Section 21.2.7).

If antennas are used at transmit and receive, the M transmit and N receive antennas both contribute to diversity. It can be shown that if simple weighting is used without additional bandwidth or time/memory processing, then maximum diversity gain is obtained if the transmitter and receiver use the left and right singular vectors of the $M \times N$ channel matrix, respectively. However, to approach the maximum $M \times N$ order diversity order will require the use of additional bandwidth or time/memory-based methods.

21.2.2 Polarization Diversity

In mobile radio environments, signals transmitted on orthogonal polarizations exhibit low fade correlation and therefore offer potential for diversity combining. Polarization diversity can be obtained by either explicit or implicit techniques. Note that with polarization only two diversity branches are available compared to space diversity, for which several branches can be obtained using multiple antennas.

In explicit polarization diversity, the signal is transmitted and received in two orthogonal polarizations. For a fixed total transmit power, the power in each branch will be 3 dB lower than if single polarization is used. In the implicit polarization technique, the signal is launched in a single polarization but is received with cross-polarized antennas. The propagation medium couples some energy into the cross-polarization plane. The observed cross-polarization coupling factor lies between 8 and 12 dB in mobile radio [3, 4]. The cross-polarization envelope decorrelation has been found to be adequate. However, the large branch imbalance reduces the available diversity gain.

With handheld phones, the handset can be held at random orientations during a call. This results in energy being launched with varying polarization angles ranging from vertical to horizontal. This further increases the advantage of cross-polarized antennas at the base station since the two antennas can be combined to match the received signal polarization. This makes polarization diversity even more attractive. Recent work [5] has shown that with variable launch polarization, a cross-polarized antenna can give comparable overall (matching plus diversity) performance to a vertically polarized space diversity antenna.

Finally, we should note that cross-polarized antennas can be deployed in a compact antenna assembly and do not need the large physical separation necessary in space diversity antennas. This is an important advantage in the PCS (personal communication services) base stations, where low profile antennas are needed.

21.2.3 Angle Diversity

Where the angle spread is very high, such as indoors or at the mobile unit in urban locations, signals collected from multiple nonoverlapping beams offer low fade correlation with balanced power in the diversity branches. Clearly, since directional beams imply use of antenna aperture, angle diversity is closely related to space diversity. Angle diversity has been utilized in indoor wireless local-area networks (LANs), where its use allows substantial increase in LAN throughputs [6].

21.2.4 Frequency Diversity

Another technique to obtain decorrelated diversity branches is to transmit the same signal over different frequencies. The frequency separation between carriers should be larger than the coherence bandwidth. The coherence bandwidth, of course, depends on the multipath delay spread of the channel. The larger the delay spread, the smaller the coherence bandwidth and the more closely we can space the frequency diversity channels. Clearly, frequency diversity is an explicit diversity technique and needs additional frequency spectrum.

A common form of frequency diversity is multicarrier (also known as multi-tone) modulation. This technique involves sending redundant data over a number of closely spaced carriers to benefit from frequency diversity, which is then exploited by applying interleaving and channel coding/forward error correction across the carriers. Another technique is to use frequency hopping, by which the interleaved and channel-coded data stream is transmitted with widely separated frequencies from burst to burst. The wide frequency separation is chosen to guarantee independent fading from burst to burst.

21.2.5 Path Diversity

The path diversity type of implicit diversity is available if the signal bandwidth is much larger than the channel coherence bandwidth. The basis for this method is that when the multipath arrivals can be resolved in the receiver, since the paths fade independently, diversity gain can be obtained. In CDMA systems, the multipath arrivals must be separated by more than one *chip* period and the RAKE receiver provides the diversity [7]. In time-division multiple-access (TDMA) systems, the multipath arrivals must be separated by more than one *symbol* period, and the maximum likelihood sequence estimation (MLSE) receiver provides the diversity.

21.2.6 Time Diversity

In mobile communication channels, the mobile motion together with scattering in the vicinity of the mobile causes time-selective fading of the signal with Rayleigh fading statistics for the signal envelope. Signal fade levels separated by the *coherence time* show low correlation and can be used as diversity branches if the same signal can be transmitted at multiple instants separated by the coherence time.

The coherence time depends on the Doppler spread of the signal, which in turn is a function of the mobile speed and the carrier frequency.

Time diversity is usually exploited via interleaving, forward error correction (FEC) coding, and automatic repeat request (ARQ). These are sophisticated techniques to exploit channel coding and time diversity. One fundamental drawback with time diversity approaches is the delay needed to collect the repeated or interleaved transmissions. If the coherence time is large, as for example when the vehicle is slow moving, the required delay becomes too large to be acceptable for interactive voice conversation.

The statistical properties of fading signals depend on the field component used by the antenna, the vehicular speed, and the carrier frequency. For an idealized case of a mobile surrounded by scatterers in all directions, the autocorrelation function of the received signal $x(t)$ [note that this is not the envelope $r(t)$] can be shown to be

$$E[x(t)x(t+\tau)] = J_0\left(\frac{2\pi\tau v}{\lambda}\right)$$

where j_0 is a Bessel function of the 0th order, and v is the mobile velocity.

21.2.7 Transformed Diversity

In transformed diversity, the space diversity branches at the transmitter are transformed into other forms of diversity branches exploitable at the receiver. This is used when the forward channel is not known and shifts the responsibility of diversity combining to the receiver, which has the necessary channel knowledge.

21.2.7.1 Space to Frequency

- *Antenna delay.* Here the signal is transmitted from two or more antennas with delays on the order of a chip or symbol period in CDMA or TDMA, respectively. The different transmissions simulate resolved path arrivals that can be used as diversity branches by the RAKE or MLSE equalizer.
- *Multicarrier modulation.* The data stream after interleaving and coding is modulated as a multicarrier output using an inverse discrete Fourier transform (DFT). The carriers are then mapped to the different antennas. The space-selective fading at the antennas is now transformed to frequency-selective fading, and diversity is obtained during decoding.

21.2.7.2 Space to Time

- *Antenna hopping/phase rolling.* In this method, the data stream after coding and interleaving is switched randomly from antenna to antenna. The space-selective fading at the transmitter is converted into a time-selective fading at the receiver. This is a form of "active" fading.

■ *Space–time coding.* The approach in space–time coding is to split the encoded data into multiple data streams, each of which is modulated and simultaneously transmitted from different antennas. The received signal is a superposition of the multiple transmitted signals. Channel decoding can be used to recover the data sequence. Since the encoded data arrive over uncorrelated fade branches, diversity gain can be realized.

21.3 Diversity Combining Techniques

Several diversity combining methods are known. We describe three main techniques: selection, maximal ratio, and equal gain. They can be used with each of the diversity schemes discussed above.

21.3.1 Selection Combining

Selection combining is the simplest and perhaps the most frequently used form of diversity combining. In this technique, one of the two diversity branches with the highest carrier-to-noise ratio (C/N) is connected to the output (see Figure 21.2a).

The performance improvement due to selection diversity can be seen as follows: Let the signal in each branch exhibit Rayleigh fading with mean power δ^2. The density function of the envelope is given by

$$p(r_i) = \frac{r_i}{\sigma^2} e^{\frac{-r_i^2}{2\sigma^2}} \tag{21.1}$$

where r_i is the signal envelope in each branch. If we define two new variables

$$\gamma_i = \frac{Instantaneous\ signal\ power\ in\ each\ branch}{Mean\ noise\ power}$$

$$\Gamma = \frac{Mean\ signal\ power\ in\ each\ branch}{Mean\ noise\ power}$$

then the probability that the C/N is less than or equal to some specified value s is

$$\mathrm{Prob}[\gamma_i \le \gamma_s] = 1 - e^{-\gamma_s/\Gamma} \tag{21.2}$$

The probability that γ_i in all branches with independent fading will be simultaneously less than or equal to γ_s is then

$$\mathrm{Prob}[\gamma_1, \gamma_2, \ldots \gamma_M \le \gamma_s] = (1 - e^{-\gamma_s/\Gamma})^M \tag{21.3}$$

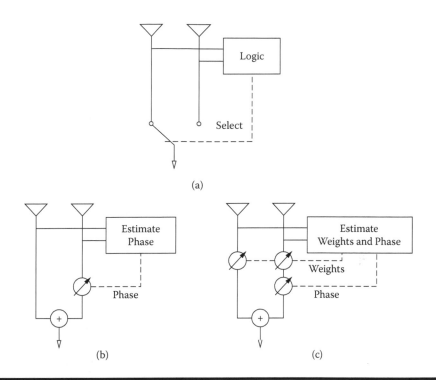

(a)

(b)   (c)

Figure 21.2   Diversity combining methods for two diversity branches. (a) Selection combining, (b) maximum ratio combining, (c) equal gain combining.

This is the distribution of the best signal envelope from the two diversity branches. Figure 21.3 shows the distribution of the combiner output C/N for $M = 1, 2, 3$, and 4 branches. The improvement in signal quality is significant. For example, at the 99% reliability level, the improvement in C/N is 10 dB for two branches and 16 dB for four branches.

Selection combining also increases the mean C/N of the combiner output and can be shown to be [1]

$$\text{Mean } (\gamma_s) = \Gamma \sum_{k=1}^{M} \frac{1}{k} \tag{21.4}$$

This indicates that with four branches, for example, the mean C/N of the selected branch is 2.08 better than the mean C/N in any one branch.

21.3.2 Maximal Ratio Combining

In the maximal ratio combining technique, the M diversity branches are first cophased and then weighted proportionally to their signal level before summing

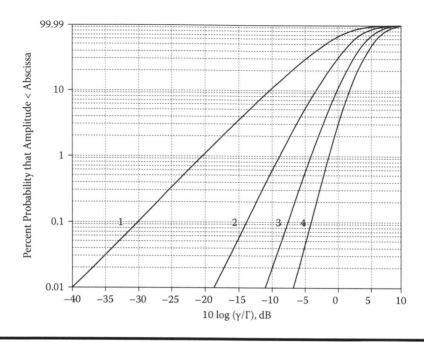

Figure 21.3 Probability distribution of signal envelope for selection combining.

(see Figure 21.2b). The distribution of the maximal ratio combiner

$$\text{Prob}[\gamma \le \gamma_s] = 1 - e^{(-\gamma_m/\Gamma)} \sum_{k=1}^{M} \frac{(\gamma_m/\Gamma)^{k-1}}{(k-1)!} \tag{21.5}$$

The distribution of output of a maximal ratio combiner is shown in Figure 21.4. Maximal ratio combining is known to be optimal in the sense that it yields the best statistical reduction of fading of any linear diversity combiner. In comparison to the selection combiner, at 99% reliability level, the maximal ratio combiner provides a 11.5-dB gain for two branches and a 19-dB gain for four branches, an improvement of 1.5 and 3 dB, respectively, over the selection diversity combiner.

The mean C/N of the combined signal may be easily shown to be Mean .*m*/D *M0*:

$$\text{Mean}(\gamma_m) = M\Gamma \tag{21.6}$$

Therefore, combiner output mean varies linearly with *M*. This confirms the intuitive result that the output C/N averaged over fades should provide gain proportional to the number of diversity branches. This is a situation similar to conventional beam forming.

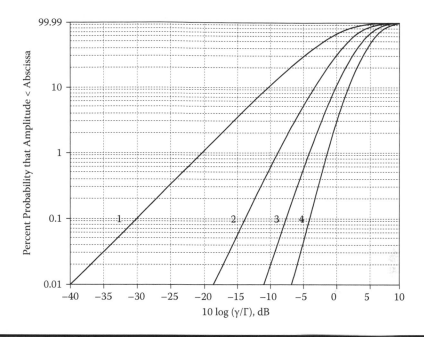

Figure 21.4 **Probability distribution of signal envelope for maximal ratio combining.**

21.3.3 Equal Gain Combining

In some applications, it may be difficult to estimate the amplitude accurately, the combining gains may all be set to unity, and the diversity branches merely summed after cophasing (see Figure 21.2c). The distribution of equal gain combiner does not have a neat expression and has been computed by numerical evaluation. Its performance has been shown to be very close to within a decibel to maximal ratio combining. The mean C/N can be shown to be [1]

$$\text{Mean}(\gamma_e) = \Gamma\left[1 + \frac{\pi}{4}(M-1)\right] \tag{21.7}$$

As with maximal ratio combining, the mean C/N for equal gain combining grows almost linearly with M and is approximately only 1 dB poorer than maximal ratio combiner even with an infinite number of branches.

21.3.4 Loss of Diversity Gain Due to Branch Correlation and Unequal Branch Powers

The analysis assumed that the fading signals in the diversity branches were all uncorrelated and of equal power. In practice, this may be difficult to achieve, and as we discussed, the branch cross-correlation coefficient $\rho = 0.7$ is considered to be acceptable. Also, equal mean powers in diversity branches are rarely available. In such cases, we can expect a certain loss of diversity gain. However, because most of the damage in fading is due to deep fades and the chance of coincidental deep fades is small even for moderate branch correlation, one can expect a reasonable tolerance to branch correlation.

The distribution of the output signal envelope of maximal ratio combiner has been shown to be [8]

$$\text{Prob}[\gamma_m] = \sum_{n=1}^{M} \frac{A_n}{2\lambda_n} e^{-\gamma_m/2\lambda_n} \tag{21.8}$$

where λ_n are the eigenvalues of the $M \times M$ branch envelope covariance matrix, whose elements are defined by

$$R_{ij} = E[r_i r_j] \tag{21.9}$$

and A_n is defined by

$$A_n = \prod_{\substack{k=1 \\ k \neq m}}^{M} \frac{1}{1 - \lambda_k/\lambda_n} \tag{21.10}$$

21.4 Effect of Diversity Combining on Bit Error Rate

So far, we have studied the distribution of the instantaneous envelope or C/N after diversity combining. We now briefly survey how diversity combining affects BER performance in digital radio links; we assume maximal ratio combining.

To begin, let us first examine the effect of Rayleigh fading on the BER performance of digital transmission links. This has been studied by several authors and is summarized in [9]. Table 21.1 gives the BER expressions in the large $E_b = N_0$ case for coherent binary phase-shift keying (BPSK) and coherent binary orthogonal

Table 21.1 Comparison of BER Performance for Unfaded and Rayleigh Faded Signals

Modulation	Unfaded BER	Faded BER
Coh BPSK	$\frac{1}{2}\text{erfc}\left(\sqrt{E_b/N_0}\right)$	$\frac{1}{4(\bar{E}_b/N_0)}$
Coh PSK	$\frac{1}{2}\text{erfc}\left(\sqrt{\frac{1}{2}E_b/N_0}\right)$	$\frac{1}{2(\bar{E}_b/N_0)}$

Table 21.2   BER Performance for Coherent BPSK and FSK with Diversity

Modulation	Post Diversity BER
Coherent BPSK	$\left(\dfrac{1}{4\overline{E}_b/N_0}\right)^L \left(\begin{array}{c} 2L-1 \\ L \end{array}\right)$
Coherent FSK	$\left(\dfrac{1}{2\overline{E}_b/N_0}\right)^L \left(\begin{array}{c} 2L-1 \\ L \end{array}\right)$

frequency shift keying (FSK) for unfaded and Rayleigh faded AWGN (additive white Gaussian noise) channels. $EN_b = N_0$ represents the average $E_b = N_0$ for the fading channel.

Observe that error rates decrease only inversely with signal-to-noise ratio (SNR) as against exponential decreases for the unfaded channel. Also, note that for fading channels, coherent BPSK is 3 dB better than coherent binary FSK, exactly the same advantage as in unfaded case. Even for the modest target BER of 10^{-2} that is usually needed in mobile communications, the loss due to fading can be very high: 17.2 dB.

To obtain the BER with maximal ratio diversity combining, we have to average the BER expression for the unfaded BER with the distribution obtained for the maximal ratio combiner given in Equation 21.5.

Analytical expressions have been derived for these in [9]. For a branch SNR greater than 10 dB, the BER after maximal ratio diversity combining is given in Table 21.2.

We observe that the probability of error varies as $1/\overline{E}_b/N_0$ raised to the Lth power. Thus, diversity reduces the error rate exponentially as the number of independent branches increases.

21.5 Concluding Remarks

Diversity provides a powerful technique for combating fading in mobile communication systems. Diversity techniques seek to generate and exploit multiple branches over which the signal shows low fade correlation. To obtain the best diversity performance, the multiple-access, modulation, coding, and antenna design of the wireless link must all be carefully chosen to provide a rich and reliable level of well-balanced, low-correlation diversity branches in the target propagation environment.

Successful diversity exploitation can have an impact on a mobile network in several ways. Reduced power requirements can result in increased coverage or improved battery life. Low signal outage improves voice quality and handoff performance. Finally, reduced fade margins directly translate to better reuse factors and hence increased system capacity.

Glossary

Automatic request for repeat: An error control mechanism in which received packets that cannot be corrected are retransmitted.

Channel coding/forward error correction: A technique that inserts redundant bits during transmission to help detect and correct bit errors during reception.

Fading: Fluctuation in the signal level due to shadowing and multipath effects.

Frequency hopping: A technique by which the signal bursts are transmitted at different frequencies, separated by random spacing, that are multiples of signal bandwidth.

Interleaving: A form of data scrambling that spreads bursts of bit errors evenly over the received data, allowing efficient forward error correction.

Outage probability: The probability that the signal level falls below a specified minimum level.

PCS: Personal communications services.

RAKE receiver: A receiver used in direct-sequence spread-spectrum signals. The receiver extracts energy in each path and then adds the energies together with appropriate weighting and delay.

References

1. Jakes, W.C., *Microwave Mobile Communications*. New York: Wiley, 1974.
2. Lee, W.C.Y., *Mobile Communications Engineering*. New York: McGraw-Hill, 1982.
3. Vaughan, R.G., Polarization diversity system in mobile communications. *IEEE Trans. Veh. Technol.*, VT-39(3), 177–186, 1990.
4. Adachi, F., Feeney, M.T., Williason, A.G., and Parsons, J.D., Crosscorrelation between the envelopes of 900 MHz signals received at a mobile radio base station site. *Proc. IEE*, 133(6), 506–512, 1986.
5. Jefford, P.A., Turkmani, A.M.D., Arowojulu, A.A., and Kellet, C.J., An experimental evaluation of the performance of the two branch space and polarization schemes at 1800 MHz. *IEEE Trans. Veh. Technol.*, VT-44(2), 318–326, 1995.
6. Freeburg, T.A., Enabling technologies for in-building network communications— Four technical challenges and four solutions. *IEEE Trans. Veh. Technol.*, 29(4), 58–64, 1991.
7. Viterbi, A.J., *CDMA: Principle of Spread Spectrum Communications*, Reading, MA: Addison-Wesley, 1995.
8. Pahlavan, K., and Levesque, A.H., *Wireless Information Networks*, New York: Wiley, 1995.
9. Proakis, J.G., *Digital Communications*, New York: McGraw-Hill, 1989.

Chapter 22

Multiple-Input Multiple-Output Systems in Wireless Communications

S. Plevel, S. Tomažič, T. Javornik, and G. Kandus

Contents

22.1 Introduction

Multiple-input multiple-output (MIMO) systems in wireless communications refer to any wireless communication systems in which at both sides of the communication path more than one antenna is used. Systems utilizing multiple transmit and multiple receive antennas are commonly known as MIMO systems. This wireless networking technology greatly improves both the range and the capacity of a wireless communication system. MIMO systems pose new challenges for digital signal processing given that the processing algorithms are becoming more complex, with multiple antennas at both ends of the communication channel. Overviews of MIMO systems can be, found for example, in [1–3].

MIMO systems constructively explore multipath propagation by using different transmission paths to the receiver. These paths can be exploited to provide redundancy of transmitted data, thus improving the reliability of transmission (diversity gain) or increasing the number of simultaneously transmitted data streams and increasing the data rate of the system (multiplexing gain). The multiple spatial signatures can also be used for combating interference in the system (interference reduction). A general model of a MIMO system is shown in the Figure 22.1.

A seminal information theory article by Foschini and Gans of Lucent Technologies [4] showed that the capacity of these systems can increase linearly with the number of transmit antennas as long as the number of receive antennas is greater than or equal to the number of transmit antennas. Since an increase in capacity means

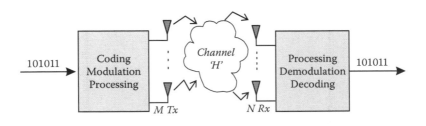

Figure 22.1   A general block diagram of multiple-input multiple-output wireless communication system.

capability of faster communication, this unmatched capacity improvement over regular one-antenna systems has fueled a huge interest in MIMO techniques, thus leading to the development of many forms of MIMO systems.

Traditional wireless communication systems with one transmit and one receive antenna are denoted as single-input single-output (SISO) systems, while systems with one transmit and multiple receive antennas are denoted as single-input multiple-output (SIMO), and systems with multiple transmit and one receive antenna are called multiple-input single-output (MISO) systems. Conventional smart antenna systems have only a transmit side or only a receive side equipped with multiple antennas, so they fall into one of last two categories. Usually, the base station has the antenna array since there is enough space and it is cheaper to install multiple antennas at base stations than to install them in every mobile station. Strictly speaking, only systems with multiple antennas at both ends can be classified as MIMO systems, although it may sometimes be noted that SIMO and MISO systems are referred to as MIMO systems. In the terminology of smart antennas SIMO and MISO systems are also called antenna arrays.

22.2 Capacity of MIMO Systems

From a mathematical point of view, the MIMO communication is performed through a matrix and not just a vector channel. So, it is possible to simultaneously transmit multiple parallel signal streams in the same frequency band and thus increase spectral efficiency. This technique is called *spatial multiplexing* and is shown in Figure 22.2. The data stream is encoded in a vector encoder and transmitted concurrently by M transmitters. The MIMO radio channel introduces distortion to the signal. The receiver has N antennas. Each antenna receives the signals from all M transmit antennas; consequently, the received signals exhibit interchannel interference. The received signals are downconverted to the baseband and sampled once per symbol interval. The MIMO processing unit estimates the

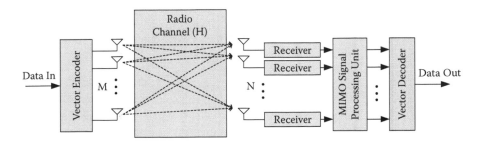

Figure 22.2   Block diagram of a MIMO system utilizing spatial multiplexing for capacity maximization.

transmitted data streams from the sampled baseband signals. The vector decoder is a parallel-to-serial converter, which combines the parallel input data streams to one output data stream.

In a system with M transmit and N receive antennas, there exist M*N subchannels between transmitter and receiver. In general, each subchannel exhibits selective fading; consequently, it is modeled as a linear discrete time finite impulse response (FIR) filter with complex coefficients. In the case of flat fading, the signal in each subchannel is only attenuated and phase shifted due to different propagation times between each receive and transmit antenna. The subchannel is reduced to one tap FIR filter (i.e., one complex coefficient). When the channel is constant during the whole time slot, the channel is quasi-static. For capacity investigation, we have assumed that the radio channel is quasi-static and fading flat.

In the aforementioned case, the received signal on the *j*th receive antenna can be expressed as

$$y_j = \sum_{i=1}^{M} h_{ij} x_i + n_j, \tag{22.1}$$

where x_i is the transmitted signal from the *i*th antenna, and y_j is the received signal at the *j*th antenna. Variable n_j denotes samples of circularly symmetric complex Gaussian noise with variance σ_n^2 at the *j*th receiver. The fading channel is described as a sum of complex paths h_{ij} between receive and transmit antennas. The complex gain coefficient h_{ij} complies with Gaussian distribution. The matrix form of Equation 22.1 is

$$\mathbf{y} = \mathbf{Hx} + \mathbf{n}, \tag{22.2}$$

where $\mathbf{y}$ is the column vector of the received signals, $\mathbf{H}$ is the channel matrix, $\mathbf{x}$ is the column vector of the transmitted signal, and $\mathbf{n}$ is a column vector of the additive white Gaussian noise.

The capacity of the system depends only on the transmitted signal power, noise, and channel characteristics. The channel capacity for a flat-fading deterministic channel can be expressed as

$$C = \log_2 \left[\det \left(\mathbf{I} + \frac{\rho}{M} \mathbf{HH}^* \right) \right], \tag{22.3}$$

where $\rho = P/\sigma_n^2$, P is the cumulative power transmitted by all antennas, and σ_n^2 is the noise power at each receive antenna. $\mathbf{H}$ is the matrix describing quasi-static channel response, and the superscript * denotes the transpose conjugate of the channel matrix $\mathbf{H}$ [5].

In the extreme case when we can assume uncorrelated paths, all eigenvalues of the product $\mathbf{HH}^*$ are nonzero and approximately equal. The capacity is then expressed as

$$C = \sum_{i=1}^{M} \log_2 \left(1 + \frac{\lambda_i}{M} \rho \right) \approx A_{min} \log_2 \left(1 + \frac{N}{A_{min}} \rho \right), \tag{22.4}$$

where $A_{min} = (M,N)$.

The amount of available capacity in an idealized MIMO channel increases linearly with A_{min} without an increase in transmit power. If the channel is time variant, the expression holds true for only one instance of the channel. Telatar [5] extended the expression for ergodic (mean) capacity in a random time-varying Gaussian channel. He found that ergodic capacity grows linearly with the number of receive antennas for a large number of transmit antennas. However, if the number of receive and transmit antennas is comparable, the benefit of adding a single antenna is much smaller.

An increase of MIMO system capacity can be achieved by multiplexing data streams into parallel subchannels (pipes) in the same frequency band. The pipes can be viewed as independent radio channels. The column vectors of flat-fading channel matrix **H** are usually nonorthogonal. However, by singular value decomposition (SVD), the channel matrix can be decomposed into diagonal matrix $\Lambda^{1/2}$ and two unitary matrices **U** and **V**:

$$\mathbf{H} = \underset{M_R \times M_R}{\mathbf{U}} \underset{M_R \times M_T}{\Lambda^{1/2}} \underset{M_T \times M_T}{\mathbf{V}^*}. \tag{22.5}$$

The diagonal entries of $\Lambda^{1/2}$ are in fact the nonnegative square roots of the eigenvalues of $\mathbf{HH}^*$. The number of nonzero eigenvalues $\lambda_1, \lambda_2, ..., \lambda_K$ of $\mathbf{HH}^*$ is equal to the rank of the channel matrix and to the number of independent subchannels. The global capacity could be expressed as the sum of the subchannel capacities. In fact, the singular values of channel matrix determine the gains of the independent parallel channels.

With the knowledge of the gain of the independent channels at the transmitter, we can determine the optimum power distribution at each transmit antenna to achieve maximum capacity. The MIMO channel capacity is determined by the water-filling theorem:

$$C = \sum_{i}^{K} \log_2 \left(1 + \lambda_i \frac{P_i}{\sigma_n^2} \right), \tag{22.6}$$

where P_i is the power allocated to the channel i, calculated by water-filling power allocation [5].

If the channel matrix at the transmitter is unknown, the uniform power distribution among transmitters is assumed for channel capacity calculation:

$$C = \sum_{i=1}^{K} \log_2 \left(1 + \frac{\lambda_i}{M} \rho \right). \tag{22.7}$$

When there is no knowledge of the channel state at the transmit side, the capacity is described by Equation 22.6, while the capacity with perfect channel knowledge at the transmit side can be calculated with Equation 22.7. In the ideal rich scattering (Rayleigh) channel knowledge at the transmit side is beneficial at a low signal-to-noise ratio (SNR), while at high SNR there is no significant difference in the capacities of Equations 22.6 and 22.7. However, the channel knowledge at the transmit side can help increase the reliability of practical systems substantially since we must not forget that the capacity is just the theoretical upper bound, which can only be achieved with codes of infinite length.

The capacity analysis of different realistic propagation environments has a significant influence on a design of communication systems. It was shown that for low-rank channels or low SNRs the usage of multiple transmit and receive antennas has much lower gain. In low-rank channels, capacity grows only logarithmically with the number of receive antennas. In these cases, usage of diversity techniques is recommended.

22.3 Benefits of Multiantenna Systems

The most important advantages of multiple-antenna systems are array gain, interference reduction, and diversity gain. MIMO systems not only can exploit the transmit and receive multiantenna benefits simultaneously but also can offer something new compared to the traditional antenna array systems (i.e., multiplexing gain). However, a compromise between diversity and multiplexing has to be made since it is not possible to exploit both maximum diversity gain and maximum multiplexing gain at the same time [6]. Ideally, adaptive systems would adapt the exploitation of multiple antennas to current conditions and thus simultaneously increase both the throughput and the reliability of the communication system.

22.3.1 Array Gain

Array gain indicates improvement of SNR at the receiver compared to traditional systems with one transmit and one receive antenna. The said improvement can be achieved with correct processing of the signals at the transmit or at the receive side, so the transmitted signals are coherently combined at the receiver. To achieve array gain at the transmitter antenna array, the channel state information (CSI) has to be known at the transmit side, while for the exploitation of antenna array gain at the receiver, the channel has to be known at the receive side. Receive array gain is achieved regardless of the correlation between the antennas.

22.3.2 *Interference Reduction*

Interference in the wireless channel appears due to frequency reuse. It decreases the performances of the communication systems. With multiple antennas, it is possible to separate the signals with different spatial signatures and thus decrease interchannel interference. When traveling through wireless medium, each signal is marked with the path that it has traveled. For the interference reduction, it is necessary to know the channel state information.

At the transmit side, the transmitted signal can be directed to the chosen users. With this, the interferences to the other users are decreased, and more efficient frequency planning is thus possible, which in turn increases the capacity of cellular systems. This technique is also called *beam forming* and is a very common spatial processing technique. A beam former can be seen as a spatial filter that separates the desired signal from interfering signals given that all the signals share the same frequency band and originate from different spatial locations. It essentially weights and sums the signals from different antennas in the antenna array to optimize the quality of the desired signal. In addition to interference rejection and multipath fading mitigation, a beam former also increases the antenna gain in the direction of the desired user.

Common beam-forming criteria include minimum mean square error (MMSE), maximum signal-to-interference-and-noise ratio (MSINR), maximum signal-to-noise ratio (MSNR), constant modulas algorithm (CMA), and maximum likelihood (ML). Beam forming is typically implemented using adaptive techniques. The adaptive array algorithms are broadly classified as trained algorithms and blind algorithms. Trained algorithms use a finite set of training symbols to adapt the weights of the array and maximize the SINR (signal-to-interference-plus-noise ratio). Blind algorithms do not require training signals to adapt their weights. As a result, these algorithms save bandwidth efficiency since all time slots can be used for transmission of useful data. A comprehensive review of adaptive antenna array systems can be found in [7].

22.3.3 *Diversity Gain*

Diversity in wireless communications is used to combat signal fading. Several techniques exist, but they are all based on the same principle: They transmit the signal through several independently fading paths. The more independently fading channels exist, the higher is the probability that at least one of them is not in deep fade.

Three types of diversity have been known for quite some time in wireless communications and have been used widely: time diversity, frequency diversity, and space diversity. For space diversity, there is no need for extra bandwidth or for extra time; however, the price to be paid is an increased complexity of the system since multiple antennas with radio-frequency chains and some processing are needed. Antennas must be separated sufficiently; otherwise, the signals are correlated, and diversity gain is reduced. The separation of the antennas needed for independent fading is called *coherence distance.* Coherence distance depends mostly on the departure and arrival angles of the signals. If the multipath is very rich, meaning that the signals arrive at the receiver from all directions, then a separation of approximately half of the wavelength is sufficient. If the angles are smaller, then the distance needed for independent fading is larger. Measurements have shown that regarding the base station, the height of the base station and the coherence distance are strongly correlated: The higher the base station antennas are, the larger is the coherence distance. For the mobile stations in an urban environment, separation of more than half of the wavelength is usually sufficient to achieve low correlation and thus high space diversity gain. If several antennas are used only at the receive side, we obtain receive diversity; if several antennas are used only on the transmit side, we obtain transmit diversity.

In MIMO systems, there are several antennas at both ends, which offer the potential of very high diversity gains. Diversity gain is equal to the number of independent channels in the system, which depends on the position of the antennas and the environment. If we have M transmit and N receive antennas, then we have $M*N$ subchannels, and the maximum diversity gain equals $M*N$. The higher the diversity gain is, the lower the probability of erroneous detection of the received signal is. The diversity gain indicates how fast the probability of error is decreasing with an increase in the signal strength.

When multiple antennas are used for reception, the received signals can be weighted and summed together. The phase shift of the received signals has to be taken into account, or the signals from different antennas would not necessarily be added together coherently at the combiner. The output signal would still have large fluctuations because of sometimes constructive and sometimes destructive combining. The method by which the weighting coefficients are chosen in such a way that the average quality of the signal (SNR) is maximized is called maximum ratio combining or MRC. With this method, the coefficients are equal to the conjugate complex value of the channel coefficients. This means that all received signals are shifted to the same phase, and the signals with higher strengths are getting a proportionally more important role at the signal combiner. The SNR at the output of the combiner is equal to the sum of the SNRs on all antennas. In addition to the array gain, MRC detection also achieves maximal diversity gain. An advantage of using receive diversity is that it is seamless to the transmitter, so it does not need to be defined in the standards to be used. Most modern communication systems are used with receive diversity if it is thus required.

Besides the receive diversity, it is also possible to use transmit diversity, which became a topic of studies in 1990. The transmit diversity is very suitable for cellular systems as more space, power, and processing capability is available at the base stations. Systems with transmit diversity differ in regard to the knowledge of the channel state information at the transmitter. When the channel state information is known to the transmitter, the system is dual to the receive diversity, the only difference is that the signal from each antenna is multiplied with the weight prior to the transmission, so that they automatically add together coherently at the receiver. When the channel state information is not known at the transmitter, it is a common practice to combine the space diversity with the time diversity. The technique is known as space–time coding (STC) and can achieve maximal diversity gain but, unfortunately, no array gain.

22.3.4 Multiplexing Gain

To exploit multiplexing gain, one needs to have several antennas at both ends of the communication system. In the MIMO system with a rich scattering environment, several communication channels in the same frequency band can be used. As shown, the capacity of the spatial multiplexing system can be increased with the minimum number of transmit and receive antennas. Such an increase in spectral efficiency of the system is especially attractive since there is no need for additional spectrum or for increasing transmit power. However, multiple antennas are needed at both ends to employ spatial multiplexing, while for other multiple-antenna benefits just an antenna array at one end is needed. The decoding of spatially multiplexed signals is very demanding, as explained later, and spatially multiplexed systems are less reliable given that in addition to low signal strength, high correlation between antennas can also cause erroneous detection.

22.4 Space–Time Codes

Coding the information across transmit antennas and time slots in a way that the receiver can reliably extract the information and exploit spatial diversity (possibly while providing coding gain) is called STC. Just one receive antenna and no channel knowledge at the transmit side are needed. The STC coder generates as many symbols as there are transmit antennas. These symbols are transmitted simultaneously, each through a different antenna. The goal of STC is to code the symbols at the transmitter in such a way that the highest diversity gain is achieved after decoding. Two main categories of STC are space–time trellis code (STTCs) and space–time block codes (STBCs). An in-depth introduction of STC and its applications in wireless communication systems can be found, for example, in [8].

22.4.1 Space–Time Trellis Codes

STTC is an extension of trellis-coded modulation (TCM) to multiple transmit antennas. It combines the advantages of transmit diversity and TCM in an ingenious way to obtain reliable, high-data-rate transmission in wireless channels without feedback from the receiver. STTC was first introduced by Tarokh et al. in 1998 [9]. They defined design criteria for STTC over slow flat fading, fast flat fading, and spatially correlated channels assuming high SNRs. They constructed codes that provided a good trade-off among data rate, diversity advantage, and trellis complexity. STTC can be illustrated in a trellis diagram, in which vertexes are defined with a diagram of state transitions. Besides maximal diversity gain, coding gain can be achieved.

STTC decoding can be done with the Viterbi algorithm. First, the branch metrics for each vertex in the trellis diagram is calculated, than the Viterbi algorithm finds the path through the trellis diagram for which the cumulative metrics are the smallest. The complexity of this decoding is considerable, which is why STBCs are more attractive for the implementation.

22.4.2 Space–Time Block Codes

STBCs map a block of input symbols into a space and time sequence. The receiver usually uses ML detection. The greatest benefit of block codes over trellis codes is that the optimal decoding is much simpler. Instead of joint detection of all of the transmitted symbols, the transmitted symbols can be separated with STBC. The class of codes that allow separation is called orthogonal STBC and is especially important for the implementation. STBC can achieve maximal diversity gain for a given number of transmit and receive antennas; however, they cannot achieve any coding gain. The first STBC with two transmit antennas was discovered by Alamouti [10] and is now widely known as the Alamouti code. Later, it was, with some limitations, generalized to different numbers of transmit antennas [11].

22.4.3 Alamouti Code

The Alamouti scheme can be compared with the MRC scheme for receive diversity exploitation; the main difference is that the Alamouti scheme is used when the antenna array is at the transmitter (MISO and MIMO), which is especially important for the downlink from the base stations. With the Alamouti scheme, two data symbols are transmitted in two transmission times, so the transmission rate (data throughput) is the same as with traditional systems with one transmit antenna. A diagram of the communication systems with the Alamouti scheme and two receive antennas is presented in Figure 22.3. In the first symbol period, symbol 1 is transmitted from the first transmit antenna, and symbol 2 is transmitted from the second antenna. In the second symbol period, symbol 2, multiplied by –1 and complexly conjugated, is transmitted from the first antenna, and the complexly conjugated symbol 1 is transmitted from the second antenna. Due to orthogonality, optimal decoding of each transmitted symbol can be done independently using

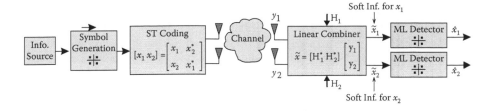

Figure 22.3 **Space–time coding with Alamouti scheme for two transmit and two receive antennas exploiting maximum diversity gain.**

simple linear decoding. In this way, the maximal diversity gain can be achieved for any number of receive antennas.

Since STBCs do not give any coding gain, they are usually combined with external forward error correction (FEC) coding, so the quality of transmission increases even further. Given that STBC can give soft output information (exact real value and not just a decision on the symbol), it is possible to combine the FEC code with advanced iterative decoding, known as *turbo coding*.

22.5 Spatial Multiplexing

In spatial multiplexing MIMO systems, independent data streams are transmitted through different antennas, which maximizes the data throughput of the MIMO systems. This type of communication is often called V-BLAST (vertical Bell Laboratories layered space–time architecture) [4]. V-BLAST systems divide the input data stream into as many independent data streams as there are transmit antennas. Then, the signals are modulated and simultaneously sent through all M transmit antennas, as shown on Figure 22.2.

In the case of spatial multiplexing, the processing at the transmitter is quite simple, but the processing at the receiver can be very complex, depending on the complexity of the receiver decoding algorithm. The performance of an MIMO spatial multiplexing system depends highly on the receiver quality since all N receive antennas receive signals from all M transmit antennas, and they have to be separated sufficiently.

22.5.1 Maximum Likelihood Decoding

The ML decoding of spatially multiplexed substreams maximizes the probability of correct detection and is therefore the optimal decoding possible. It can be denoted with the following equation:

$$\mathbf{x}_{ML} = \arg\min_{\mathbf{x}_j \in \{\mathbf{x}_1, \ldots, \mathbf{x}_K\}} \| \mathbf{y} - \mathbf{H}\mathbf{x}_j \|^2. \tag{22.8}$$

The problem of ML decoding is that it usually requires an extensive search for all possible combinations of transmitted symbols. The time of computation is exponentially proportional to the number of transmit antennas and the number of bits coded in each spatially multiplexed substream. Since in most cases this decoding is too complex to be implemented in communication systems, other faster methods of decoding are usually applied. However, ML decoding is important as a measurement of the proximity of other decoding algorithms to the optimum. It is also used in combination with other decoding methods on a limited subset of possible solutions.

22.5.2 Matrix Pseudoinversion Decoding

The most simple, but also the least-efficient, decoding method is matrix inversion. Since matrix inversion exists only for square matrices, pseudoinversion is used. Interference is removed by multiplying the received signal $\mathbf{y}$ with the pseudoinverse of the channel matrix. This method is also called zero forcing (ZF) since the interferences are zeroed by multiplication with the matrix inverse. When the channel matrix is badly conditioned (antennas are correlated), multiplication of the received signal with the matrix inverse significantly increases the noise. A bit better performance is achieved with a similar method called ZF-MMSE, by which the SNR is taken into account when calculating the matrix inversion to achieve MMSE. Due to their simplicity, these linear methods are sometimes used as a basis for other detection methods. The diversity gain achieved with this detection method is just $M - N + 1$; however, it is worth noting that this simple linear decoding can give very good results in adaptive MIMO systems when a number of spatially multiplexed substreams and the set of used transmit antennas are selected carefully based on the current channel state [12,13].

22.5.3 Successive Interference Cancellation Decoding

Successive interference cancellation (SIC) is an iterative detection method that consists of three steps: zeroing, quantization, and interference cancellation [14]. These steps are iteratively repeated until all transmitted symbols are detected. In each iteration, one symbol is detected with the ZF or ZF-MMSE method, then it is quantized (truncated to the nearest possible transmitted values), and the influence of this symbol is subtracted from the received vector $\mathbf{y}$.

This decoding is a good compromise between the complexity (which is much lower than with ML decoding) and efficiency (which is much better than a simple linear ZF decoding). The drawback of iterative cancellation decoding is error propagation. If one symbol is detected erroneously, then it is very likely that all the remaining symbols will also be decoded erroneously. That is why ordered SIC (OSIC) is usually used where the transmitted symbols are decoded in the order of the most likely correct decoding [14].

22.5.4 Sphere Decoding

The idea of sphere decoding (SD) is to search for the solution of ML decoding just for those points **Hx** that are inside a defined M-dimensional hypersphere with defined radius and center in **y**. The size of the radius defines the compromise between computational time and efficiency of decoding. The details of SD are beyond the scope of an encyclopedia, but a good review can be found in [15]. Let us just conclude that SD, if properly implemented, can give results very close to optimal ML decoding with huge complexity reduction.

22.6 Adaptive MIMO Systems

Communication systems can adapt the throughput rate to the current conditions of the channel if channel knowledge is available at the transmit side. This is known as adaptive coding modulation (ACM) techniques. In a MIMO system, ACM can be extended to the selecting mode of operation. It is commonly known that adaptive communication systems are more reliable and robust.

Knowing the channel state at the transmitter is much more difficult than knowing the channel state at the receiver, where the channel can be estimated by the training sequence. Feedback information from the receiver to the transmitter is needed, except in the case of a reciprocal channel like time-division duplex systems, for which the same parameters can be used for both sides, but precisely calibrated equipment is needed for that. Several schemes of adaptation were proposed in the literature; one of them has been mentioned (beam forming). Since the channel state information in MIMO systems is quite substantial, it is a good idea to select only from a set of predefined modes of operation. In spatial multiplexing systems, the number of spatially multiplexed streams can be selected based on the SNR and correlation between antennas. Transmit antenna selection can be performed to select a set of transmit antennas to be used to meet certain conditions [13,16].

In adaptive MIMO systems, it would be ideal to be able to achieve any trade-off between speed and reliability. An example of such a code, which includes V–BLAST and the orthogonal design STBCs as special cases, was proposed by Hassibi and Hochwald [17] and is called linear dispersion code (LDC). LDC can be used for any number of transmit and receive antennas and can be decoded with V–BLAST-like algorithms. The most important property of LDCs is that they satisfy an information–theoretic optimality criterion.

22.7 Applications of MIMO Systems

As shown, MIMO systems can significantly increase the reliability or capacity of communication systems, but many problems are to be faced. Capacities in a real environment are much lower than the capacities mentioned at the beginning of this chapter; since the antennas are not always uncorrelated, the channel estimation is not always accurate, there can be Doppler shift of frequency, and there might be synchronization problems. It is important to determine how many antennas should be used and how large the distance should be between them. The limitations are physical dimensions of the equipment, processing capabilities, and in the case of mobile stations, power consumption. The price of the equipment grows with the number of antennas used since extra amplifiers, filters, and processing power are needed; therefore, it is better to use multiple antennas only at the base station side.

On the other hand, the base stations are located higher than the subscriber stations; therefore, the propagation properties of the channel are less favorable for the exploitations of the MIMO systems. To ensure a satisfactory level of independency, they should be separated by around 10 wavelengths of the signal. For example, at 2-GHz frequency this distance is around 1.5 meters. For the subscriber stations, which are usually quite low and in many cases there is no line of sight to the base station, a distance of half of the wavelength is usually sufficient.

It is expected that in most of the modern wireless standards multiple-antenna techniques will be enabled, at least as an optional feature [18]. Let us have a closer look at three examples of the standardized MIMO solutions.

22.7.1 WiMax: IEEE 802.16

The usage of multiple antennas is foreseen in the standard for broadband wireless access (BWA) from the Institute of Electrical and Electronics Engineers (IEEE) 802.16. In the *d* version of the standard, known also as fixed WiMax (802.16-2004), multiple antennas may be used for beam forming (AAS, adaptive antenna system) or for exploiting diversity in downlink for two transit antennas with the described Alamouti scheme (the standard regards it as STC).

A newer version of the WiMax standard, also known as mobile WiMax (802.16e) or the *e* version, was ratified in December 2005. Besides beam forming and Alamouti STC schemes, it also provides spatial multiplexing of 2×2 MIMO in the downlink, which has the potential to double the speed of communication [19]. In the uplink, there is also a special feature available, called collaborative multiplexing, that provides the ability to use spatial multiplexing in the uplink even with mobile stations with just one antenna. The said operation is performed by collaborative transmission of two mobile stations on the same frequency at the same time. Although these two mobile stations transmit independent data, they are spatially well separated; therefore, the base station can decode both of them if at least two receive antennas are available at the base station. This does not increase the data throughput per user but increases overall data throughput in the sector. Mobile WiMax also supports adaptive switching between these options to maximize the benefit of smart antenna technologies under different channel conditions. This adaptive MIMO switching (AMS) is done between multiple MIMO modes to maximize spectral efficiency with no reduction in coverage area.

22.7.2 Wireless Local-Area Network: IEEE 802.11n

The MIMO systems have also a great potential in wireless local-area networks (WLANs). IEEE formed a new 802.11 Task Group (TGn) in January 2004 to develop the 802.11n standard for WLAN. In July 2005, competitors TGnSync and WWiSE agreed to merge their proposals as a draft and sent it to the IEEE in September 2005. It was expected that the standardization process could be completed by the second half of 2006, but unfortunately ratification of the standard was delayed until 2009.

The *n* version of WLAN is supposed to offer great improvement in both capacity and reliability. With more efficient use of orthogonal frequency division multiplexing (OFDM), an increase in the bandwidth from 20 to 40 MHz, and the use of spatial multiplexing with up to four simultaneously spatially multiplexed substreams, data with throughput speeds of over 500 Mbps can theoretically be achieved. The Alamouti scheme can be used for the exploitation of transmit diversity and MRC for the receive diversity.

Even before the 802.11n standard was approved, some manufacturers were already developing and testing equipment. An example is Airgo Networks, which was based on MIMO solutions produces WLAN devices. Airgo's True MIMO™ technology is based on drafts of IEEE 802.11n; however, the question remains whether it will be compatible with the standard.

22.7.3 Third Generation Partnership Project

The Third Generation Partnership Project (3GPP) is intensively engaged to implement MIMO systems in the broadband code-division multiple-access (CDMA) systems [20]. The use of spatial diversity with two antennas was proposed in the R99 version of the CDMA system. Multiple-antenna systems were divided into those working in open loop (no channel state information at the transmit side) and those working in closed loop (channel state information available at the transmit side). Examples of open loop are again the Alamouti scheme (here called space–time transmit diversity, STTD) and time switch transmit diversity (TSTD). In the closed-loop systems, transmit beam forming can be used; it is called transmit adaptive array (TAA). In later versions of the standard, larger numbers of transmit antennas were introduced, and they were divided into subgroups hierarchically weighted to exploit transmit diversity and beam forming. Several documents examining the multiantenna use on both the base station side and on the terminal (mobile) station side are available on the Internet page of 3GPP. There were several STC schemes proposed for one, two, or four antennas on both sides.

22.8 Summary

In the field of wireless communications, MIMO systems have enabled a huge step forward since they can significantly increase both the coverage and the capacity of cellular systems. The technology is developing very fast and is already present in several standards. Standardization difficulties can appear in supporting the compatibility for previous versions of standards, which is why it is easier to incorporate MIMO in completely new standards like WiMax. However, no standardization will resolve all issues. Standards help improve product efficiency, but the actual design and manufacturing issues alone will decide on the performance of the final product.

In the future, it is expected that several antennas will be included in many laptop computers or mobile devices. Massive use of multiple antennas will decrease the prices of such devices, which in turn can make the technology available to a wider range of users. It is hard to predict which standard or technology will continue the fourth-generation (4G) wireless systems, but it will almost certainly incorporate MIMO systems. Those systems will have to include the ability to adapt to the time-changing nature of the wireless channel using some form of at least partial feedback to make a fine compromise between rate maximization (spatial multiplexing) and diversity (STC) solutions.

References

1. Gesbert, D., Shafi, M., Shiu, D.S., Smith, P.J., and Naguib, A., From theory to practice: An overview of MIMO space-time coded wireless systems. *IEEE J. Sel. Areas Commun.*, 21, 281–302, April 2003.
2. Paulraj, A.J., Gore, D.A., Nabar, R.U., and Bölcskei, H., An overview of MIMO communications—A key to gigabit wireless. *Proc. IEEE*, 92, February 2004.
3. Gershman, A.B., and Sidiropoulus, N.D., *Space-Time Processing for MIMO Communications*. New York: Wiley, 2005.
4. Foschini, G.J., and Gans, M.J., Layered space–time architecture for wireless communication in a fading environment when using multiple antennas. *Bell Labs. Syst. Tech. J.*, 1, 41–59, 1996.
5. Telatar, E., Capacity of multi-antenna Gaussian channels. *Eur. Trans. Telecomm. ETT*, 10, 585–596, November 1999.
6. Zheng, L., and Tse, D., Diversity and multiplexing: A fundamental tradeoff in multiple antenna channels. *IEEE Trans. Inform. Theory*, 49(5), 1073–1096, May 2003.
7. Allen, B., and Ghavami, M., *Adaptive Array Systems, Fundamentals and Applications*. New York: Wiley, 2005.
8. Vucetic, B., and Yuan, J., *Space–Time Coding*. New York: Wiley, 2003.
9. Tarokh, V., Seshadri, N., and Calderbank, A.R., Space–time trellis codes for high data rate wireless communication: Performance criterion and code construction. *IEEE Trans. Inform. Theory*, 44(2), March 1998.
10. Alamouti, S., A simple transmitter diversity technique for wireless communications. *IEEE J. Sel. Areas Commun.*, Special Issue, 16(8), 1998.

11. Tarokh, V., Jafarkhani, H., and Calderbank, A.R., Space–time block codes from orthogonal designs. *IEEE Trans. Inform. Theory*, 45(5), 1456–1467, July 1999.
12. Plevel, S., Javornik, T., and Kandus, G., A recursive link adaptation algorithm for MIMO systems. *AEU Int. J. Electron. Commun.*, 59(1), 52–54, March 2005.
13. Heath, R.W., Jr., and Paulraj, A.J., Switching between diversity and multiplexing in MIMO systems. *IEEE Trans. Sig. Proc.*, 53(6), 962–968, June 2005.
14. Wolaniansky, P.W., Foschini, G.J., Golden, G.D., and Valenzula, R.A., V-BLAST: An architecture for realizing very high data rates over the rich-scattering wireless channel. *Proc. Int. Symp. Adv. Technol.*, September 1998.
15. Hassibi, B., and Vikalo, H., On sphere decoding algorithm. I. Expected complexity. *IEEE Trans. Sig. Proc.*, 53(8), 2806–2818, August 2005.
16. Plevel, S., Javornik, T., Kandus, G., and Jelovcan, I., Transmission scheme selection algorithm for spatial multiplexing MIMO systems with linear detection. *WSEAS Trans. Commun.*, 6(5), June 2006.
17. Hassibi, B., and Hochwald, B., High rates codes that are linear in space and time. *IEEE Trans. Inform. Theory*, 48, 1804–824, July 2002.
18. Hottinen, A., Kuusela, M., Hugl, K., Zhang, J., and Raghothaman, B., Industrial embrace of smart antennas and MIMO. *IEEE Wireless Commun.*, 13(4), 8–16, August 2006.
19. Wimax Forum, Mobile WiMAX—Part I: A technical overview and performance evaluation. *WiMAX forum*, August 2006.
20. Soni, R.A., Buehrer, R.M., and Benning, R.D., An intelligent antenna system for cdma2000. *IEEE Sig. Proc. Mag.*, 19(4), 54–67, July 2002.

Links

1. http://userver.ftw.at/~zemen/MIMO.html
2. http://www.ece.utexas.edu/~rheath/research/mimo
3. http://www.wimaxforum.org
4. http://www.oreilly.com/catalog/802dot112/chapter/ch15.pdf
5. http://www.3GPP.org
6. http://www.airgonetworks.com
7. http://www.ieee.org

Index

Printed and bound by CPI Group (UK) Ltd, Croydon, CR0 4YY

21/10/2024

01777084-0017